W9-DGL-789

CREATING VALUE THROUGH SKILL-BASED STRATEGY AND ENTREPRENEURIAL LEADERSHIP

TECHNOLOGY, INNOVATION, ENTREPRENEURSHIP AND COMPETITIVE STRATEGY SERIES

Series Editors: John McGee and Howard Thomas

Published

STEFFENS
Newgames: Strategic Competition in the PC Revolution

BULL, THOMAS & WILLARD
Entrepreneurship: Perspectives on Theory Building

SANCHEZ, HEENE & THOMAS
Dynamics of Competence-Based Competition

DAI
Corporate Strategy, Public Policy and New Technologies

BOGNER
Drugs to Market

SUSMAN & O'KEEFE
The Defense Industry in the Post-Cold War Era

NORDBERG & VERBEKE
The Strategic Management of High Technology Contracts: The Case of CERN

Forthcoming titles

PORAC & VENTRESCA
Constructing Industries and Markets

Other titles of interest

DAEMS & THOMAS
Strategic Groups, Strategic Moves, and Performance

DENNING
Making Strategic Planning Work in Practice

DOZ
Strategic Management in Multinational Companies

McNAMEE
Developing Strategies for Competitive Advantage

TSE
Marks and Spencer

Related journals—sample copies available on request

European Management Journal
International Business Review
Journal of Retailing and Consumer Services
Long Range Planning
Scandinavian Journal of Management

CREATING VALUE THROUGH SKILL-BASED STRATEGY AND ENTREPRENEURIAL LEADERSHIP

William C. Schulz III and Charles W. Hofer

1999

Pergamon

An Imprint of Elsevier Science

Amsterdam – Lausanne – New York – Oxford – Shannon – Singapore – Tokyo

ELSEVIER SCIENCE Ltd
The Boulevard, Langford Lane
Kidlington, Oxford OX5 1GB, UK

© 1999 Elsevier Science Ltd. All rights reserved.

This work is protected under copyright by Elsevier Science, and the following terms and conditions apply to its use:

Photocopying
Single photocopies of single chapters may be made for personal use as allowed by national copyright laws. Permission of the publisher and payment of a fee is required for all other photocopying, including multiple or systematic copying, copying for advertising or promotional purposes, resale, and all forms of document delivery. Special rates are available for educational institutions that wish to make photocopies for non-profit educational classroom use.

Permissions may be sought directly from Elsevier Science Rights & Permissions Department, PO Box 800, Oxford OX5 1DX, UK; phone: (+44) 1865 843830, fax: (+44) 1865 853333, e-mail: permissions@elsevier.co.uk. You may also contact Rights & Permissions directly through Elsevier's home page (http://www.elsevier.nl), selecting first 'Customer Support', then 'General Information', then 'Permissions Query Form'.

In the USA, users may clear permissions and make payments through the Copyright Clearance Center, Inc., 222 Rosewood Drive, Danvers, MA 01923, USA; phone: (978) 7508400, fax: (978) 7504744, and in the UK through the Copyright Licensing Agency Rapid Clearance Service (CLARCS), 90 Tottenham Court Road, London W1P 0LP, UK; phone: (+44) 171 631 5555; fax: (+44) 171 631 5500. Other countries may have a local reprographic rights agency for payments.

Derivative Works
Tables of contents may be reproduced for internal circulation, but permission of Elsevier Science is requireed for external resale or distribution of such material.
Permission of the Publisher is required for all derivative works, including complications and translations.

Electronic Storage or Usage
Permission of the publisher is required to store or use electronically any material contained in this work, including any chapter or part of a chapter.

Except as outlined above, no part of this work may be reproduced, stored in a retrieval system or transmitted in any form or by any means, electronic, mechanical, photocopying, recording or otherwise, without prior written permission of the publisher.
Address permissions requests to: Elsevier Science Rights & Permissions Department, at the mail, fax and e-mail addresses noted above.

Notice
No responsibility is assumed by the Publisher for any injury and/or damage to persons or property as a matter of products liability, negligence or otherwise, or from any use or operation of any methods, products, instructions or ideas contained in the material herein. Because of rapid advances in the medical sciences, in particular, independent verification of diagnoses and drug dosages should be made.

Although all advertising material is expected to conform to ethical (medical) standards, inclusion in this publication does not constitute a guarantee or endorsement of the quality or value of such product or of the claims made of it by its manufacturer.

First edition 1999

Library of Congress Cataloging-in-Publication Data
A catalog record from the Library of Congress has been applied for.

British Library Cataloguing in Publication Data
A catalogue record from the British Library has been applied for.

ISBN: 0-08-043444-4

Typeset by The Midlands Book Typesetting Company, United Kingdom

♾ The paper used in this publication meets the requirements of ANSI/NISO Z39.48-1992 (Permanence of Paper).
Printed in The Netherlands.

CONTENTS

To date, there have been three generations of improvements in Andrews' SWOT analysis paradigm, one dealing with each of the three basic components of Andrews' model.

The first major improvement in Andrews' model was the development of contingency theory models of effective strategy first proposed by Hofer in 1975. Basically, contingency theory models changed the nature of the functional relationship (f) between the independent variables (Strengths, Weaknesses, Opportunities, and Threats) and the dependent variable (Strategy) in Andrews' model. More specifically, contingency theory proposed that the nature of the functional relationship between Strategy and Strengths, Weaknesses, Opportunities, and Threats would vary in both magnitude and in functional form (linear/non-linear; continuous/discontinuous, etc. etc.) depending on the basic characteristics of the organization's skills and resources and on its external environment.

Put differently, contingency theory argued that the basic components of different strategy models would differ based on the circumstances involved, i.e., that Hofer's (1980) model of effective turnaround strategies would differ from Harrigan's (1980) model of effective strategies for declining industries; that both would differ from Sandberg's model of effective strategies for new ventures; and that all three would differ from Rumelt's model of effective corporate strategies—a result which all the strategy research conducted to date has confirmed.

Two of the major implications of contingency theory are: (1) that situation matters, e.g., industry knowledge is important, and (2) that few universal principles of strategy exist since a principle, to be effective, must apply to most situations.

The second major improvement in Andrews' SWOT Analysis model was the development of Porter's "Five Forces" model of competitive strategy (1980,1985). Porter's model represented a quantum improvement in the ways that a strategist should analyze any organization's environment in order to identify potential future Opportunities and Threats. While both Hofer (1975) and Hofer & Schendel (1978) have suggested that stage of industry evolution/life cycle was an important determinant of environmental Opportunities and Threats, Porter described in detail the particular variables that would influence both the number and magnitude of the Opportunities and Threats that an organization might face at the three most critical industry life cycle stages (Emergence, Maturation, and Decline). In addition, Porter described in depth two unique industry environments (fragmented and global industries) in which the organization's Opportunities and Threats would be unique because of the unusual characteristics of these environments. Porter also developed a generic business strategy classification system.

Two major implications of Porter's "Five Forces" model are : (1) that businesses can influence the opportunities and threats that they will face through proactive strategic action; and (2) that no competitive advantage is infinitely sustainable because of the changing nature of the organization's external environment.

The third major improvement in Andrews' SWOT analysis model has been the development of "Resource-based models of Strategy." One of the major contributions of these models has been the specification of the properties that a firm's resources and skills must have in order to build and maintain sustainable competitive advantages. In particular, they must be: (1) valuable (one cannot build significant advantages on resources that have little value); (2) unique (without uniqueness no differential advantage is possible); and (3) rare (one cannot build enduring advantages on resources that are so commonplace that most firms could acquire them); (4) difficult-to-imitate (otherwise any advantage gained would be of limited duration). The second primary contribution of Resource-based models of Strategy has been to argue that in, most situations, resources are a more important determinant of performance than opportunities—an assertion that has been confirmed by the research of Hansen and Wernerfelt (1989).

One of the major implications of the Resource-based models is that the genesis of significant, sustainable competitive advantage must come from Specific knowledge of a firm's capabilities and the opportunities emerging in its environment, not from Generic knowledge of these factors because generic knowledge, by its nature, cannot produce the valuable, unique, rare, and difficult-to-initiate characteristics required to build such advantages.

What then is Skill-based Strategy? Skill-based Strategy is the term we have created to describe the way that highly successful, independent, entrepreneurs and corporate intrapreneurs actually develop and craft effective strategies for their organizations. Put differently, the concept of Skill-based Strategy has been developed from research on the real world practices of highly successful entrepreneurs. At the same time, we also show that Skill-based Strategy concepts successfully and effectively integrate the three generations of theoretical improvements in Andrews' SWOT analysis model of Strategy Formulation.

While a full discussion of the concept of Skill-based Strategy must be covered in the text, there are four aspects of the concept that we would highlight here. They are the need to focus : (1) first on Skills and Resources and secondly on Opportunities; (2) first on Customers and then on Competitors; (3) first on Customer Benefits and then on Price, and (4) first on Dynamic Capabilities and then on Static Assets. Each is a critical aspect of the Skill-based Strategy formulation process, and each will be examined in slightly more depth here.

First, like Andrews' original SWOT Analysis model, Skill-based Strategy fully recognizes the need to understand both Skills and Resources and environmental Opportunities and Threats in order to identify effective matches on which enduring competitive advantages can be built. Nonetheless,

Skill-based Strategists understand their organization's skills and resources first, then they examine their organization's environment using this skill and resource "profile" in order to identify the Opportunities that will be most attractive to their organization. The rationale for this process is quite simple. While SWOT Analysis requires one to identify both Strengths & Weaknesses and Opportunities & Threats, these two sets are not of equal size. As Figure 2 illustrates, the set of attractive environmental Opportunities & Threats is far, far larger than the set of Strengths & Weaknesses possessed by the organization. This reality implies that the most efficient search procedure is to first understand organizational Strengths & Weaknesses, and then to search for Opportunities that these might be applied to—which is exactly what Skill-based Strategists do.

Second, while both Porter and Resource-based Strategy theorists correctly point out the need for sustainable competitive advantages, Drucker is nonetheless correct when he states that the first task of business is "to create a "customer". The first focus of Strategy must be on the creation (and maintenance) of customers by satisfying the customers' needs and wants. Only after one has secured a customer can one then ask: "What sustainable advantages can I create that will enable me to keep this customer in the future?" If one analyzes an organization's set of Skills and Resources (a) in order to try to identify some subset of these Skills and Resources (b) on which enduring competitive advantages might be built, it may be that technology, tacit knowledge, uniqueness, rarity, and duplication difficulty will be the key components of the finial selection process. It is also true, however, that it is the Customer who will decide which of the elements in subset b have value from the *Customer's* perspective—which is ultimately the only perspective that counts, since it is the only one the Customer will pay for. Once again, our research shows that Skill-based Strategists understand both the primacy of the Customer and the concurrent

Fig. 2. SWOT Analysis in the Real World

need to establish enduring competitive advantage, and that they have built this understanding into their strategic decision-making processes.

Third, Skill-based Strategy formulation fully recognizes the need to use both Customer Benefits and Price as long-term strategic weapons. Every new innovation that has ever been brought to market—ranging from automobiles to bicycles, cameras, micro-waves, personal computers, radios, television, and Xerox copiers—has been sold initially on the basis of the unique functions the product performs and the benefits it provides to its customers. At the same time, few such products have reached the mass market until lower-priced models of the same products have been produced. Skill-based Strategists have developed the ability to position their organization's products and/or services on the Price-Benefit continuum in niches that uniquely fit the skill profile of their organization and that secure not only initial sales, but also enduring competitive advantages.

Fourth, Skill-based Strategists deeply understand not only the unique Assets and Resources that their organizations' possess, but also the dynamic Skills and Capabilities that their organizations can develop in the future, and they use both to screen potential future Opportunities in order to select those that they will seek to exploit in the world of tomorrow. While their competitors are planning how to overcome the Skill-based Strategist's current advantages, the Skill-based Strategist is thinking carefully about how to create an entirely new, but equally effective set of advantages for the future.

The world is always changing. Consequently, there is no way to *guarantee* future success because no one can foresee the future. All that one can do is to build a strong set of Skills and Resources that provide significant customer benefits and competitive advantages, which one can use to prepare for the future. As noted earlier, Strategic Management research has shown that effective Strategy is the primary determinant of long-term organizational success. And, our research has shown that Skill-based Strategic Thinking is one of the better ways to develop effective Strategies, i.e., that Skill-based Strategy can contribute to the long-term ability of an organization to create wealth and economic value.

ENTREPRENEURIAL LEADERSHIP

If Skill-based Strategy is the key to creating Economic Value and Wealth, then Entrepreneurial Leadership is the key to developing and crafting effective Skill-based Strategies.

Leadership is important for two reasons. First, and most importantly, all significant organizational activity involves human beings. Ultimately, it is people, not computers, machines, or robots, that make things happen. Leadership is essential to coordinated, motivated human activity. Second, the future is about change, and leadership is an essential component of the "change process".

There are, of course, numerous factors on which leadership can be based, including charisma, position, and power. Every observer of the political scene remembers and understands the charismatic leadership of Presidents John F. Kennedy and Ronald Reagan, as well as the power-based leadership of Mayor Richard F. Daley of Chicago and President Lyndon B. Johnson. Charisma and power are two of the most important bases of leadership in the political arena. Neither is sufficient in leading business organizations, however, because of businesses' unique role of providing the primary goods and services demanded by society.

Two other skills are necessary for effective business leaders in the 21st Century. First, such leaders must have a "Vision" of the future of their industry so that they can develop Strategies for seizing the Opportunities provided by that future. Second, they must understand how to develop the organizational systems and processes that will be needed to effectively implement such Strategies. Put differently, they need to understand how to make their Vision of the future actually happen in the real world. We call this set of capabilities Entrepreneurial Leadership, and it is one of the basic concepts and themes that this text will examine in depth.

CONCLUSION

This book provides a detailed description of the understanding and insights which we have developed about how individual entrepreneurs and corporate intrapreneurs use "Skill-based Strategy" and "Entrepreneurial Leadership" to create significant new "Economic Value" and "Wealth" for the organizations they lead.

This book is based on an in-depth research study of four highly successful entrepreneurs and/or intrapreneurs, as well as on a comprehensive analysis of the latest concepts and theories in the fields of Entrepreneurship, Industrial Organization Economics, Organization Theory, and Strategic Management.

The book itself is organized into four different parts. Part One reviews the Foundations of Resource-based approaches to Strategic Management Thinking. Part Two describes the four companies which we have studied in depth. Part Three weaves the insights and understanding we have gained from these companies into a practitioner-oriented model of "Value Creation". Lastly, Part Four covers various important aspects of this model, such as the "Skill Life Cycle" that are not covered either in depth or at all in most traditional texts in the field. Finally, since we are "explorers" in the ever evolving world of Strategic Management, we both welcome and encourage our readers to share their experiences and insights with us.

William C. Schulz III
Charles W. Hofer
July, 1999

THE AUTHORS

William C. Schulz is Associate Professor of Business and Economics at Oglethorpe University in Atlanta, Georgia. He teaches courses in strategic management, entrepreneurship, total quality management and international business competitiveness. Dr Schulz worked for McDonnell Douglas and the Eaton Corporation before entering the University. His research interests are in the areas of entrepreneurship and strategy, and he has won international recognition and two awards for the research on which this book is based.

Charles W. Hofer is Regents Professor of Strategy and Entrepreneurship at the University of Georgia's Terry College of Business. During his career, Dr Hofer has consulted with numerous large and small companies and not-for-profit organizations, including AT&T, Exxon, Ford, and General Electric; has helped his students successfully start over a dozen new businesses; and has taught in executive programs around the world. He is internationally recognized as one of the founding scholars of the fields of Entrepreneurship and Strategic Management, and according to a vote of the Business Policy and Strategy Division of the Academy of Management, Dr Hofer has written three of the top thirty works in the field of Strategic Management. In addition, Dr Hofer is one of only two scholars in the field of Strategic Management who have written two or more books and articles that have reached the 'one-in-a-thousand citation count minimums' for the Social Sciences. To date, Dr Hofer's PhD students have won ten of the thirty-six major dissertation awards given in the field of Entrepreneurship, including the two awards won by Dr Schulz's dissertation, upon which this book is based.

PART ONE: FOUNDATIONS

1. FROM RESOURCES TO ADVANTAGE
2. RESEARCH OPPORTUNITIES, RATIONALE AND OBJECTIVES
3. RESOURCES FROM THE GROUND UP: THE FOUNDATIONS OF PERFORMANCE

1

FROM RESOURCES TO ADVANTAGE

INTRODUCTION

At their foundation, the fields of entrepreneurship and strategic management address the fundamental issues and processes that affect the ability of entrepreneurs and general managers to create and sustain organizations that generate value in the economy. In general, both theory and research in the field of strategy suggest that the successful formation, survival, and prosperity of a firm are linked to the ability of its leaders and managers to recognize and establish a "fit" between their firm's current and future internal resources, including the skills of the workforce, and external opportunities and threats in the environment (Hofer and Schendel, 1978; Andrews, 1987; Venkatraman and Prescott, 1990).

However, some of the important ingredients in this "fit" equation, such as the internal management of the skills and specific "knowledge-based" resources of a firm, have not been studied as much, or as systematically, as have other more external aspects of the equation, such as market characteristics and position (Wernerfelt, 1984; Chrisman 1986; Ansoff, 1987; Barney, 1991; Grant, 1996; Teece *et al.*, 1997).

While recent theories have been introduced that explore the management of "core competencies" (Sanchez *et al.*, 1996; Gorman and Thomas, 1997), "dynamic capabilities" (Teece *et al.*, 1997), and the management of "knowledge as a competitive resource" (Hamel and Heene, 1994; Grant, 1996; Spender, 1996), as of 1998 there remains little in-depth empirical evidence and analysis to help refine and advance theory in these areas, as Teece *et al.* (1997), note:

> Further theoretical work is needed to tighten the framework [of a dynamic capabilities theory], and empirical research is critical to helping us

> understand how firms get to be good, how they sometimes stay that way, why and how they improve, and why they sometimes decline (p. 530).

And, as Grant (1997) notes:

> While making some progress in integrating prior research on organizational learning and organizational resources and capabilities, much remains to be done at both the empirical and the theoretical level, especially in relation to understanding the organizational processes through which knowledge is integrated . . . . Further progress is critically dependent upon closer observation of the processes through which tacit knowledge is transferred and integrated (p. 384).

Clearly, as of early 1998, there is a need for in-depth empirical evidence and research that explores and describes the actual processes by which basic human and material resources are converted into potential advantages, and which can enhance both the practice and theory of entrepreneurship and strategic thinking.

OVERVIEW OF THE BOOK

This book, *Creating Value Through Skill-Based Strategy and Entrepreneurial Leadership*, takes a comprehensive look at the process by which leaders, as entrepreneurs and strategists, attempt to build and craft the skill-bases of their firms to best create long-term value for their customers. Through rigorous qualitative analysis of longitudinal case histories, both the practice and theory of skill-based strategic thinking and entrepreneurial leadership are discussed and examined in detail. The book consists of four major sections:

(1) Part One examines the foundations of resource-based approaches to management and strategic thinking. The relationship between human and other resources is discussed, and a detailed process-typology of organizational resources that serves as a basis for understanding the detailed processes of how resources can be leveraged into sustainable strategic advantage is presented.
(2) Part Two of the book presents the case histories of four very different firms, including a high-tech chemicals research company, a custom cabinet manufacturer, a large corporate industrial engineering firm, and a craft-oriented surgical instrument maker. Each case provides a unique setting from which to gain insights into the processes of skill and competence development, and also the value creation process, as discussed by exemplary practitioners.
(3) Part Three of the book compares and contrasts the insights discussed from the four case analyses, and integrates the findings into a "practitioner-based" model of value creation. This model is then further integrated with current academic theory, and a more formal theory of the value creation process is presented.

to recognize and adapt to any emergent phenomenon that affects the firm and its intended strategy (Mintzberg and Waters, 1985); and, (3) the skills and abilities of the managers and other employees of the firm to implement the intended strategy and to respond to ongoing changes in both the internal and external environments (Galbraith and Kazanjian, 1986; Grant, 1996; Spender, 1997; Teece *et al.*, 1997).

FOUR MAJOR APPROACHES TO STRATEGIC MANAGEMENT

The Focus of Traditional Strategy Models

Traditional models of strategy, such as those proposed by Andrews (1971, 1987) and Hofer and Schendel (1978) suggest that a firm's performance is a function of its general managers' abilities to match their firm's internal strengths and weaknesses with the opportunities and threats in its external environment. The primary focus of these models has been on the strategy formulation process, and on the creation of deliberate or intended strategies (Mintzberg, 1991). Figure 2.2 illustrates the focus of the traditional models and identifies some of their strengths and weaknesses.

In the traditional strategy models, a firm's strategy is an output of the strategic thinking process and reflects, as Hofer and Schendel (1978, p. 25) note, "the fundamental pattern of present and planned resource deployments and environmental interactions that indicates how the organization will achieve its objectives".

The strengths of the traditional strategy models include: (1) they take a holistic view of the "fit" equation and examine both firm and environmental

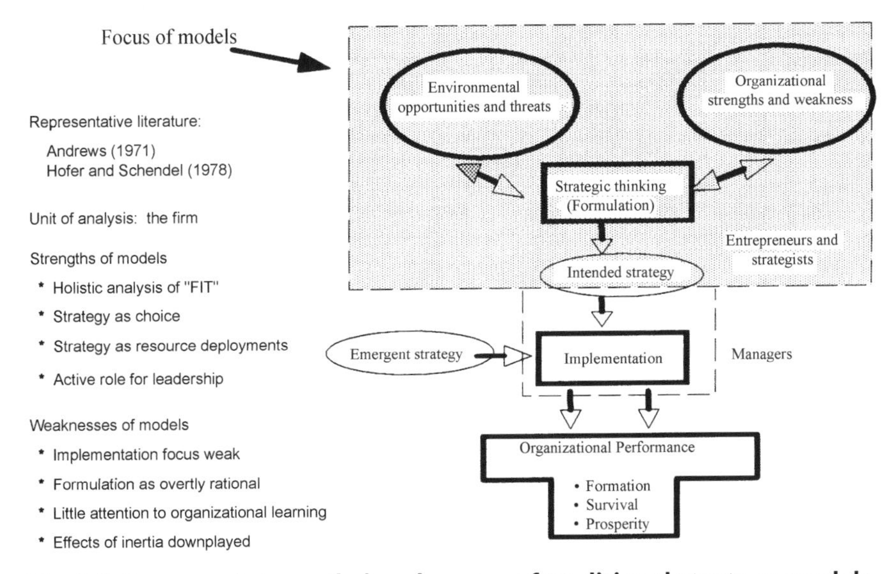

Fig. 2.2. Focus and strengths/weaknesses of traditional strategy models.

factors; (2) they recognize strategy as a choice variable and argue that managers can actively help to determine their firm's future; (3) they view strategy as a pattern of resource deployments and interactions with the environment, which recognizes the importance of the firm to be able to implement its strategy; and finally, (4) they recognize the important role that leaders play in shaping the culture and context for action in the firm (Bower, 1970).

The weaknesses of the traditional strategy models include: (1) a relatively weak focus on implementation issues, particular as they relate to *how* the firm's strengths and weaknesses are developed from the firm's resource base to meet future requirements (Ansoff, 1979); the strategy formulation process they discuss is primarily based on rational decision-making models and does not adequately account for emergent and bounded-rationality phenomenon (Mintzberg, 1990); (3) they do not systematically address issues of organizational learning and interpretation (Daft and Weick, 1984); and finally, (4) they do not adequately model the processes of organizational inertia and change (Miller and Friesen, 1980; Ginsberg, 1988).

The Focus of Competition-Based Strategy Models

Competition-based models are a second major approach to examining strategic management issues. These models, based on Porter's 1980 book *Competitive Strategy*, focus on the competitive environment of firms. Strategy, in these models, helps the leaders and managers of a firm to establish and defend their position, relative to competitors, in a particular market environment. These competition-based models suggest that the firm's performance is primarily a function of the firm's scope and positioning within a given industry structure. Figure 2.3 illustrates the focus and identifies some of their strengths and weaknesses.

The strengths of the competition-based models include: (1) they focus on strategy as position; (2) they provide systematic tools for identifying and analysing the competitive environment, which can be used in the strategy formulation process; (3) they examine the role of relative power and bargaining position in the environment (Pfeffer, 1987); and, (4) they examine the structural elements of the competitive environment which might limit the mobility of a firm to change its strategy.

The weaknesses of the competition-based models include: (1) they assume that all firms in a given industry have access to the same resource base and that they possess homogeneous resources over the long run (Barney, 1991); (2) they provide virtually no direction for the analysis of firm-specific attributes that might serve as a source of performance differences among firms, since, in these models, "the sources of competitive advantage lie at the level of the industry, or possibly groups within an industry" (Teece *et al.*, 1992, p. 36); and (3) as a result of their focus on industry-level phenomena, they do not focus on implementation and organizational

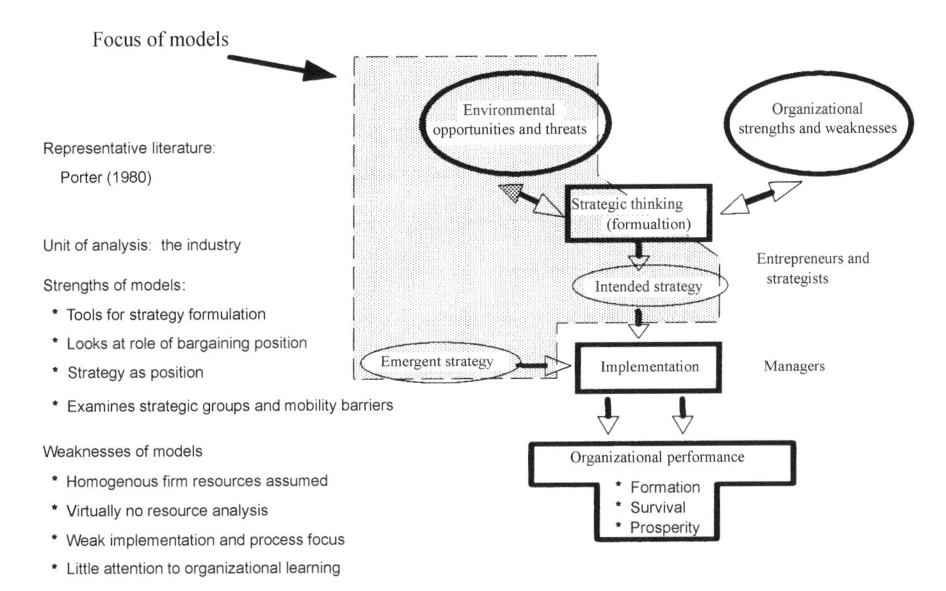

Fig. 2.3. The focus and strengths/weaknesses of competition-based models.

processes such as (4) organizational learning and systems development (Teece *et al.*, 1992).

The Focus of Structural Resource-Based Models

It has been argued that the emerging resource-based set of management concepts and models have roots in, and complement, traditional strategy models and competition-based approaches by focusing on the performance benefits to firms that are able to acquire and build unique, rare, valuable, and not easily imitated resources (Wernerfelt, 1984; Rumelt, 1987; Perteraf, 1993; Barney, 1991; Conner, 1991; Teece *et al.*, 1997).

Kathleen Conner (1991) suggests that a resource-based approach to organization theory and strategic management is "reaching for a new theory of the firm" (p. 143). As Schulze (1992) notes, however, there are different or complementary perspectives on this emerging resource-based theory of the firm. One of these perspectives is the "structural" class of resource-based theory, which is rooted in the industrial economics tradition and which

> emphasizes Ricardian[1] and/or Monopoly Rents and is common in strategy content research (e.g. Chatterjee and Wernerfelt, 1991, 1990). Principal authors include Barney (1986a; 1991), Wernerfelt (1984; 1989) and Dierickx and Cool (1989), among others (Schulze, 1992, p. 5) . . . . the structural model describes economic activity as occurring within efficient markets whose important parameters of behavior are presumed to be known . . . the presence of efficiency rent yielding resources is unlikely and of limited value . . . . Managers of firms operating in this type of economic system should primarily be interested

> in resources whose unique qualities persist in equilibrium (Lippman and Rumelt, 1982; Peteraf, 1993) . . . . An objective of the structural model is to identify a set of rules through which potentially under-utilized [and idiosyncratic] resources can be identified and their capacity to generate rents evaluated (p. 9)

This form of resource-based theory focuses on the impact that idiosyncratic firm attributes may have on a firm's ability to achieve a viable competitive position (Barney, 1991, p. 100), and sustain a competitive advantage through the appropriation of Ricardian rents (Rumelt, 1987). Figure 2.4 illustrates the focus and strengths and weaknesses of the structural resource-based models.

The strengths of the structural resource-based models include: (1) they examine issues related to how firms can sustain their advantages over time through the acquisition and management of unique and scarce resources; (2) they provide a means to analyze the attributes of different resources, which can be used to formulate resource-based strategies; (3) they acknowledge the impact that firm history has on the potential set of future strategic directions, and argue that imitability is a key competitive variable; and (4) competition for resources and factor supplies is considered every bit as important to the long-run potential performance of a firm as is competition for customers.

The weaknesses of the structural resource-based models include: (1) they do not examine the organizational processes by which resources are developed and converted into organizational services (Teece *et al.*, 1992); and (2) they do not identify how resource analysis fits into the larger strategic

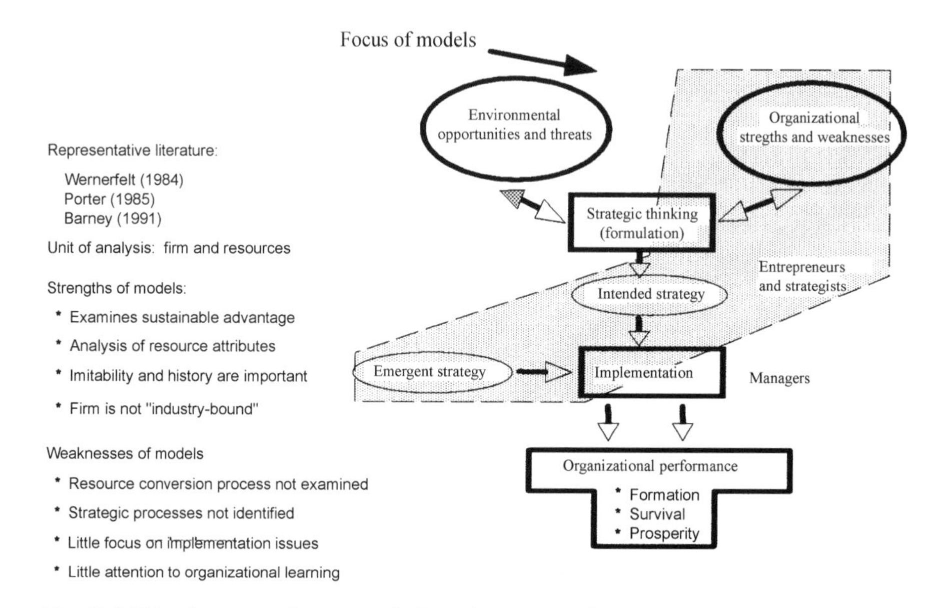

Fig. 2.4. The focus and strengths/weaknesses of structural resource-based models.

management process, particularly with respect to (3) organizational learning and (4) implementation processes (Teece *et al.*, 1992).

The Focus of Process-Oriented Resource-Based Models

Schulze (1992) notes that in addition to the structural, economics-oriented, resource-based theories of the firm, there is a process-based approach that embraces dynamic analysis and focuses on the role of the firm as a creator of resources and unique value, as opposed to just the owner of scarce or valuable resources (Conner, 1991, p. 139). Teece *et al.* (1992) note that the general resource-based perspective, and the process-based focus in particular:

> focuses on strategies for exploiting existing firm-specific assets. However, the resource-based perspective also invites consideration of strategies for developing new capabilities. Indeed, if control over scarce resources is the source of economic profits, then it follows that such issues as skill acquisition and learning become fundamental strategic issues.
>
> It is in this second dimension, encompassing skill acquisition, learning and capability accumulation, that we believe lies the greatest potential for the [*process-based*] resource-based theory to contribute to strategy. (pp. 8–9) [our emphasis]

Schulze (1992) describes the characteristics of the process-based resources theories. They:

> describe economic action as occurring in markets that are less efficient. The parameters which influence the behavior of the economic system are subject to both endogenous and exogenous influence . . . managers of firms operating in this type of economic system are primarily interested in the *process* through which resources can be employed to generate efficiency rents (Reed and DeFillipi, 1990)[2] . . . . it is recognized that wealth can be created through the ongoing process of creating quasi-rents (Grant, 1991) as well as through the process of seeking to maximize returns from idiosyncratic resources (p. 10).

Figure 2.5 illustrates the focus and strengths and weaknesses of the process resource-based models.

The strengths of the process resource-based models, such as those of Penrose (1959), Itami & Roehl (1987) and Prahalad and Hamel (1991) include: (1) they are a blend of structural resource-based models and traditional strategy models. That is, they acknowledge the importance of idiosyncratic resources to firm performance, *and* they focus on the holistic process of ensuring that there is a “fit” between the firm’s resources and its opportunities and threats; (2) they recognize that firms can choose their strategies, and can prepare themselves for change. They recognize that, “strategy analysis must be situational and that there is no algorithm for creating wealth for an entire industry” (Teece *et al.*, 1992, p. 38); (3) they

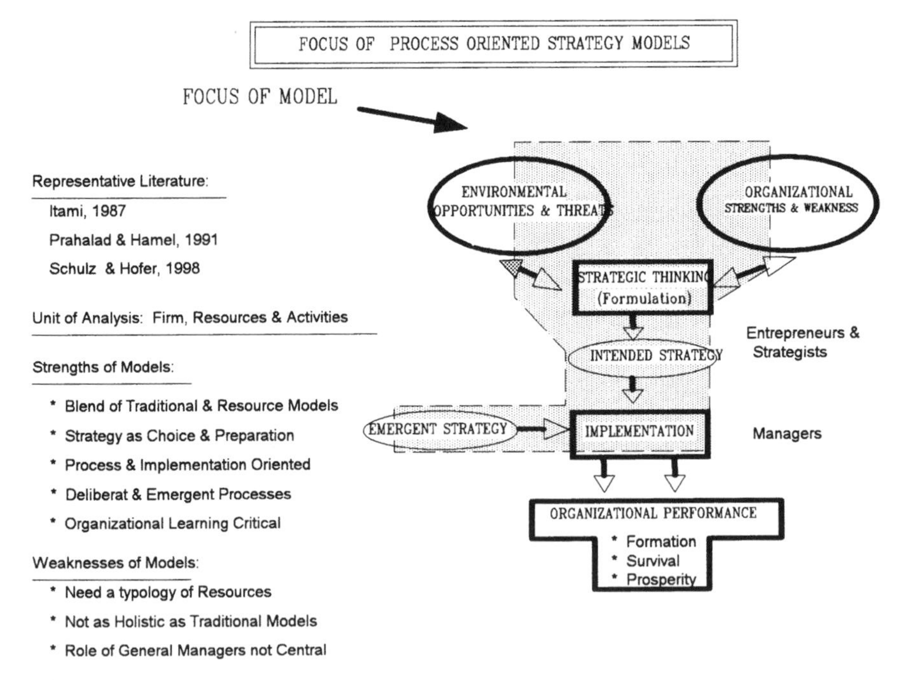

Fig. 2.5. The focus and strengths/weaknesses of process resource-based models.

accommodate both deliberate and emergent strategy processes, and are process and implementation oriented, and do recognize the importance of organizational learning to firm performance (Teece *et al.*, 1992); and finally (4) they have a strong emphasis on the importance of human expertise. The weaknesses of the process resource-based models include: (1) the basic vocabulary for identifying and classifying resources as they are transformed in the value-added process is weak (Conner, 1991); (2) they do not examine external and environmental elements of the strategic "fit" equation as much as either the traditional models or competition-based models; and (3) they do not focus on the role of general managers or leadership as much as the traditional models.

THE IDENTIFICATION OF OPPORTUNITIES FOR RESEARCH

In the previous section, four general approaches to strategic management were identified and their strengths and weaknesses were identified in terms of the elements of a generic model of the strategy process. Overall, from analysis of the extant approaches, it appears that there is a strong need to examine the *process elements* of strategic management, particularly as they relate to: (1) the development process of the firm's resource base (including its skills), which is the foundation for the establishment of a firm's present and future strengths and weaknesses; and (2) the area of *implementation*, particularly as it relates to role of general management and leadership in supporting organizational learning and the development of organizational flexibility. This analysis is consistent with Teece *et al.*'s (1992; 1997) judgment that a process-oriented resource-based approach to strategy, which, "encompasses skill acquisition, learning and capability accumulation", has great potential to contribute to the field of strategy (Teece *et al.*, 1992, p. 9).

A PREVIEW OF THE STRATEGY MODEL DEVELOPED IN THIS BOOK

Figure 2.6 presents a simplified diagram of the strategy model that was developed as a result of the research carried out in this book. This is a resource-based process-oriented "flow" model that examines the roles and relationships among three primary processes that are believed to be fundamental aspects of the ability of a firm to establish sustainable competitive advantage. This model was developed using systematic grounded theory-building methods (Glaser and Strauss, 1967; Eisenhardt, 1989; Yin, 1989) that are discussed in detail in Appendix Two. The model reflects the practitioner-based "working theory", and also integrates more formal academic theory.

The three processes identified in the model include: (1) an *Entrepreneurial "outside-in" process*, which involves the elements by which a firm identifies

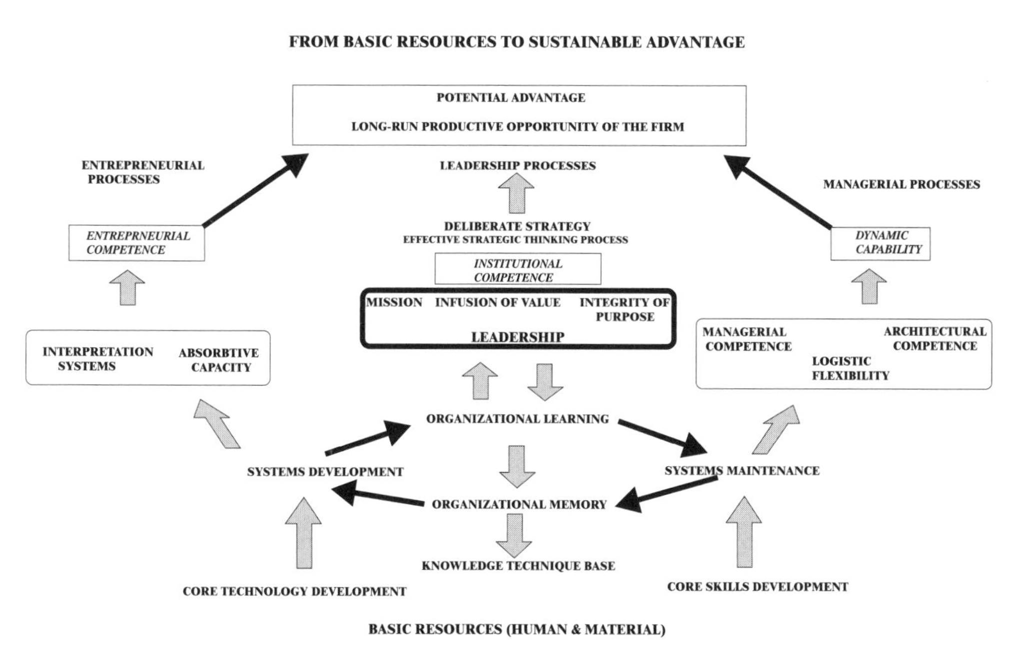

Fig. 2.6. Resource-based model developed in the book.

and interprets opportunities to create value in its current environment or in new environments; (2) a *Managerial "inside-out" process* which involves the elements by which the firm develops the ability to transform its basic skills and material resources into organizational services that can be used to create competitive goods and services for customers (and hence create value); and (3) a *Leadership "integrative" process* which involves the elements by which a firm establishes the ability to integrate its entrepreneurial and managerial competencies, and develops a shared set of organizational values and norms that help it protect its key resources.

This model addresses the weaknesses of many extant resource-based models in that it is implementation oriented and looks at both structural and process attributes of resource management. The model developed in the book supplements traditional models of strategy by acknowledging the key role of "fit" with the environment, and also complements the competition-based models by examining issues related to the development of organizational competencies and the implementation of resource-based strategies.

In the remainder of this book, we will systematically present evidence and analysis that is the substance behind the summary diagram in Fig. 2.6. This summary is intended, at this point, to give the reader a rough "outline" of where we are going with the research.

DETAILED RATIONALE FOR THE RESEARCH

In this section we examine the specific rationale for conducting research in the general area of resource management, entrepreneurship and skill-based strategy in particular.

Research Rationale

There are four major rationales for conducting theory-building research that explores the relationships between the processes of resource development and entrepreneurial choice:

(1) *Prior research suggested that resource-based management issues were an increasingly important factor in understanding the determinants of firm success.*

> Research suggests that resource-based management issues are an increasingly important factor in determining firm success at *both the corporate and business levels of analysis*. Given these findings, it is important that the general topic of resource-based management be explored in further detail.

(2) *Various business theorists and practitioners also suggested that resource management and skill-based strategic thinking were key determinants of firm success.*

> The specific focus on resource management and skill-based strategy processes may help improve business practice. A focus on *how and why* general managers and their firms build and maintain critical skill-based resources over time may be an important factor in the effort to help companies improve their abilities to uniquely create value for current and new customers, thus leading to better long-term performance.

(3) *The mixed results of empirical research in the specific area of distinctive competence suggested that more theory development is needed.*

> Research in the specific area of distinctive competence (which is an element of the resource and skill-based focus of this book) has yielded mixed results on the assessment of the relationship between a firm's areas of competence and performance. The most likely cause of these mixed findings is that the organizational resource aspects of the theories that were evaluated were incomplete or misspecified, or both. A more systematic *theoretical development* approach is required.

(4) *Academic theorists also recognized that there was a need for more of a focus on resource-development and skill-base issues.*

> Theory on the topic of resource and skill development and utilization remains incomplete. Current business-level strategy models, in particular, under-emphasize skill development and utilization issues and do not adequately examine the processes by which organizational skills and competencies are developed, maintained and deployed in an effort to build and sustain competitive advantage.

Each of these major rationales is discussed in more detail below.

(1) *Prior research suggested that resource-based management issues were an increasingly important factor in understanding the determinants of firm success.*

Corporate Level Research

Previous strategic management research at the corporate level of analysis has found that the deployment pattern of skills and resources across the corporation has had an impact on corporate performance. Rumelt's 1974 work on strategy, structure and economic performance suggested that a firm's, "economic performance is more closely associated with the type rather than the extent of its product-market scope, and with the way in which businesses are related to one another rather than their number" (p.

123). Rumelt defined diversification strategy as, "(1) the firm's commitment to diversity per se; together with (2) the strengths, skills or purposes that span this diversity, demonstrated by the way new activities are related to old activities" (p. 11). He found that:

> The Dominant-Constrained and Related-Constrained groups were unquestionably the best overall performers, and both strategies were based upon the concept of controlled diversity . . . *these companies have strategies of entering only those businesses that build on, draw strength from, and enlarge some central strength or competence*. While such firms frequently develop new products and enter new businesses, they are loath to invest in areas that are unfamiliar to management. (pp. 150–151) [our emphasis]

Schmalensee (1985) found that, "knowing a firm's profitability in market A tells nothing about its likely profitability in *randomly selected market 'B'*. This is consistent with the conglomerate bust of the past decade and with a central prescriptive thrust of Peters and Waterman (1982, Ch. 10, p. 349): wise firms do not diversify beyond their demonstrated spheres of competence" (1985, p. 349). This also supports Rumelt's findings.

Farjoun's (1990) research supports and helps further explain the findings of Rumelt (1974, 1982, 1991), Salter and Weinhold (1979), Lecraw (1984), and Palepu (1985) who all found that firms which diversified into businesses that could leverage established competencies outperformed firms that attempted to broaden outside their competence base through acquisition.

Farjoun (1990) examines the concept of "corporate relatedness" from a resource-based perspective, and shows that significant diversification patterns, based on corporate resource and skill attributes, can be identified and are complementary to product diversification patterns. The study also demonstrates that human expertise, or "the types of knowledge, skills, ideas and experience possessed by individuals, teams or larger units in organizations [*skill-based resources*]" (1990, p. 87), is an important factor that can explain diversification, and that human resources are indeed generators of economies of scope (Farjoun, 1990, p. 217).

Business Level Research

At the business level, Hansen and Wernerfelt (1989) found that the internal characteristics of a firm account for nearly twice the variance in the performance of firms than external, market-oriented variables. They argue that, "if our findings of the relative importance can be generalized, it would suggest that the critical issue in firm success and development is not primarily the selection of growth industries or product niches, but it is the building of an effective, directed, human organization in the selected industries" (p. 409).

Chrisman's (1986) findings are consistent with those of Hansen and Wernerfelt. He found that the determinants of business success are different from the determinants of failure. He found that successful firm's tend, through strategy, to match their critical skills with industry key success factors; while firms that fail tend to suffer from organizational problems that hinder their ability to match skills with success factors.

Sousa de Vasconcellos and Hambrick's (1989) research supports Chrisman's findings. They found a "very strong relation between how a firm rates on the industry key success factor index (which they developed in their study), and how it performs". The ability of firms to develop appropriate internal strengths, given environmental conditions, is an important aspect of business level strategy.

Venkatraman and Prescott's (1990) analysis of PIMS data suggests that firms that are able to co-align their resources with environmental conditions, or to find a "fit", outperformed other firms. Within the context of their multi-time-period study (1976–1979 and 1980–1983), they also found evidence that strategies which leverage a firm's resources, in terms of production capacity, employee productivity and ability to produce relatively high quality products, are positively related to return on investment.

(2) *Various business theorists and practitioners also suggested that resource management and skill-based strategic thinking were key determinants of firm success.*

The specific focus on resource management and skill-based strategy processes may help improve business practice. A focus on *how and why* general managers and their firms build and maintain critical skill-based resources over time may be an important factor in the effort to help companies improve their abilities to uniquely create value for current and new customers, thus leading to better long-term performance. Irvin and Michaels (1989), from the McKinsey and Company Consulting Firm, note that, "in many businesses, opportunities to develop novel, and sustainable, sources of competitive advantage are few and becoming fewer. Skill-based competitive advantages, however, generally have significant impact on economic performance and are extremely difficult to replicate" (p. 4). Likewise, McKinsey and Company's Coyne (1986) and others suggest that sustainable competitive advantage can be achieved only if the firm has the ability to generate a "capability gap", rooted in its skill-based resources, that cannot be easily eroded by competitors (Peters and Waterman, 1982; Wernerfelt, 1984; Drucker, 1985; Rumelt, 1987; Barney, 1991).

Klein *et al.* (1991), in a theoretical article in the *Journal of General Management*, based on their years of work in consulting, note that,

> on a practical level, we have increasingly found our high technology clients to be asking, 'What skills, capabilities and technologies should be build up?', whereas traditional formal techniques attempt to answer the question, 'what markets should we enter, and with what products?'

As Joseph Badarocco, among others, observes in his 1990 book, *The Knowledge Link*, "*executives must rethink basic assumptions about strategy*, acknowledging that successful strategies depend upon learning, creating, adapting, and commercializing knowledge and skills" (1991, p.14) (Warner, 1989; Naugle and Davies, 1987; MIT Commission, 1989; Pascale, 1991).

(3) *The mixed results of empirical research in the specific area of distinctive competence suggests that more theory development is needed.*

Extant research studies that have examined the specific relationships between skills and distinctive competencies and firm performance have found mixed results. Table 2.1 identifies seven of the more important and relevant empirically oriented research efforts that explicitly evaluate the role of competence as a variable in explaining business activity and performance. Six of the seven studies directly examine the relationship between a firm's skills and competencies and its performance.[3] Of these six studies, all of them found some evidence of a positive relationship between competency and performance. However, a more careful examination of these studies reveals that nearly 40% of their empirical tests either failed to find any statistical relationship (when one was expected), or found a significant relationship that was opposite to the one predicted by the theories they were attempting to assess.[4]

These mixed findings suggest that the relationship between competency areas and performance are not that well understood. For example, in the 1985 Hitt and Ireland article, engineering (process and value analysis) and research and development competencies were either not related to firm performance or were negatively related to performance across all the general industry and strategy types identified. Clearly, this is an anomaly in light of the success of firms that implement total quality management programs and of the general results of PIMS studies that show that R&D as a percentage of sales is positively related to return on investment for mature industries (Buzzell, 1987).

Likewise, the findings of Hitt *et al.* (1982) that production competencies do not seem to have a significant or positive role in firm performance is suspect in light of the success of Japanese firms in exploiting manufacturing skills to their advantage in many industries.

Further, the general findings by Venkatraman and Prescott (1990), which show that capacity utilization and employee productivity are both significantly and positively associated with performance seem to conflict with the findings of Hitt and Ireland (1985; 1986), who found either no relationship, or negative relationships between engineering R&D (process innovation) and performance. It would seem that both these competence areas play an important part in a firm's ability to improve its capacity utilization and productivity, yet the results indicate otherwise.

how each area of business activity uses these capabilities and skills". Yet, as of early 1998, very few strategic management theorists have systematically addressed the questions of: (1) how and why these skills should be identified, (2) at what level of detail should firm skills be understood, and why; (3) how and why are these skills nurtured and built into competencies and capabilities; and (4) how broadly should the skills and competencies be applied from a strategic standpoint? In short, there is little theoretical guidance in the literature to address such questions about essential skill development and competition.

THE CURRENT RESEARCH SITUATION, OBJECTIVES AND QUESTIONS

The Current Research Situation

Overall, then, the state of academic research and theory at the end of 1997, as well as practitioner needs, all suggest that there is a need for research and in-depth empirical study in the areas of organizational resources and skills, distinctive competence, and skill-based strategy. As of the beginning of 1998, there is little theoretical guidance in the literature to address questions about organizational skill development and the role that entrepreneurs and general managers have in the strategic process of skill development; therefore, there is a need for exploratory process-oriented theory-building research that is grounded in the practice of exemplary leaders and managers, and which seeks to better understand what organizational skills and resources are and how they are developed and managed.

The grounded model and related theory proposed in this research project will hopefully improve and supplement previous models in the field of strategy by: (1) investigating the role of entrepreneurs and general management in making choices that guide which resources, particularly which skill-based resources, a firm will develop and which markets it will use those resources to serve; (2) examining the processes over time by which leaders and managers transform their resources into activities that create value for customers; and (3) identifying the important characteristics and relationships of exemplary resource and skill-based management processes.

Research Objectives, Questions and Tasks

There are *three basic objectives* of this book:

(1) *To develop a more complete understanding of skill-based development issues that is grounded in the experience and understanding of adroit entrepreneurs and general managers.*

This "grounded understanding", in the form of idiographic-level case histories and models, seeks to identify and explain the central processes by which firms convert skill-based resources into sustainable competitive advantage. The in-depth empirical work should serve as a good foundation for improving the practice of entrepreneurial leadership and skill-based strategic thinking.

(2) *To enfold the grounded, idiographic-level understanding developed in the field research with relevant extant literatures and in the process generate an integrated and grounded, substantive-level theory of skill-based strategy.*

This enfolded substantive-level theory should serve as a base from which to conduct future empirical research that examines the theorized relationships between process-based resource management issues and firm performance.

(3) *To discuss the implications of the empirical research presented in this book, as they relate to the improvement of the practice of entrepreneurship and strategic thinking, and as they relate to academic scholarship and government policy.*

A set of research questions and tasks was created for each of the first two primary research objectives. These questions and tasks were used to guide the research design and execution of the project. These questions and tasks are presented and discussed below.

Research Questions and Tasks for Objective One

Q1-A: What language do practitioners use in describing and understanding skills and resource-based phenomena?

Q1-B: What processes are involved in how managers build, nurture, change and deploy skill-based resources?

Q1-C: How are these processes related to one another and to the long-term performance of the organization?

T1-A: To develop an improved vocabulary and set of constructs for describing and thinking about organizational resources and skills, and to generate a process-oriented typology of organizational resources that could be used as a basis for effective strategic thinking with regards to resource deployment, particularly in terms of a firm's skills and competencies.

T1-B: To develop several grounded idiographic-level models of skill-based strategies.

Research Questions and Tasks for Objective Two

Q2-A: What are the primary resource-based process determinants of firm performance and potential sustainable competitive advantage?

Q2-B: How are these determinants related to performance and one another, and why are they important?

T2-A: To develop an integrated, grounded, substantive-level theory of skill-based strategy.

T2-B: To develop an inventory of propositions that specifies the theory in more precise form.

T2-C: To develop a series of diagrams and a hierarchical spreadsheet summary of the theory.

If these objectives are met, the book will contribute to our understanding of: (1) how general managers conceptualize and identify the skills that their firms have and need; (2) how they invest in and build relevant skills; and (3) how they convert these skills into strategic capability and deploy them in an attempt to achieve sustainable competitive advantages. These research questions have served as the primary guides during the development of the research design and the subsequent choice of research methods.

CHAPTER SUMMARY

In this chapter four major approaches to strategic management were identified and their strengths and weaknesses with respect to the elements of a generic strategy process model were discussed. Opportunities for theory-building research were then identified, based on a summary analysis of the strengths and weaknesses of the four strategic management approaches.

It was argued that there appears to be a strong need to examine the *process elements* of strategic management and entrepreneurship further, particularly as they relate to: (1) the development process of the firm's resource base (including its skills), which is the foundation for the establishment of a firm's present and future strengths and weaknesses; and (2) the area of *implementation*, particularly as it relates to the role of general management in supporting organizational learning and the development of organizational flexibility.

A preview of the model that was developed in this book, using systematic, grounded, theory-building techniques, was presented, and four general rationales for conducting theory-building research in the resource management and skill-based strategy area were identified and then discussed at length. The rationales were that:

(1) *Research suggests that resource issues are an increasingly important factor in determining firm success.*

(2) *The mixed results of empirical research in the specific area of distinctive competence suggests that more theory development is needed.*

(3) *Theorists recognize that there needs to be more of a focus on resource development and skill-base issues.*

(4) *The focus on resource management and skill-based strategy may help improve business practice.*

It was argued in the rationale section that, until recently, the resource, skill and competence aspect of the "fit" equation has been relatively ignored in the development of the field of strategic management. As of 1992, very little research has systematically addressed four key issues: (1) how various types of organizational resources are related to one another (Conner, 1991); (2) how leaders and managers responsible for their firms' strategies identify what resources, including skills and competencies, their firms have (Wernerfelt, 1984; Burgleman and Sayles, 1986); (3) how and why leaders and managers build and invest in relevant skills (Schultz, 1961; 1990); and (4) how leaders and managers convert these skills into more general strategic capabilities and deploy them in an attempt to create value in the market-place and achieve sustainable competitive advantages (Lenz, 1980).

Given the basic rationale for the research, the primary research objectives, questions and tasks were then identified. The remainder of this book has been designed to contribute to the understanding of resource management issues in general, and skill-based strategy issues in particular. Specifically: (1) it develops a typology of organizational resources, based on extant academic literatures, that identifies four distinct levels of organizational resources, shows how they are related to one another in a generic resource transformation process, and discusses the concepts of skills and competence as strategic resources; (2) it explores how exemplary entrepreneurs and general managers "in the field" actually conceptualize, identify, invest in, develop, and protect skills and competencies within their organizations (which are an important class of organizational resources) and explores how they provide leadership in developing and implementing *skill-based strategies* which may lead to potential value creation opportunities; and (3) it strives to augment and clarify current theory on the subject of skill-based strategy through the systematic comparison, analysis and integration of academic and practitioner perspectives.

NOTES

1. Ricardian rents reflect the potential above-normal differences in payments received by the owners of a particular resource or service (Penrose, 1959) compared with others that possess similar, but not identical, resources.
2. Efficiency rents reflect above-normal profits received by a firm due to persistent cost advantages relative to competitors.
3. Stevenson's research examined the process by which strengths and weaknesses were defined by managers, and looked at whether such information on strengths and weaknesses was useful in managing the firm.
4. The average percentage of total performance variance explained by the studies that reported such figures was 27%.

3

RESOURCES FROM THE GROUND UP: THE FOUNDATIONS OF PERFORMANCE

> Surely, extensive questionnaires are not required to convince us that the able businessmen are well aware that the more they can learn about the resources with which they are working and about their business the greater will be their prospects of successful action.
>
> Penrose (1959, p. 77)

INTRODUCTION: THINKING STRATEGICALLY ABOUT RESOURCES

If a firm is to survive and grow over time, its general managers must have a clear understanding of what resources are and how they can create or acquire them, and they must also be able to *deploy those resources strategically* with respect to customer preferences and competitive pressures. There are two primary elements of this chapter: first, the authors review what is meant by the concepts of strategy and strategic thinking within the context of the book, particularly as they relate to the deployment of resources. Second, given the argument that was presented in Chapter 2 that there was a need for an inclusive typology of organizational resources that identifies the fundamental types and levels of resources that may exist within an organization, a typology of organizational resources is presented that attempts to fill the holes in resource-based theory that were identified by Yuchtman and Seashore (1967), and others, and which can provide a base for strategic thinking with respect to resources.

The typology examines both content (e.g. what different types of resources exist?) and process issues. This resource typology is considered a "process typology" because in addition to providing a basic layout of resource types, it also shows how resources are transformed from one type to another during the exercise of business activities.

The emphasis on the *process dimensions* of resources and managing resources differentiates this research project from many of the more structural approaches to resource-based theory, which, as Schulze (1992, p. 12) notes:

> emphasizes Ricardian and/or Monopoly Rents and is common in strategy content research . . . the structural model describes economic activity as occurring within efficient markets whose important parameters of behavior are presumed to be known . . . the presence of efficiency rent yielding resources is unlikely and of limited value.

The perspective that guided the development of the resource typology presented in this book was based on what Schulze (1992) describes as the more "process" oriented perspective of the resource-based theory of the firm which embraces dynamic analysis and focuses on the role of the firm as a creator of resources and unique value.

Teece *et al.* (1992) note that the process-oriented resource-based perspective:

> invites consideration of strategies for developing new capabilities. Indeed, if control over scarce resources is the source of economic profits, then it follows that such issues as skill acquisition and learning become fundamental strategic issues. It is in this second dimension, encompassing skill acquisition, learning and capability accumulation, that we believe lies the greatest potential for the [*process-based*] resource-based theory to contribute to strategy. (pp. 8–9) [our emphasis]

STRATEGY: VISION IN ACTION

Strategy and Strategic Thinking as Integrative Tools

It has been nearly twenty years since a group of scholars met at a conference to discuss how business policy could be more effective as an *integrative tool* for general managers. The proceedings of that conference were published as a book edited by Schendel and Hofer (1979). In this book the editors note that:

> New concepts or paradigms are important to scientific progress . . . . the new paradigm we propose for the policy field is 'Strategic Management', and it rests squarely on the concept of strategy . . . . [and] entrepreneurial choice is at the heart of the concept of strategy, and it is good strategy that insures the formation, renewal, and survival of the total enterprise (Schendel and Hofer, 1979, pp. 2–6).

Thus the core element of traditional strategic management thinking, as it initially began, revolved around *entrepreneurial choice*. The essence of entrepreneurial choice revolves around the creation and renewal, response and redirection of a firm's resources in pursuit of opportunity or avoidance of threat (Rumelt, 1987; Ginsberg, 1988). Such choices are often made at critical junctures of a firm's history and require frame-breaking changes in the way leaders and organizations work.

Strategy

As Rumelt (1979, p. 198) notes, "critical or strategic problems . . . lie outside the limits of behavior and may require a redesign of the system. In our biological analogy, strategic issues are ones that lead to extinction unless mutation (or its behavioral equivalent) occurs". Ideally, this "mutation" results from the exercise of entrepreneurial choices, which presupposes that "intended strategies" (Mintzberg, 1987a) are being enacted. Winter (1987, pp. 160–161) emphasizes this point in noting that:

> An organizational strategy . . . is a *summary account* of the principal characteristics and relationships of the organization and its environment – an account developed for the purpose of informing decisions affecting the organization's success and survival. This formulation emphasizes the *normative intent of strategic analysis* and rejects the notion that there are strategies that have "evolved implicitly" (Porter 1980, xiii) or that strategy is a "non-rational" concept.

Note that strategy, in the context of Winter's comments, is a tool that is used to inform decisions affecting the long-term health of the firm: decisions in which leadership commits to a vision (or acquiesces to an inertia) of where the firm should go—and then relies on strategy to get the firm from where it is to where it wants to be.

Strategic Thinking

The actual decisions that are made as to where the firm *should go* are not strategy, *per se*, but can be informed by sound strategic thinking. Strategic thinking is a thinking process by which one attempts to understand, with some certainty, what activities, actions, resources, opportunities, threats, skills etc. are really the most important in a given situation. Rumelt (1979, p. 200) elaborates, and suggests that "a principal function of [strategic thinking] is to structure a situation—to separate the important from the unimportant and to define the critical subproblems to be dealt with". Good strategic thinking can help break complex and messy problems into an array of subproblems that can be understood and solved (Rumelt, 1979; Siffin, 1983).

A firm's strategy is a content and process output of the strategic thinking process—it reflects, as Hofer and Schendel (1978, p. 25) note, "the fundamental pattern of present and planned resource deployments and environmental interactions that indicates how the organization will achieve its objectives".

The comments above reflect the fact that in this book, like in Venkatraman's (1989a) work, strategies are separate from goals and are the outputs of decision-making processes. This is consistent with the views of Hofer and Schendel (1978), Schendel and Hofer (1979) and Mintzberg and Waters (1985). Thus, when the authors speak of strategy in this book, their focus is on the means a firm adopts (i.e. resource deployment patterns) to achieve its desired objectives. The pursuit of the firm's multiple objectives occurs within the context of the overall goals or mission of the organization's leadership (Hofer and Schendel, p. 200).

General Management: Integrating the Parts

In terms of the role that general managers play in the formation and implementation of strategy, it is assumed here that the general managers, as leaders and strategists of a firm, are responsible for providing the "vision" and context for action (Bower, 1970) and values (Selznick, 1957) for the firm, and for being the "creators, or shapers, or keepers of skills" (Peters, 1984), or the "architects of purpose" (Andrews, 1971, 1987) for the firm. As Andrews (1971, 1987, p. 1) remarks, "general management is, in its simplest form, the management of a total enterprise or of an autonomous subunit. Its diverse forms in all kinds of businesses always include the *integration of the work of functional managers or specialists* (our emphasis)".

Strategic thinking and strategy provide general managers with the tools they need to better understand the complex internal and external environments of their firm, and to help them integrate the various parts of the firm into a viable, coordinated whole.

Strategy and Operations: From the Whole to the Parts and Back

This book will focus primarily on how firms have executed strategies, particularly from a skilled resource perspective, and how the skill base and resources of the firm affect future entrepreneurial choice activities. The interrelationship between strategy and operations will be central to the discussion. Benjamin Tregoe and John Zimmerman (1989) provide a lucid insight on this relationship as a continuum:

> We define ... strategy as *the framework which guides those choices that determine the nature and direction of an organization*. We define operations as the day-to-day planning and decision making which guide the processes of development, manufacturing, distribution, marketing, sales, and servicing of an organization's products or services along the way to its customers. It is *how* an organization is run. The continuum begins with getting vision articulated and ends when that vision is an integral part of day-to-day operations.

Strategy is vision in action, according to Tregoe and his associates; it is a process that helps translate entrepreneurial choices, which affect the long-term development of the firm, such as the selection of a "driving force", into action and potentially to competitive advantage (Hofer and Schendel, 1978). The driving force is the governing principle, according to Tregoe and associates, that provides focus in making strategic choices; it is the "central hook for strategic vision" (Tregoe *et al.*, 1989, p. 46) and is the "primary determiner of the scope of future products and markets", as well as firm skills and capabilities (Tregoe and Zimmerman, 1980, p. 40). According to Tregoe and Zimmerman (1988), a firm can define and re-define itself with respect to one of the following eight primary *driving forces*:

(1) **Products offered**: The firm seeks to meet a long-term need in the market by offering leading-edge, differentiated products that add value for the customer. The product scope is typically narrow and the number of customer segments served is typically large and grows concentrically.
(2) **Markets served**: The firm builds unique long-term relationships with a well-defined customer base and attempts to meet the various needs of that customer base through a broad and innovative product mix.
(3) **Technology**: The firm builds its strategy around a unique body of knowledge and technical capabilities from which it develops distinctive competence in exploiting applications of its technologies.
(4) **Low-cost operations**: The firm focuses on process improvements, economies of scale, and other unit cost reduction techniques; it targets markets that are price sensitive and offers the low-price product in those markets.
(5) **Operations capability**: The firm develops a distinctive competency in achieving economies of scope through operations flexibility. The firm targets a broad number of diverse markets and is generally a make-to-order producer.
(6) **Method of sale/distribution**: The firm develops extensive capability in distribution and sales, and attempts to offer either low-cost or differentiated products. It targets markets selectively, given its particular strengths.
(7) **Return/profit**: The firm attempts to realize return and profit measures through diversification and portfolio management.
(8) **Natural resources**: A firm is able to control and develop products based on the ownership of a unique natural resource.

The Tregoe and Zimmerman strategy framework is important because it is implementation oriented and because it is one of the early, and perhaps better, resource-based strategy perspectives in the field. The driving force concept requires that the strategists in a firm recognize the primary role that functional activities and resources play in enabling their firm to produce value, and then build their strategy around the core activities.

RESOURCES DRIVE SCOPE AND MAY BE AT THE ROOT OF COMPETITIVE ADVANTAGE

Ultimately, within the Tregoe and Zimmerman framework, the resources that a firm can acquire and learn to master determines who the firm can serve and how to best serve them. Their perspective focuses on the internal characteristics of a firm, and particularly the skills of the firm (which then determines the firm's abilities to understand external markets), and on the integration processes that are necessary to carry out the challenge of creating value. According to this resource-based process strategy approach, which is shared by others (Penrose, 1959; Itami and Roehl, 1987; Teece *et al.*, 1992), *a firm establishes competitive advantage through its ability to assemble and coordinate its resources and create value for its customers, on a day-to-day level, more effectively than its competitors.*

Interestingly, the definition of "resource" itself indicates both the strategic and skill-oriented nature of resources. *Webster's Third New Unabridged International Dictionary* (1981) contains the following:

> **Resource**–from the Latin, *resurgere*, to rise again; from the Old French, *ressourse*, relief; (1) a new or reserve source of supply or support; a fresh or additional stock or store available at need . . . . (5) capability of or skill in meeting a situation (1981, p. 1934).

Resources are, by definition, those assets which are important and available at times of need, of which all strategic situations, again by definition, are. The skills of the human beings of a firm are the principal resource for a firm, both in terms of enabling it to "meet a situation" and to draw on an "additional stock at need".

Hofer and Schendel (1978) contend that, at the business level, the "level and patterns of [an] organization's past and present resource and skill deployments" (resource deployments), may be more important than the "extent of the organization's present and planned interactions with its environment (scope—including market/product domain) in determining the unique position [the firm] develops *vis-à-vis* its competitors" (p. 25). This "unique position" is called competitive advantage.

This book is designed to explore issues pertaining to competitive advantage that arise from the development, maintenance and changing of the specific skills that are the roots of Porter's "value chain activities" or McKinsey's "business system concept" (Porter 1985, p. 36). We are seeking to establish a more detailed understanding of the process by which firms are limited or augmented by the discrete skills they possess. We are attempting to understand the fundamental resource-based origins of competitive advantage, for as Tilles (1983) notes,

> the essential attribute of resources is that they represent action potential . . . a resource may be critical in two senses: (1) as a factor limiting the

> achievement of corporate goals; and, (2) as that which the company will exploit as the basis for its strategy (p. 85).

In most previous models of strategic management, however, the role of resource analysis has been relatively underemphasized, as was noted in Chapter 2 by Ansoff (1979), Hofer and Schendel (1978), Wernerfelt (1984) and Chrisman (1986). The resource analysis element of the strategic management process, according to Hofer and Schendel (1978), consists of the following activities:

> First, the business should develop a profile of its principal resources and skills. Next, it should compare this resource profile to the key success requirements of the product/market segments in which it competes [or might compete] in order to identify the major strengths on which it can build a viable economic strategy and the critical weaknesses which it must overcome to avoid failure (this pattern of strengths and weakness is called its competence profile). Finally, it should compare its strengths and weaknesses with those of it competitors in order to identify those areas in which it has sufficiently superior resources and skills to create economically meaningful competitive advantages in the marketplace (p. 145).

Hofer and Schendel (1978) note that, "one reason that many strategy formulation models skip the resource profile step in the resource analysis process is the fact that *resources have no value in and of themselves*. They gain value only when one specifies the ways in which they are to be used" (p. 148).

Indeed, in addition to the reason mentioned by Hofer and Schendel above, another more fundamental reason that resource analysis is the weak link in most strategic management models is that no systematic means, or typology of organizational resources, exists by which to understand what resources are and what constitutes meaningful differences among them. It is to the challenge of building an organizational resource typology that can be useful to strategic thinking and management that we now turn.

TOWARDS AN ORGANIZATIONAL RESOURCE TYPOLOGY

It is important to have a clear understanding of the types and levels of organizational resources because they in large part define the *productive opportunity* of a firm, which comprises all of the productive possibilities that its "entrepreneurs" see and can take advantage of (Penrose, 1959, p. 31; Tilles, 1983; Itami, 1987). Further, rare and valuable resources are as much a part of the basis for competition and cooperation among organizations as are the products firms produce (Yuchtman and Seashore, 1967; Pfeffer and Salancik, 1978). Yet, as Wernerfelt (1984) notes, "nothing is known, for example, about the practical difficulties involved in identifying resources (products are easy to identify), nor about to what extent one in practice can combine capabilities across operating divisions . . . " (p. 180).

In the next subsections a brief summary table that identifies important citations that discuss organizational resource *types* and **attributes** will be presented, and the elements of an organizational resource typology that builds upon and integrates the previous literature will be identified.

Synopsis of Resource Types and Attributes Identified in Literature

Table 3.1 provides a synopsis of the *types* of organizational resources [printed in *italic*] that may exist in organizations, and the **attributes** of those resources [printed in **bold**] as they have been identified by authors in the fields of economics, human resource management, organization theory and strategic management. Each citation provides some unique element to our understanding of organizational resource types and attributes. Most of the works were written to achieve a distinctive objective, and hence there is little cross-referencing or integration among the citations. In total, however, they provide an excellent base from which to construct an integrative resource typology.

Overview of the Resource Typology

The resource typology that is presented in this section uses the resource types and attributes identified in Table 3.1 as "raw materials" and organizes them in a theoretically and practically useful and coherent fashion. Four hierarchical and interrelated, process-oriented levels of resources can be identified in the literature. The hierarchical relationships mean that as one moves "downstream" towards more complex bundling of basic resources, there is still a "dependency" on the basic resources that precede. The four levels include: (1) a basic resource level, (2) a systemic resources level, (3) a services resource level, and (4) a stock resources level. A simplified illustration of the four resource levels and their relationship to one another is presented in Fig. 3.1.

If one considers that most resources are "indivisible" in one way or another, then it is appropriate to define the *basic resource level as that level that contains the "least reducible human and material resource elements"*. In the case of human-based resources this defines individual human beings and their skills. In the case of material-based resources, this includes all financial resources and any single "least reducible" material items that might fall under the categories of "realty", or "equipment".

When "basic level" resources are combined into aggregates, then one is no longer dealing with "least reducible components", but rather with systems. *The "systemic resource level" includes systems aggregates of the basic resources*. Most basic resources are amenable to packaging and assembly into larger, systems applications. Individuals join teams and equipment is assembled into coordinated facilities.

Table 3.1. Synopsis of literature on resource types and attributes.

Authors	Primary *types* and **attributes** of resources
Ansoff (1965)	*Facilities and equipment*; *personnel*; *organizational capabilities* (standards, procedures); *management capabilities* (experience, decision-making)
Barney (1991)	*Physical capital* (plant and equipment, geographic location, access to raw materials); *human capital* (training, experience, judgement, insight of humans in firm); *organizational capital* (reporting structure, planning, coordination, control systems); **resource mobility**, **homogeneity**, **availability**, **value**, **imitability**, **causal ambiguity**, **social complexity**, **substitutability**
Becker and Gordon (1966)	**General**, **specific** resources. A resource is specific if it is earmarked for a particular use; general if its use is undetermined.
Dierickx and Cool (1989)	*Stock* and *flow* assets or resources. Important attributes include: **time compression diseconomies; asset mass efficiencies; interconnectedness of asset stocks; asset erosion; causal ambiguity**
Fossum *et al.* (1986)	*Human resources (capital)*; **general human capital**, knowledge and skills of equivalent value to many employers; **specific human capital**, consists of knowledge and skills which are of value to a single (or select set) of employers
Hofer and Schendel (1978)	*Financial* (cash flow, debt capacity, new equity); *physical* (office, plant and equipment inventories, warehouse and distribution facilities; *human* (scientists, engineers, sales, analysts); *organizational* (quality control, cash management systems); *technological* (high-quality, low-cost plants)
Itami and Roehl (1987)	*Financial*, *people*, *physical*, *invisible resources*, information or knowledge-based
Penrose (1959)	**Resources**: bundles of potential services that can be defined independent of their use. *Services*: specific applications of resources. *Physical resources* (plant, equipment, land, natural resources, raw materials, semi-finished goods, by-products and finished products); *human resources* (unskilled and skilled labour, clerical, administrative, financial, legal, technical and managerial staff)
Pfeffer (1987)	*Political* resources (skills) become critical if a firm wishes to build inter-organizational power.
Wernerfelt (1984)	**Attractive resources**. *Machine capacity, customer loyalty, production experience, technological leads*
Yuchtman and Seashore (1967)	*Human energies*. **Scarce** and **valuable** resources. Resource **liquidity**; **stability**, or storage without depreciation; **relevance**, to the extent resources are capable of transformation and exchange; **universality**, degree of demand; **substitution**

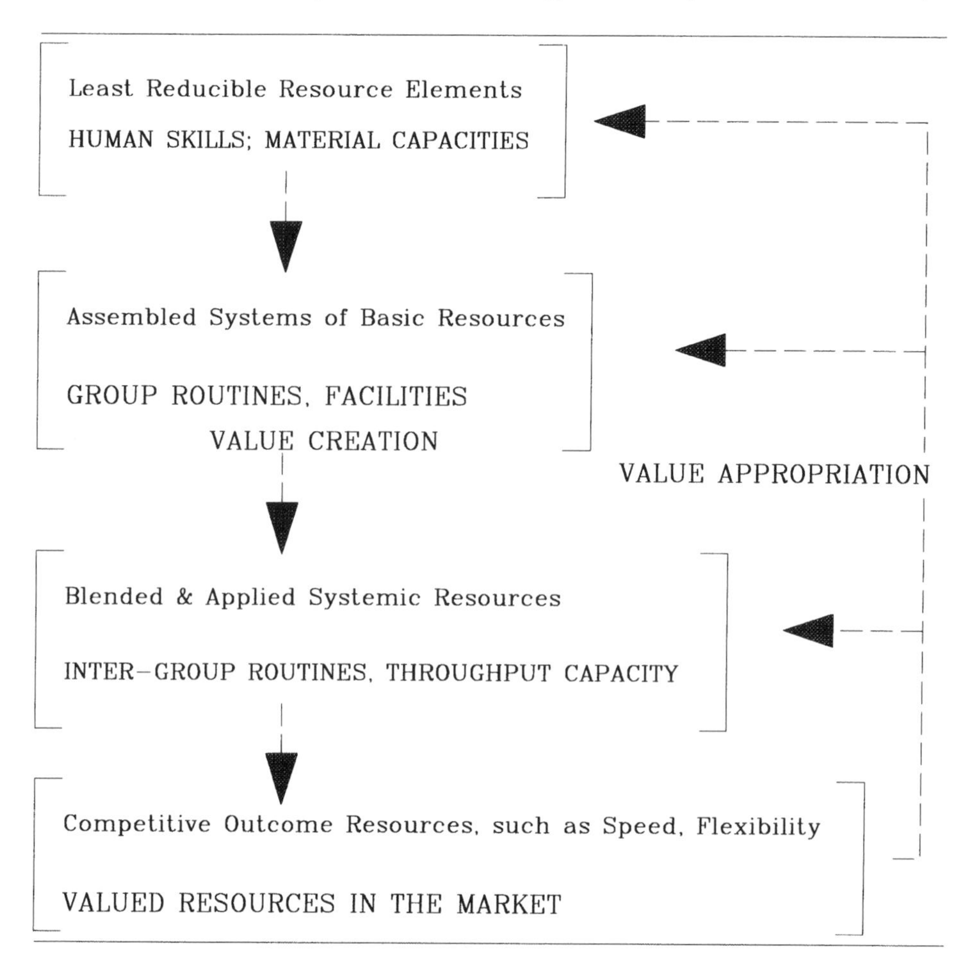

Fig. 3.1. A simple process and resource levels typology.

Taken together, "basic" and "systemic" resources are the building blocks that any organization must have in order to function. Yet, as Edith Penrose (1959) recognized, they provide only the base from which the "productive opportunity" of a firm is ultimately defined. Penrose outlined a theory of the growth of the firm that, in large part, was built around the concept that the firm was a "collection of productive resources, the disposal of which, between different uses and over time, is determined by administrative decision" (p. 24). For Penrose it is critical for the "entrepreneurs" and "managers" of a firm to be able to *translate basic resources into viable and needed organizational services*, which could then be used to produce specific stock assets. For:

> Strictly speaking, it is never resources themselves that are the 'inputs' in the production process, but only the services that the resources can render . . . . the important distinction between resources and services . . . lies in the fact that resources consist of a bundle of potential services and

> can, for the most part, be defined independently of their use, while services cannot be so defined, the very word 'service' implying a function or activity. As we shall see, it is largely in this distinction that we find the source of the uniqueness of each individual firm. (1959, p. 25)

Organizational services, are really a class of organizational resource—they are "value added activity" (Porter, 1985), or "process" resources that are part of the transformation of "basic resources" into "stock asset resources" (Dierickx and Cool, 1989). Services are basic and systemic resources in action. As Penrose notes, "although the 'outputs' in which the firm is interested are productive services, it is [basic and systemic] *resources* that, with few exceptions, must be acquired in order to obtain services" (1959, p. 67). The productive services of an organization really determine the "essence" of the firm. They are where value is created and customer needs addressed.

Finally, organizational services, when appropriately managed, generate "stock asset resources". Dierickx and Cool (1989) suggest that "*'stock resources', like Itami and Roehl's (1987) 'invisible assets', are critical resources that are generally accumulated rather than acquired in "strategic factor markets"* (Barney, 1986b). General stock asset resources include: commitment (Ghemawat, 1991), shared value (Pascale and Athos, 1981), vision, trust, goodwill, experience, organizational memory (Walsh and Ungson, 1991), productivity, flexibility, speed, creativity, quality (Garvin, 1988), brand recognition, patents, absorptive capacity (Cohen and Levinthal, 1990), liquidity, proximity, capacity, finished goods inventory, organizational slack (Bourgeois, 1980a), and organizational throughput (Goldratt, 1990). The accumulation of stocks occurs through the investment in basic and systemic resource "flows", which are then converted by the service-level resources into stocks.

The level of a firm's stock asset resources depends on a firm's commitment to acquiring and deploying more elementary resources, such as those described at the basic and systemic resource level. Stocks cannot be built instantaneously, and Dierickx and Cool argue that "appropriate time paths of relevant flow variables must be chosen to build required stock assets".

DETAILS OF THE ORGANIZATIONAL RESOURCES TYPOLOGY

In the next subsections, the details that underlie the typology of resources that was just introduced are presented.

Basic Level Resources

One of the fundamental tasks of management is to ensure that the organization is properly staffed and that it has the requisite basic resources to execute its strategy. If one considers that most resources are "indivisible" in one way or another, then it is appropriate to define the *basic resource level as that level that contains the "least reducible human and material resource elements"*. In the case of human-based resources this defines individual human beings and the skills they have embodied. In the case of material-based resources, this includes all financial resources, and any single "least reducible" material items that might fall under the categories of "realty" or "equipment".

Human-Based Resources

Basic human-based resources derive from the skills of individuals and are often referred to as human capital. Most economic theories of the firm, and opportunity-based strategy models (Porter, 1980), slight the role of human capital in the analysis of firm activities and performance (Teece *et al.*, 1997). Theodore Schultz (1961), however, has been an outspoken economist on the critical role that human resources ultimately play in the economy. He comments:

> Although it is obvious that people acquire useful skills and knowledge, it is not obvious that these skills and knowledge are a form of capital, that this capital is in substantial part a product of deliberate investment, that it has grown in Western Societies at a much faster rate that conventional (nonhuman) capital, and that its growth may well be the most distinctive feature of the economic system. (Schultz, 1961, p. 1)

Two primary kinds of human capital investment are possible: (1) investments in general human capital, where the knowledge and skills of an individual are of equivalent value to many employers; or investments in (2) specific human capital, where the knowledge and skills of an individual are of value to a single or very select set of employers (Fossum *et al.*, 1986). Skills, knowledge and abilities in humans, which are valued in the economy, are the stuff of human capital.

Skills: Basic Yet Strategic Resources

Resources at the basic level are considered "indivisible". In the case of human resources this means that the basic human resource element is the individual human being, who has skills borne of natural aptitude, practice and experience. Ultimately the skills of a productive organization's members determine the success or failure of the firm to survive and grow. Yet few organizational scholars have *critically inspected the strategic role* that skills play in the development and effectiveness of a firm, although some theory and research does exist (Tregoe and Zimmerman, 1980, 1988; Pascale and Athos, 1981; Peters and Waterman, 1982; Peters, 1984; Nonaka and Johansson, 1985; Itami & Roehl, 1987; Naugle and Davies, 1987; Butler, 1988; Lengnick-Hall and Lengnick-Hall, 1988; Klein *et al.*, 1992).

Most work in the area fails to make connections between how individual skills and abilities are ultimately translated into strategic capability (Lenz, 1980). Rumelt (1974), for example, claims to use the "range of skills possessed corporately by a firm" (p. 11) as a point of departure in identifying the existence and type of firm diversification:

> The concept of "diversification strategy" . . . is defined here as (1) the firm's commitment to diversity per se; together with (2) the strengths, skills or purposes that span this diversity, demonstrated by the way new activities are related to old activities. (p. 11)

Yet despite having said this, Rumelt goes on to operationalize his diversification concept, through the refinement of a "specialization ratio" and a "related ratio", in terms of end-product revenues, which reflect a bias towards market definitions of the firm rather than skill definitions. Rumelt does not systematically define or identify specific skill-based connections that define how and why "new activities are related to old". This weakness, in part, stems from the fact that there was, and still is, a thin base of research and theory on organizational skills from which Rumelt could have built upon; and skill definition was not his primary task.

In fact, in the field of strategic management "skill" as a concept has received virtually no explicit attention, with the exception of Hofer and Schendel's (1978) brief discussion of "skill profiles", Pascale and Athos's inclusion of "skill" as one element in the "Seven-S" managerial framework, Peters' (1984) work on the development of "distinctive skills", Itami's inclusion of "skill" as an "invisible asset", and Naugle and Davies' (1987) work on "strategic-skill pools [*skill-sets*] and competitive advantage". There is still a need for an explicit analysis of skill as an organizational concept within the strategic management context (Ansoff, 1978; Wernerfelt, 1984; Chrisman, 1986).

A Definition of Skill

In order to begin understanding the link between these concepts, let us turn to an etymological analysis of the word skill:

> D1: *Skill: From the Old Norse (before 1500) word SKIL, meaning "discernment in knowledge". Later, it became "proficient in a specific art, trade, technique". From the Latin PERITIAE, meaning experience, practical knowledge, familiarity with.*

Three points need to be emphasized:

(1) The appropriate referent, or level of analysis, in measuring and discussing skill, *per se*, is the individual—in relation to a specific art, trade or technique. "Marketing", for instance, is a general technique, and as such is not a skill: it is a service-level resource that reflects the skills of many individuals working in concert with material resources. "To video edit local television advertising", on the other hand, is a far more

specific technique, and may be considered an appropriate skill referent. In general, skills are described by specific action verbs, as they apply to individuals.

(2) "Proficiency", which comes from the Latin *PROFICERE* (to advance, to gain, which is the same root of "profit"), implies that skill results from "expert facility in a given art or branch of learning". Learning and improvement in skills become potentially important organizational investment tools, as expertise is valued in the economy.

(3) Skill is associated with experience and practice—it is manifest in the timely application of specific tasks by individuals (Barnard, 1938, pp. 127–128). Skill can be cumulative, and its application is dynamic. Itami & Roehl (1987) refers to skill as an "invisible asset" that resides, as we have noted also, in the personnel of the firm. It is "invisible" because its source is knowledge, information and experience, which is not apparent:

> Much of the invisible assets of the firm are embodied in people . . . it is of course impossible to separate people from the invisible assets they carry. Engineers store technical knowledge in their brains; workers acquire skills and savvy on the job. People are important resources, not just as participants in the labor force, but as *accumulators and producers of invisible assets*. (p. 14)

As an "invisible asset", Itami argues that skills (embodied in people) provide the fundamental platform for competitive advantage:

> Invisible assets are the real source of competitive power and are the key factor in corporate adaptability for three reasons: they are hard to accumulate [and thus imitate], they are capable of simultaneous uses, and they are both inputs and outputs of business activity (p. 12–13).

Types of Skill

Figure 3.2 illustrates the categories and types of skill that help define the human capital component of basic human resources. The list of the specific skills in the figure comes from an integration of Katz's (1974) skill typology, the skill classification system employed by the US Department of Labor's *Dictionary of Occupational Titles* (1991), and the perspectives of the practitioners studied in this project.

An individual embodies a mixture of two basic skill types: (1) knowledge skills that are borne from "knowing, or seeking to know" (gnostic skills), which include spiritual and conceptual skills (Maier, 1965; Katz, 1974; Herron, 1990); and (2) skills that are borne from "action and habit" (practic skills), which include administrative skills, political skills (Pfeffer, 1987) and technical skills (Katz, 1974), which are related to a specific "technique or art".

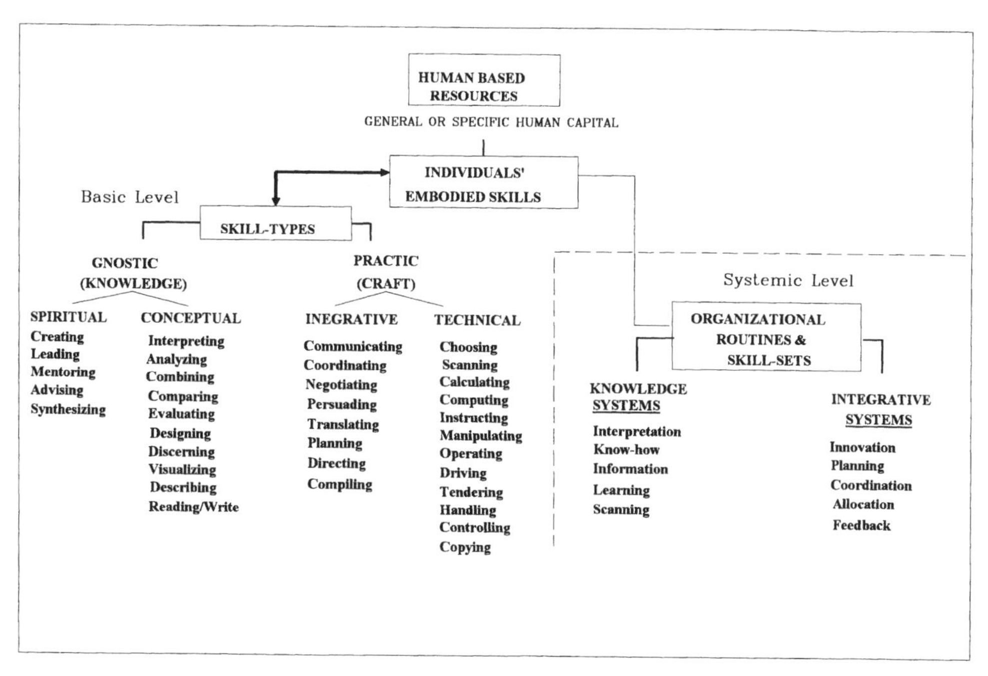

Fig. 3.2. Detailed explosion of basic human resource elements.

Gnostic skills relate to those skills inherent in "seeking to know and discover new things". The "spiritual" aspect of these types of skill relates to the "breathing of life" meaning of the word spiritual. Thus, "creating, leading, mentoring, advising and synthesizing" are all skill-types that involve creating and being part of the creative process. In the case of mentoring and advising, what is being created is a new organizational personality. Gnostic skills are the key to the innovation process in firms and to entrepreneurial leadership, and reside only in human resources, as Amendola and Bruno (1990) recognize:

> the human resource is the key of innovative outlooks: it is the *reservoir of potential capabilities* which can create new and different productive options. In this light, it does not appear as an input to be passively associated with given productive structures, and whose relevant aspect is then availability in the right amount and proportion, but as an asset made up of *skills* which are the end result of a sequential process of specialization. *As an asset rather than an input, labor is no longer something to minimize or get rid of, but to maintain and possibly enrich.* (p. 428) [our emphasis]

Practic skills, or "craft skills", on the other hand, are far more rooted in "action and habit", and are the fruits of experience, training, practice and "learning by doing" (Dutton and Thomas, 1985). The key to the creation and maintenance of practic skills in an organization is the recognition that organizational learning must first begin with learning by individuals. As

Garratt (1987) says, "the only resource capable of learning in an organisation are the people that comprise it" (p. 42).

The potential of human capital, though recognized by Human Resource Management scholars, has not been central to, or well integrated into, many of the strategic management models in use today (Ohmae, 1982). Like Ohmae (1983), who discusses the central focus of Japanese managers on its key resources: the first is "Hito" (people), and then "Kane" (money) and "Mono" (plant), Garratt (1987) and Warner (1986) emphasize that investments in human resources that bear human capital may be the most critical strategic issue for firms, because investments in human capital exploit economies of scope and make physical capital more effective than it might otherwise be. Garratt (1987) notes that:

> future investment in plant will need to be balanced, *or subservient to, investments in to skilled designers, thinkers, engineers, artisans, and administrators who will be the "learners" of the organisation* . . . acceptance of this idea transforms fundamentally the position of people in the previously accepted hierarchy of organisational values and investments (pp. 42–43).

And Warner (1989) highlights the fact that "human resources investment supplements the physical capital cost. There is, however, the following *paradox of human resource investment*, for the full effectiveness of physical capital investment may not be realized if skilled labor is not deployed and trained appropriately . . . (p. 285)".

Material-Based Resources

Material-based resources include two primary types, as illustrated in Fig. 3.3: (1) financial, which includes monetary resources such as cash, receivables and marketable securities, and credit resources; and (2) physical resources, which can be broken down into "realty" and "equipment" categories. Realty refers to the resources an organization owns or controls that are related to location and access of land properties and individual buildings, rights to minerals and energy, and to any resources that the organization has purchased or built that are the raw materials for the firm's value added processes. Purchased parts, for example, are items that will go into an assembly or manufacturing process and will become part of a finished product. Until they are converted into a finished product, they are basic, physical, realty resources.

Equipment refers to any item that may be purchased or built as part of the value-added assembly, production or support systems of the organization. Equipment resource items, such as machine tools, are used to alter the nature of a product or service as it moves through the value-added chain of the organization. They may also be used, like many computers, to support the primary value-creation activities. In all cases, equipment resources

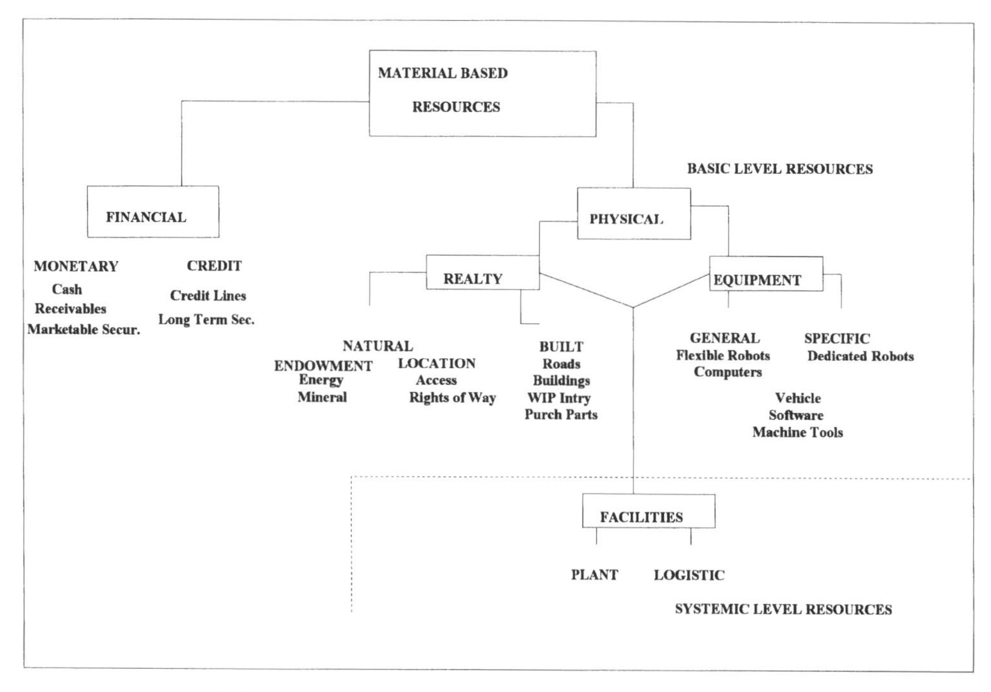

Fig. 3.3. Detailed explosion of basic material resources.

differ from realty in that they are resources that are used to add value to a product or service, whereas realty resources provide the material setting to, and in which, the value added process takes place.

THE SYSTEMIC RESOURCE LEVEL

When basic level resources are combined into aggregates, then one is no longer dealing with least reducible components, but rather with systems of basic resources tied together. The systemic resource level includes systems aggregates of the basic resources. Refer back to Figs. 3.2 and 3.3, which also list general systemic resources.

Aggregating Human Skills: Skill-sets and Organizational Routines

On the human-based resource side the systemic analogues to individual skills are organizational routines (Nelson and Winter, 1982; Teece, 1982) and skill-sets (Naugle and Davies, 1987), which consist of groups of individuals and the mechanisms by which they communicate and make decisions. These mechanisms are systemic resources and are manifest in integrative, administrative (Fayol, 1930; Bower, 1970) and political systems, and in organizational knowledge systems (Daft and Weick, 1984; Winter, 1987; Badarocco, 1990; Walsh and Ungson, 1991).

At the systemic level, basic resources are assembled and aggregated into more holistic forms that in turn can be applied to create organizational services. If skill, as a concept, refers to individual human resource attributes, then we must examine intermediate concepts that address group-level human resource attributes.

A Definition of Skill-Sets

Naugle and Davies (1987) explore the idea of "strategic skill-pools (SSPs)" and consider how these "groups of functional talents" can be translated into competitive advantage. They associate SSPs with "functional" skills, but here "functional" does not necessarily connote "functions" like production, finance, or manufacturing for example:

> The strategist must *look deeply into the function to identify the key skills* which could allow the manufacturing organization to provide a competitive edge to more than one business. (p. 38) [our emphasis]

Rather, these key skills reside in skill-pools, or groups of individuals. It is not clear whether Naugle and Davies consider it necessary that a skill-pool resides within the boundary of a traditional "function", so in an effort to avoid misinterpretation it is preferable to think of skill-pools as "*skill-sets*". The word "set" implies only that there are multiple people, working together, who are utilizing their skills towards some shared purpose. It remains an empirical question and task of this book to learn more about the particular nature of these sets, including how permeable and fluid the boundaries of these sets are (Hamel, 1991). Skill-sets are very similar to what Nelson and Winter (1982) and Teece (1982) call organizational routines:

> In a sense, organizations 'remember by doing'. Routine operation is the organizational counterpart of the exercise of skills by an individual [Nelson and Winter 1982]. Thus routines function as the basis of organizational memory. To utilize organizational knowledge, it is necessary not only that all members know their routines, but also that all members know when it is appropriate to perform certain routines. (Teece, 1982, p. 44)

At minimum, skill-sets are a more general concept than skill—which reflects the characteristics of an individual. Skill-sets represent a grouping of skilled individuals—which may or may not cut across traditional "functional boundaries".

Material Systems Resources

On the material-based side, facilities are classified as systemic resources. Facilities represent the systems aggregation of basic physical resources—the combination of location, groups of buildings and equipment, designed to work as the physical part of a value added system, may be a unique organizational resource—if the facility represents more than the sum of its basic resource parts.

THE ORGANIZATIONAL SERVICES LEVEL

Taken together, basic and systemic resources are the building blocks that any organization must have in order to function. Yet, as Edith Penrose (1959) recognized, they provide *only the base* from which the productive opportunity of a firm is ultimately defined. Organizational services, are really a class of organizational resource—they are value-added activity (Porter, 1985), or process resources that are part of the transformation of basic resources into stock asset resources (Dierickx and Cool, 1989). The types of organizational services that a firm may produce include: *Primary creation services*, which include all those specific activities that directly involve the design and ultimate delivery of a good to a customer, such as value analysis activities, product development and manufacturing, sales, and distribution. *Information support services*, which include those general activities that involve the generation and dissemination of data and information within the organization, and include scheduling activities, environmental scanning activities, and boundary spanning activities; while *technical support services* include those general activities that are needed to support the primary value-added processes, such as procurement, logistics, maintenance and testing.

Competence: Measuring the Effective Deployment of Systemic Resources to Create Services

Within the context of the process-oriented resource typology discussed so far, we have examined the concepts of skill and skill-sets as components of human resources and human capital. Skills reside within an individual, who is a basic resource, while skill-sets reside within a group of individuals, which are a systemic resource.

At the organizational services level of resource definition there is also a critical human capital resource—competence—that must be addressed more fully, as it reflects the higher order abilities of the organization's personnel to blend and coordinate basic and systemic resources into potentially valuable organizational services and stock level resources.

Extant Uses of the term "Competence"

The terms "*competence*" (Selznick, 1957; McKelvey, 1982; Reimann, 1982; Naugle and Davies, 1987; Cleveland *et al.*, 1980; Reed and DeFillipi, 1990), "*core competence*" (Prahalad and Hamel, 1990;

Waddock, 1999; Collis, 1991; Hamel, 1991), and "*distinctive competence*" (Selznick, 1957; Andrews, 1971, 1987; Hofer and Schendel, 1978; Snow and Hrebiniak 1980; Hitt *et al.*, 1982, 1985, 1986; Chrisman, 1986; Stoner, 1987; Fahey, 1989; Curran and Goodfellow, 1990), have all been used in a myriad of strategic management contexts, and each term, and how it relates to the others, has been given a variety of meanings by different authors. The progress of the field, in terms of how we understand and define these terms, appears to be at a pre-paradigmatic stage (Schendel and Hofer, 1979; Freeman and Lorange, 1985; Ansoff, 1987). There is little theoretical or conceptual integration, as Cleveland *et al.* (1990) note:

> theoretical knowledge about production competence as an element in a broader concept (such as operations strategy) *or as a concept in and of itself* is tenuous and inconclusive. (p. 658) [our emphasis]

Take the definitions and relationships between distinctive competence and competence, for example: Hitt and Ireland's 1985 definition of distinctive competence represents the mainstream approach to the concept, and reflects Snow and Hrebiniak's (1980) interpretation of Selznick's 1957 introduction and use of the term:

> A distinctive competence represents those activities in which a firm, or one of its units, does better relative to its competition. (p. 273)

This definition is incomplete in that it fails to connect backwards to inform one what "competence" is, and does not provide one with a sufficient basis to distinguish it from the concept of "competitive advantage".

Fahey and Christensen's 1986 discussion of how firms can build distinctive competencies into competitive advantage builds on Hitt and Ireland's theme, improving upon it, and yet differs in that it emphasizes that competence is a capability rather than an activity:

> Every successful company has found at least one way to out-perform its competitors. This capability, whatever it may be is called a *distinctive competence*: that is, the ability of a firm to do something better than competitors .

Fahey and Christensen's definition also fails to integrate backwards and inform one as to what "competence" means in the term "distinctive competence".

Cleveland *et al.* (1989) is one of the few works that does examine "competence" as a standalone concept, emphasizing that competence is more related to the effective execution of strategy than it is a set of activities a firm performs better relative to competition:

> We see competence as a variable rather than a fixed attribute . . . . As the research progressed, competence emerged as a highly complex issue. It is not a simple, outstanding strength or attribute, but a manufacturer's overall ability to support and prosecute the business strategy. (p. 658)

Cleveland *et al.*'s discussion of "competence" leads to a fundamentally different perspective on the concept than one gets from Fahey. Competence is more an internal measure of effectiveness, linked directly to strategy, than it is a simple comparison of attributes.

Overall, there is a disconcerting inconsistency in how organizational scholars define and utilize the general term "competence", and how this term relates to others, particularly "distinctive competence". On the one hand competence is seen as a set of activities or skills, while on the other hand it is the outcome of a process of learning and coordination of such skills. Some scholars define competence independent of strategy, while others think that it is a reflection of the ability of a firm to pursue some given end. Most work in the area has not explicitly identified the conceptual links among the terms, nor has it attempted to connect them theoretically to other concepts such as skill or capability.

The remainder of this section explores the definitions of the concepts "competence", "core competence", "distinctive competence", and "capability", and lays a general foundation for their theoretical interconnection and for understanding their role in a process-oriented resource typology.

Competence: Further Review and Synthesis

In order to understand a concept or term at a general level it is often helpful to go outside the standard lexicon of those who use it for specific purposes; and it is also helpful to look backwards at the root meanings of a word, to trace its etymology. So that is where we begin:

> D2: ***Competence****: Middle French (1400–1600), from COM (together) and PETERE (to go, to seek or strive): a coming together to seek. From the Latin COMPETENS (principle part of COMPETERE, to be suitable). Coming together suitably. From the Classic Latin period (80 BC to 200AD), FACULTUS (opportunity, feasibility, capacity).*

From the brief etymology above, competence reflects a coming together of qualities, or resources, in such a way that it is adequate for a given purpose. If one examines the word competence in light of the Classic Latin *FACULTUS*, to which it is related, it also means being facile in the pursuit of opportunity.

This suggests that competence may be defined generally as:

> D3: ***Competence***: *the adroit blending of resources and skills to adequately pursue a particular opportunity.*

This is in the spirit of the definition provided in the Second College Edition of *The American Heritage Dictionary* (1982):

> D3a: ***Competence***: *1. the state or quality of being competent (properly or well qualified), 2. Sufficient means for a comfortable existence.*

It is important to recognize at least four facets of the definition (D3) of competence, particularly as it relates to organizations:

(1) Competence can only be properly identified and evaluated with respect to the specific purposes or opportunities a firm is employing its resources to pursue. That is, competence cannot exist without reference to some intendedly purposeful direction.

(2) Competence refers to the ability of a firm to build, maintain and coordinate resources in such a way as to "make it appear without effort (facile)". This implies that competence requires a mastery of both general organizational skills, such as communications, and planning, as well as a mastery of the specific craft-skills within each resource area. In other words, skills drive competence (Peters 1984), and, as Reed and DeFillipi (1990) have noted, the source of a competency is always internal to the firm—it is skill.

(3) If one accepts the above implications, competence is a concept that indicates how *effective* a firm's managers and leaders are in deploying specific skill-based and material assets in pursuit of a given opportunity. As such, the appropriate level at which to measure a firm's competence is at the "functional" or "operating unit" level, which is the level at which individual skills are deployed in pursuit of organizational ends (Cleveland *et al.*, 1989). Competence is a non-financial measurement of "operating unit" skill utilization or effectiveness. Similarly, McKelvey (1982) concludes that "competence" is an excellent term to convey the idea that [his organizational species or classification framework] is based on differentiating organizations in terms of how they arrange their affairs so as to compete and survive (p. 171).

(4) The issue of examining "competence" in light of "competition" leads to an interesting etymological point of comparison. The word "compete" also comes from the Latin root *COMPETERE*, to strive together. It is closely associated with the word "competence" in that "competition" means, among other things, "a contest or similar test of skill or ability". This suggests that a firm's "competitors" are any outside forces that are striving to deploy similar resources or "to be competent" with regard to a shared opportunity-set. From a strategic management point of view, emphasizing the "competence" base of "competition" directs one's attention more to the resource side of the firm than the market side, when considering the attributes of the competitive environment (Day, 1981). This conceptualization compels one to address a few strategic questions: (1) upon what base(s) of competence can or should the firm add value in its operations? (2) Should there be one skill-set that a firm is extra-competent at managing so that it dominates or paces all others? (3) How can the firm's resources best be structured in order to ensure their most effective deployment? In order to be able to address

such questions we need to expand our discussion to include the definition of "core competence".

Core Competence

In 1990 Prahalad and Hamel introduced the concept of "core competence" as part of a "re-thinking of the corporation" (p. 80) that proposes that the development and consolidation of firm skills is at the root of competitive advantage. They state that:

> Core competence is the collective learning in the organization, especially how to *coordinate diverse production skills and integrate multiple streams of technologies* . . . . if core competence is about harmonizing streams of technology, it is also about the organization of work and the delivery of value. (p. 82) [our emphasis]

and that:

> The real sources of advantage are to be found in management's ability to consolidate corporate wide technologies and production skills into competencies that then empower individual businesses to adapt quickly to changing opportunities. (p. 81)

This conception addresses the process aspects of competence, in terms of organizational learning, skill utilization and coordination, and technology management. To Prahalad and Hamel, a firm is best defined by its "core competence", or its cardinal ability to identify, build, maintain, harmonize and support key skilled-resources so that the firm will remain effective in the rapidly changing competitive environment. Like Quinn *et al.* (1990), who envision the firm as "an intellectual holding company", Hamel (1991) suggests that the competitive focus from this "core competence" conception of the firm is on resources, not necessarily products. Hamel (1991) continues, and narrows the focus further from "resource competition" to "skill competition":

> Conceiving the firm as a portfolio of core competencies and disciplines suggests that inter-firm competition, as opposed to inter-product competition, is essentially concerned with the acquisition of skills. (p. 83)

Drawing on our earlier discussion on "competence", "core competencies" are reflections of the firm's ability to *manage its critical skill areas*.

> D4: What distinguishes "core competence" from "competence" is that the *referent set of skills is considered to be of the utmost strategic importance to the identity and survival of the firm.*

This "core set of skills" is similar to J. D. Thompson's (1967) notion of a "core technology", which, according to Thompson, must be carefully identified, managed and buffered if a firm is to survive and grow. Indeed, the similar ideas of core competence and core technology provide a theoretical basis for examining how firms define and expand their domains. Thompson (1967), like Pfeffer and Salancik (1978), Prahalad and Hamel (1990) and Quinn *et al.* (1990), suggests that:

> Organizations under norms of rationality seek to place their boundaries around those activities which if left to the task environment would be *crucial contingencies*. (p. 39) [our emphasis]

McKelvey (1982) refers to a similar construct when he discusses an organization's "dominant competence" or "primary task" (Miller and Rice, 1967). He defines the latter as:

> *that set of activities that bear directly on the conversion of inputs into those outputs critical to a population's survival* . . . critical outputs may be defined as those outputs that return the resources necessary for continued survival. (p. 174) [his italics].

These definitions suggest that the development of a core competence, which reflects a firm's general ability to manage and coordinate critical resources, is essential for continued firm survival and prosperity. Effective general managers continually seek to define what particular critical skills and material resources are central to their continued existence and then attempt to internalize and manage these resources. In short, they are attempting to identify and achieve and maintain at least one "core competence" with a given set of critical skills. The strategic questions again are: (1) what skills should be the critical ones in the firm? And (2) how can these skills be acquired, organized and protected to ensure their survival and effective application?

Distinctive and Institutional Competence: Selznick Revisited

Of all the "competence-related" terms, the concept of "distinctive competence" has perhaps had the most written about it, and has had an unusually high degree of common acceptance in the strategic management and organization literature. As Snow and Hrebiniak (1980) point out, the term was introduced by Philip Selznick, in his 1957 classic, *Leadership in Administration*. Like Andrews (1971, 1987), Snow and Hrebiniak interpret Selznick's use of the term to refer to those things an organization does especially well in comparison with its competitors. Snow and Hrebiniak (1980) define "distinctive competence" as:

> An aggregate of numerous specific activities that the organization tends to perform better than other organizations within a similar environment.

This definition provided the base from which Hitt *et al.* (1982, 1985, 1986), and Stoner (1987) have conducted research on the relationship between distinctive competencies and organizational performance. The high level of acceptance and use of the Andrews (1971) and Snow and Hrebiniak

(1980) interpretation of Selznick's term "distinctive competence" by scholars in strategic management could be seen as a sign of maturity in this area of the field.

However, until Prahalad and Hamel (1990) introduced the idea of "core competence", little conceptual progress had been made since 1980 in developing and refining any of the "competence-related" terms, including distinctive competence. Further, much of the research cited above, which uses only the "base-line" definition of distinctive competence, and does not explore either "competence" as a standalone concept or "core competence", has had mixed results, as discussed in Chapter 1. In order to get a better handle on the origins and meaning of distinctive competence, let's go back to Selznick and see within what context he used the term.

Selznick's (1957) Use of the Term "Distinctive Competence"

Philip Selznick (1957) is often cited as the first organizational scholar to discuss "distinctive competence" as a concept in understanding organizational action. Selznick refers to the term at length in at least three locations in *Leadership in Administration* (1957): pp. 42, 49–54 and 139. The central theme that emerges is that "distinctive competence" is far more than just "An aggregate of numerous specific activities that the organization tends to perform better than other organizations within a similar environment (Snow and Hrebiniak, 1980).

In *Leadership in Administration*, Selznick was attempting to build a general theory of how and why organizations transform themselves from "expendable tools" into institutions or "social organisms infused with value and possessing character". His first reference to "distinctive competence" comes on p. 42, as he is discussing "organizational character":

> In studying character we are interested in the *distinctive competence* or *inadequacy* that an organization has acquired. *In doing so we look beyond the formal aspects to examine the commitments* that have been accepted in the course of adaptation to internal and external pressures. [our emphasis]

Here Selznick seems to be arguing that distinctive competence is more than the formal identification of abilities, and that it is closely related to the emergence of an overall "organizational character". This "character" is forged in the way the firm chooses to behave on a day-to-day basis with respect to both its internal and external constituents. In this sense, Selznick uses the term "distinctive competence" to mean having an adequate character in dealing with stakeholders.

On p. 49 Selznick suggests that special capabilities and limitations are developed as an organization becomes an institution, and that:

> The assessment of organizations necessarily scrutinizes this element of competence or disability (p. 49) . . . . The *distinctive competence* to do a kind of thing is in question when we ask whether an agency is well adapted

> to carrying out an "action" program. *This has little to do with administrative efficiency; rather it reflects the general* orientation of personnel, the flexibility of organizational forms, and the nature of the institutional environment to which the organization is committed. (pp. 50–51) [our emphasis]

In this instance Selznick appears to acknowledge a difference between a "competence", a special capability or limitation, and "distinctive competence", a general character of the firm to pursue its commitments. He makes this link between distinctive competence and character more explicit through the concept of "organizational integrity", which is the "institutional embodiment of purpose" (1957, p. 139). The protection of integrity is a major function of leadership and:

> . . . the defense of integrity is also a defense of the organization's distinctive competence. As institutionalization progresses the enterprise takes on a special character, and this means that it becomes peculiarly competent (or incompetent) to do a particular kind of work. (1957, p. 139)

Indeed, as Selznick continues he makes the connection even more explicit:

> The terms "institution", "organization character", and "distinctive competence" all refer to the same basic process (our emphasis)—the transformation of an engineered, technical arrangement of building blocks into a social organism. (p. 139)

To Selznick, "distinctive competence" reflects a firm's:

> D5: (1) ability to adequately build a unique and viable "organizational personality" that is not easily imitated nor destroyed.
>
> (2) ability to establish a credible institution that can add value to both its internal and external constituents in the face of continuous pressures.

Re-Casting Selznick's Concept: Institutional Competence

The above interpretation of Selznick's "distinctive competence" differs substantially brom the currently accepted interpretation of the term as codified by Andrews (1971, 1987) and Snow and Hrebiniak (1980). Rather than attempt to change the current use of the term "distinctive competence", which is a term that refers to those things an organization does especially well in comparison to its competitors, we will introduce another term, "*institutional competence*", to refer to the concepts discussed in definition D5 above.

This perspective on institutional competence implies the following:

(1) The technical building blocks of a firm, its functional areas (which have been generally considered *areas of "distinctive competence"*, following Snow and Hrebiniak, 1980), can be engineered to provide relatively efficient outputs; but without the infusion of character and their overall linking to an organization's "institutional embodiment of purpose", the

technical functions are easily imitable and won't provide sustainable competitive advantage. This is theoretically in line with the reasoning of Lippman and Rumelt (1982), Peters (1984), Barney (1986c), Fiol (1991), and Ulrich and Lake (1991).

(2) Institutional competence is a measure of the firm's overall ability to transform itself from an "expendable tool" into a viable "institution". It has little to do with individual firm capabilities, though these individual areas of competence do contribute to the definition of the institution. Institutional competence is really a reflection of the particular skills of general management and leadership to build and maintain a cohesive, sustainable, organizational "climate" or "culture" that helps the firm maintain its skill base so that it can serve its customers and maintain credibility with its external stakeholders.

Summary of the Discussion on Competence:

The term competence has had a multi-faceted and rather indefinite tenure as a concept in organizational and management theory. In the past few sections of this book, an attempt has been made to review and clarify the meaning of the term competence, and its companion terms, core competence, distinctive competence and institutional competence. A brief review is in order, given the length of the previous analysis.

First, all four terms deal with the general facility of humans to assemble, blend and coordinate various types and kinds of resources together in the pursuit of opportunity. The differences between the terms are reflected in what resources are being managed.

(1) *Competence*, which is the parent term, was defined as:

> D3: ***Competence***: *the adroit blending of resources and skills to adequately pursue a particular opportunity.*

Competence refers to the general ability of a firm to build, maintain and coordinate resources in such a way as to "make it appear without effort (facile)". The source of a competence is always internal to the firm—it is skill that resides in its people. Competence is a concept that indicates how *effective* a firm's managers and leaders are in deploying specific resources in pursuit of a given opportunity.

(2) *Core competence*

> Core competences are a special class of competence. They are reflections of the firm's ability to *manage its critical skill and material resource areas.*

> D4: What distinguishes "core competence" from "competence" is that the referent set of skills and material resources is considered to be of the utmost strategic importance to the identity and survival of the firm.

This "core set of resources and skills" is similar to J. D. Thompson's (1967) notion of a "core technology", which, according to Thompson, must be carefully identified, managed and buffered if a firm is to survive and grow.

(3) *Institutional Competence (Selznick's Distinctive Competence Recast)* *Institutional competence* reflects the firm's ability to manage its cultural resources and those that deal with what the firm stands for—its value-base. The term is derived from Selznick's (1957) discussion of "distinctive competence", which, to Selznick, reflects a firm's:

D5: 1. ability to adequately build a unique and viable "organizational personality" that is not easily imitated nor destroyed.

2. ability to establish a credible institution that can add value to both its internal and external constituents in the face of continuous pressures.

The above interpretation of Selznick's "distinctive competence" differs substantially from the currently accepted interpretation of the term as codified by Andrews (1971, 1987) and Snow and Hrebiniak (1980). Rather than attempt to change the current use of the term "distinctive competence", which is a term that refers to those things an organization does especially well in comparison to its competitors, we will introduce another term, "*institutional competence*", to refer to the concepts discussed in definition D5 above.

THE STOCK ASSET RESOURCE LEVEL

Organizational services, when appropriately managed (that is, when a firm exhibits competence), generate stock asset resources. Dierickx and Cool (1989) suggest that stock resources, like Itami and Roehl's (1987) invisible assets, are critical resources that are generally *accumulated* rather than acquired in strategic factor markets (Barney, 1986b). General stock asset resources include: commitment (Ghemawat, 1991), shared value (Pascale and Athos, 1981), vision, trust, goodwill, experience, organizational memory (Walsh and Ungson, 1991), productivity, flexibility, speed, creativity, quality (Garvin, 1988), brand recognition, patents, absorptive capacity (Cohen and Levinthal, 1990), liquidity, proximity, capacity, finished goods inventory, organizational slack (Bourgeois, 1980a), and organizational throughput (Goldratt, 1990). The accumulation of stocks occurs through the investment in basic and systemic resource flows, which are then converted by the service-level resources into stocks.

The level of a firm's stock asset resources depends on a firm's commitment to acquiring and deploying more elementary resources, such as those described at the basic and systemic resource level. Stocks cannot be built instantaneously, and Dierickx and Cool argue that "appropriate time paths of relevant flow variables must be chosen to build required stock assets. *Critical* or *strategic* assets stocks are those which are *nontradable, nonimitable,* and *nonsubstitutable*" (1989, p. 1507).

CHAPTER SUMMARY

Summary Diagram and Discussion of the Process-Oriented Resource Typology

In this chapter the concepts of strategy and strategic thinking were examined, particularly in light of the management of a firm's resources. Since no systematic typology of resources existed in the extant literature that could be used to help managers identify and understand resources, one was developed.

Working from a list of resources identified in extant literatures, the researchers proposed a four-level, interactive, process-oriented typology of organizational resources. Figure 3.4 displays a detailed diagram of the process-oriented organizational resource typology that has been presented in this chapter. Note that both the content aspects (what resources) and the process aspects (how do they change in time) are illustrated in this diagram.

The process flow in this model is top-down: it addresses the process by which basic organizational resources are transformed into systemic resources and then into organizational services, which provide the base for stock resource development. Stock resources are what are directly competed on in the market-place, and are sequentially dependent on the management of all previous resource levels.

The left-hand side of the figure traces how one might assess the

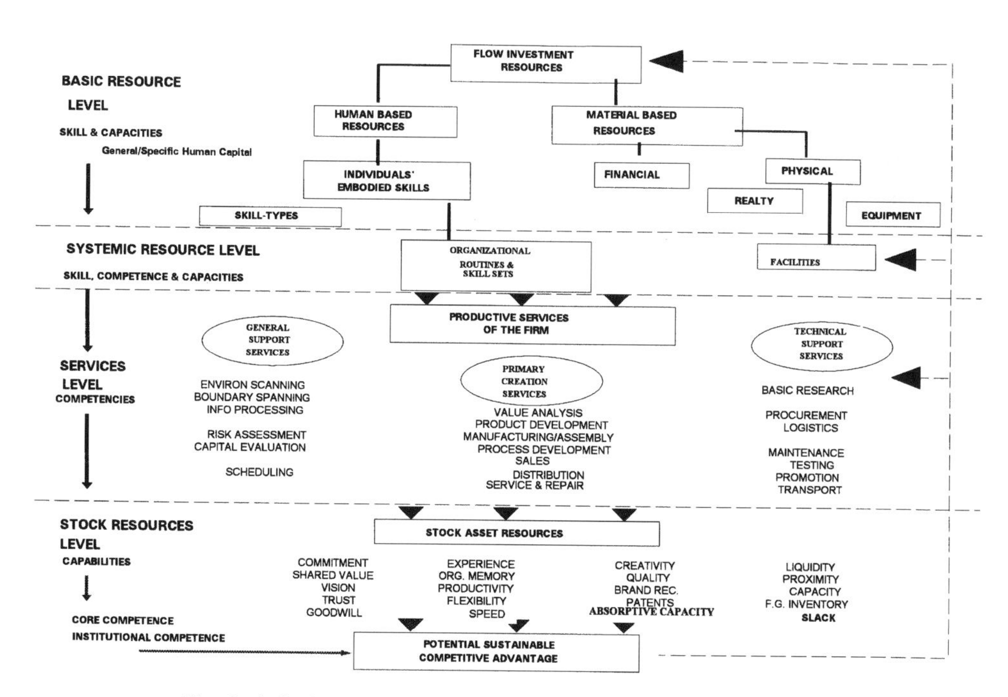

Fig. 3.4. A detailed process-oriented resource typology.

effectiveness of a firm's personnel at managing its resources at the various levels. It was argued that competence, core competence and institutional competence are all terms that deal with the general facility of humans to assemble, blend and coordinate various types and kinds of resources together in the pursuit of opportunity. The differences between the terms are reflected in what resources are being managed and at what level.

The major implication of this typology is that potential sustainable competitive advantages can arise out of the competent management of the resource transformation process, which puts the model into the general class of process-oriented resource-based theories of the firm (Schulze, 1992).

The remainder of this book will examine, based on rigorous analysis of empirical evidence (in the form of case histories), how the general resource transformation process described in this chapter, actually takes place, and will specify a theory of entrepreneurial leadership and skill-based strategic thinking that can help both the development of new theory and the practice of entrepreneurship and strategy.

SYNOPSIS OF PART ONE: FOUNDATIONS—INTRODUCTION, RATIONALE AND LITERATURE

Part One has introduced the research topic and discussed the rationale and opportunities for research. We have also reviewed the literature as to what resources are and how one might better manage these resources strategically, and have presented a process-oriented organizational typology.

Research Questions and Tasks Addressed in Part One

Part One has begun to address the first major objective of the research project, and relevant research questions.

Research Objective One

> To develop a more complete understanding of skill-base development issues that is grounded in the experience and understanding of adroit entrepreneurs and general managers.

Research Task Achieved for Objective One

> T1-A: To develop an improved vocabulary and set of constructs for describing and thinking about organizational resources and skills, and to generate a process-oriented typology of organizational resources that can be used as a basis for effective strategic thinking with regard to resource deployments.

A typology of organizational resources was presented that addresses various concerns about the tautological nature of the concept of resources, and that integrates content and process aspects of resource management. Research question Q1-A has been explicitly addressed, but at this point in the research process it is not clear whether the typology of organizational resources presented in Part One can be used as a basis for effective strategic thinking with regard to resource deployment.

PART TWO:
PRACTICE

PREVIEW OF PART TWO

In Part Two we will begin to collect and analyze data, and present theory with respect to how and why adroit practitioners manage their resources — particularly their skill-based resources—in an effort to build sustainable competitive advantages. Following a brief discussion of the "case study" research methods used to collect and analyze the field data, data from the Pilot Study site (Precision Instruments) are presented using an extant data-gathering framework as a guide.

The data from the Pilot Study is then re-cast in the form of a focused case history with an accompanying idiographic theory storyline. Then, the scope of inquiry is broadened in an attempt to provide a more adequate empirical base from which to build a substantive-level theory of skill-based strategy. Specifically, three new contexts (sites) are explored with respect to issues related to skill-based strategy and entrepreneurial leadership.

The data for each new case are presented in the format introduced in Chapter 5 for Precision Instruments. A narrative history of the skill-base and entrepreneurial leadership development of the firm is augmented with two academic storylines that discuss extant and emerging theoretical issues.

The research questions to be addressed in Part Two include:

Q1-A: What language do practitioners use in describing and understanding skills and resource-based phenomena?

Q1-B: What processes are involved in how managers build, nurture, change and deploy skill-based resources?

Q1-C: How are these processes related to one another and to the long-term performance of the organization?

4

PRELIMINARY FIELDWORK: THE PRECISION INSTRUMENTS COMPANY

> Contact either in the form of visits and observations or perhaps through descriptive case analyses provides the intellectual raw material for useful theory.
>
> Richard Daft (1982, p. 544)

INTRODUCTION

Why the Case Study Approach?

Each of the primary questions addressed in this book examines an aspect of the *process of how* entrepreneurial development occurs and is related to the formation of a high-quality human capital base and the establishment of potential competitive advantages. As Yin (1989) argues, the case study approach has a distinctive advantage over other research strategies when "a *how* or *why* question is being asked about a contemporary set of events, over which the investigator has little or no control" (p. 20). Hofer and Bygrave (1992) note that the case study method provides a strong foundation for building theory about the entrepreneurial process. They suggest that research in entrepreneurship focus on in-depth data-gathering techniques that can provide the context for analyzing holistic, dynamic and complex data.

The presentation in this book of four detailed, exemplary, case histories to examine entrepreneurial development and leadership has been chosen because cases provide a rich and dynamic context from which to look at the actual processes involved in the development of a firm's entrepreneurial environment (Feeser, 1991) and skill-base, each of which is central to the character and essence of a dynamic organization. As Philip Selznick argues in his 1957 book, *Leadership in Administration*, it is critical for organizational scholars to recognize the dynamic and evolutionary processes that affect the "character" of an organization over time. He suggests that in order to understand the development of an organization, one must "see [the character of an organization] as the product of self-preserving efforts to deal with inner impulses and external demands" (1957, p. 141). According to Selznick:

> The study of institutions is in some ways comparable to the clinical study of personality. It requires a genetic and developmental approach, an emphasis on historical origins and growth stages. There is a need to see the enterprise as a whole and to see how it is transformed as new ways of dealing with a changing environment evolve. (p. 141)

The case histories that are presented in Part Two have been written so that the reader may follow the development of the firm's skill and human asset base, from the perspective of the founder or top management team. Following a summary of the initial exploration at the pilot site, each of the main cases is presented chronologically in historical sections. At the end of each section an "*Applying Extant Concepts and Theory and Developing New Perspectives*" commentary is presented, along with an integrative diagram.

PILOT STUDY RESULTS: SEQUENCED CATEGORIZED DATA SUMMARY

In the remainder of this chapter, raw data and basic commentary on the Precision Instruments Company, which served as the pilot site, is presented using an initial data collection protocol discussed in Appendix Two as the organizing guide. Using the extant guide, data was placed into its appropriate historical chronology, and was further sorted by extant code category. Examination of the initial data, across categories, informed the primary researcher as to areas that were potentially under-represented as theory categories, and steps were taken to either augment the data or confirm why it was not central. This data was supplemented, using theoretical sampling techniques that will be discussed in Appendix Two, so that it could be re-analyzed and compared systematically with the data collected at the three follow-on sites.

THE FOUNDING YEARS: THE EARLY 1950s–1971

Business and Corporate Strategy

Under the leadership of its founders, Precision Instruments Co.'s basic business strategy was to apply the tungsten carbide bonding technologies developed in the 1940s to as many types of surgical needle holders and scissors as possible. In short, the company's strategy was one of product line expansion in the specialty surgical instrument market based on a new manufacturing technology that permitted significant improvements in product quality and durability.

Driving Force

According to Len Petrush, President and CEO of Precision Instruments, "historically the business was defined by type of metal and metal technology". This, in conjunction with the business strategy suggests that the driving force in the early years was *technology*, where the firm builds its strategy around a unique body of knowledge and technical capabilities from which it develops distinctive competence in exploiting applications of its technologies.

Organizational Resources and Resource Allocation

The primary resources in the early years revolved around the skills and knowledge of the founding members themselves.

Organizational goals and objectives
Strategy formulation
Functional goals, strategy and systems
Functional area skills and management
Employees and personnel systems
General management
Competencies
Information and evaluation systems
Leadership style

Information not available or collected in these categories for this time period.

THE CORPORATE SUBSIDIARY YEARS: 1971–1984

Business and Corporate Strategy

In 1971, the business was sold to a larger medical products firm. The business continued to grow for the next few years under the big-company umbrella. As often happens, however, both founders eventually left the firm and were replaced by "traditional" corporate middle managers who tried to make the business fit the "big company mold". By 1978, the business lost money for the first time; and by 1983, its sales had dropped to 30% of their former levels and its losses approached $1 million per year.

Under the "big company's" direction the basic strategy was to maintain/ harvest Precision Co.'s market position as profitably as possible for as long as possible. When profitable operations were no longer possible, the “big company” took a few steps to try to turn the business around, but when these did not pay off as soon or as well as expected, the business was sold.

Driving Force

Given the above information it is reasonable to assume that the driving force of the big company, with respect to the Precision Instruments Division was *Return / Profit*, where the firm attempts to realize return and profit measures through diversification and portfolio management.

Organizational goals and objectives
Strategy formulation
Organizational resources and resource allocation
Functional goals, strategy and systems
Functional area skills and management
Employees and personnel systems
General management and competencies
Information and evaluation systems
Leadership style

Information not available or collected in these categories for this time period.

THE TURNAROUND: 1984–1989

Organizational Goals and Objectives

According to Petrush, “in the mid 1980s the competition in our business was 60% German, and 40% US/UK . . . . we felt we needed to re-position ourselves vs Germany. Our goal in 1986 was to be able to make money vs the Germans with the Mark/Dollar ratio at 3/1 (it was at 2.5/1 at the time).”

In order to achieve this goal Petrush had to establish goals in the areas of both productivity and innovation. “We pursued this re-positioning by:

(1) Beginning to improve our technology and products. (We had actually considered outsourcing to the Germans and having our name put on the instruments, as our name was highly regarded.)
(2) In the long-range we wanted to broaden into specialty segments, and split production into domestic and international components.
(3) We needed to re-train and work our ‘skill-base’, in terms of both its ‘depth’ and ‘breadth’. We needed to change our bonus system to reward skill breadth and functional abilities . . . raises based on skills. We were/are looking for people with long-term interest in being in our workforce.”

Between 1985 and 1989, Len Petrush turned Precision Co. around through his entrepreneurial initiatives and an action program designed to produce world leadership, rather than just a "quick fix." Len Petrush's turnaround strategy for Precision Co., Inc. involved major entrepreneurial efforts in four areas: (1) productivity; (2) innovation; (3) strategic vision; and (4) social responsiveness. Moreover, these efforts went far beyond the ordinary in each area, and as a totality re-established Precision Co., Inc. as the world leader in the products that it makes.

While Petrush realized in 1985 that it would be possible to return Precision Co. to profitable operations solely through a combination of various productivity and product line improvements, he felt that the company's long-term survival and profitability demanded more than just correcting the mistakes of the past—especially since the basic driving force behind the company's current product line was a manufacturing technology that was now well over 40 years old!

Business Strategy and Formulation

As a result, Petrush built his new strategic vision for Precision Co. around five concepts. The first was to focus on particular groups of surgeons who demanded the highest quality products possible and who were willing to pay premium prices for these products because of their importance to the surgeon's work. More specifically, Petrush chose to focus on surgeons who performed especially critical and/or visible operations in which superior products were essential to success. Thus, in 1988, he focused primarily on plastic surgeons, with a secondary emphasis on heart surgeons and neurosurgeons.

Petrush's second concept was to develop the broadest possible line of high-quality innovative products for the most difficult parts of these surgeons' most important operations. More specifically, Petrush wanted to have superior products for every critical task that such a surgeon faced during an important operation, but to avoid any products that offered no such advantages. In this way, he could amortize both his product development efforts and his marketing costs over the largest possible volume, while simultaneously ensuring the greatest possible market acceptance of his products.

Third, he wanted to concentrate on products that required advanced metallurgical technologies and skilled labour for their manufacture, as these factors would simultaneously enhance his quality image and serve as barriers to entry to other competitors.

Fourth, Petrush wanted to be the world quality leader in whatever he did. Finally, he tried to communicate his new vision for Precision Co. to all of the company's employees—managers, production personnel, secretarial staff etc.—since he firmly believed that the company could not succeed to the degree he wanted it to without the cooperation of all its people.

Business Strategy Implementation

In order to implement this strategy, Petrush spent nearly 10% of his time actually observing new surgical procedures by leading surgeons in these three specialties in order to develop ideas and leads for new products that Precision Co. might develop. In addition, he continually encouraged the leading surgeons in each of these specialties to call him if they encountered difficulties that might be overcome with new instruments. Finally, he actively involved both the surgeons and his own manufacturing personnel in the design task, as described earlier.

Petrush also spent substantial time investigating new metallurgical developments and technologies that Precision Co. might use to develop such products. At the same time, during the past three years Petrush had subcontracted certain aspects of Precision Co.'s production operations that were not essential to the basic quality of its products in order to cut costs.

However, in selected other areas in which automated manufacturing produced products of lesser quality, he went back to hand manufacturing techniques even though this needed some moderately expensive machinery that the company had previously purchased. The basic thrust of all these efforts was to focus Precision Co.'s unique but limited resources on those specific manufacturing tasks where it could generate the greatest possible degree of customer satisfaction and/or competitive advantage.

Petrush's Innovation Efforts

The primary objective of Petrush's innovation efforts was to restore Precision Co.'s position of leadership in the specialty surgical instrument market through a series of new product developments that could be sold on the basis of superior quality and design, rather than on the basis of "price and personality".

To re-establish Precision Co.'s position of leadership, Len Petrush sought innovations in three areas: (1) new designs for surgical instruments not previously made by Precision Co.; (2) improved designs for existing Precision Co. products; and (3) the application of new technologies to the design of specialty surgical instruments. The four products described below illustrate these initiatives.

The first such improvement involved a product that had been developed just before Len Petrush joined Precision Co., namely, the company's rasp, which is used for rhinoplasty operations, i.e. "nose jobs". Traditionally in "nose job" operations, excessive bone was removed through the use of surgical chisels. Using its tungsten carbide manufacturing technology, Precision Co. developed in 1983 a rasp that enabled plastic surgeons to "file" rather than "chisel" such excess bone tissue away—a procedure that is far more accurate than the former process. This feature substantially reduced the long-term complications that accompanied this operation.

Unfortunately, with its large losses under "big company", Precision Co. never pushed this product aggressively. Consequently, the 1984 sales of this rasp were under 200 units per year. Shortly after he took over as general manager, Len chose this product to be one of the tools to re-establish Precision Co.'s image in the market. He, therefore, featured it in all the sales calls he made to surgeons and in all the company's marketing literature. The net result was that by 1989 the total sales of these rasps had increased twentyfold, and Precision Co. had become the worldwide leader in this product line.

A second product improvement, this one in scissors design, occurred when Len Petrush was informed by one of the plastic surgeons with whom he worked that, on occasion, the tissue that needed to be cut in a particular operation would sometimes slip along the edge of the scissors used in the operation, thereby making the cut less accurate than it should be. To remedy this situation, Mr. Petrush helped design a new scissors that had "serrated" grooves cut into its bottom edge in order to hold in place the tissue being cut. Not only did this innovation work, but Precision Co. was able to patent several developments associated with its manufacture. In addition, Precision Co. added a gold band to the handle shank of all such scissors to assist in their identification in the operating room. It then trademarked both this band and the trade name that it used to describe these scissors.

Another such product improvement occurred as the result of a comment made to Len Petrush by a leading plastic surgeon with whom he discusses most of Precision Co.'s products. After a particularly long operating day, the surgeon complained to Len about how tired his hands were because of the need to continuously open and close the surgical scissors used to separate skin from underlying tissues in facelift operations. (This surgical technique is called "spread dissection".) Petrush wondered what made this effort so tiring, and whether anything could be done about it.

To get some answers, he commissioned a study of the motions of the human hand while opening and closing surgical scissors as done in facelift operations. From this study, he learned that it takes two to three times as much effort to open a closed set of scissors against resistance the first inch as it does to open them further. As a consequence, he redesigned the handles of Precision Co.'s "spread dissection" scissors to have a Y-shaped flare at the base of the handle so that the fingers of the surgeon using them would already be open an inch or more when the scissors were closed. He then asked the surgeon to test the new scissors.

A more recent Precision Co. innovation involves a technology push as well as a market pull. It is the development of a new line of titanium forceps for use in an innovative new heart bypass surgery procedure. The operation involves the replacement of clogged heart arteries with another artery from the patient's chest, rather than with a leg vein as has been done in the past. The fact that an artery was used rather than a vein increased the effectiveness of the operation. The trauma to the patient was also reduced, since only one part of the body was operated on. The operation had been difficult to perform, however, because the artery in question was both "delicate" and hard to handle. Thus, with traditional forceps, great pressure was normally needed to effectively handle and control the artery, yet this same pressure often damaged the artery.

Petrush, who had sensed a need to incorporate more modern metallurgical technologies into Precision Co.'s manufacturing skill-base, decided to try to make an improved set of titanium forceps that would overcome these difficulties. Although it took some time, the company eventually developed a method for bonding tungsten carbide particles on the jaw surface of these forceps using a manufacturing process originally developed in Russia. This "rough" jaw surface contributed to both the ease and safety of the operation. So much so, in fact, that one of the leading surgeons in the world in the use of this technique called Len Petrush and volunteered to "promote" without charge these new instruments in an instructional videotape that he was making on this operation. As a consequence, Precision Co., in slightly less than a year, has become the major supplier to this market, while adding to the list of its products that are on the leading edge of current surgical procedures.

As a result of these innovations and others like them, Precision Co. has expanded its product line by over 70%, so that it is now nearly triple the size of that of its next leading competitor in carbide instruments, V-Mueller. In fact, Precision Co.'s product line of specialty needle holders, scissors, forceps and rasps now exceeds the combined total of its top three competitors by more than 40%. Equally important, these innovations have allowed Precision Co. to increase the average price of its instruments by almost 25% during the past three years, and to claim, without fear of contradiction, that it is the world's product and quality leader in the market segments in which it competes.

To help emphasize this fact and to instil pride in the company's work force, Petrush has replaced the stylized logo that was attached to Precision Co.'s products while the company was owned by the "big company" with the words "Precision Co., U.S.A." which are etched in the product's gold-plated handle with a laser beam. He explained that few of his customers knew what the old logo stood for, whereas the "Precision Co." brand name connoted the highest quality in specialty surgical instruments. He also noted that the designation "Made in USA" substantially enhanced his ability to sell his products overseas, since the USA was still considered the world leader in the development of medical technology.

Driving Force

During the turnaround era the driving force for Precision Instruments changed from technology to *markets served*, where the firm builds unique long-term relationships with a well-defined customer base and attempts to meet the various needs of that customer base through a broad and innovative product mix.

Organizational Resources and Resource Allocation

According to Petrush, "Prior to my arrival, Precision manufactured 98% of what we sold, but in order to compete today, and be full market oriented, we are beginning to out source non-critical products . . . . we are keeping the products and processes that require the skilled craftsman we have".

Functional Goals, Strategy and Systems

The Productivity Plan

The primary objective of almost all turnarounds is to restore profitable operations as quickly as possible. Petrush's productivity plan was the primary vehicle for doing this at Precision Co. It consisted of three parts. First, in the short term, attention was given to cutting costs by combining duplicate functions and/or by eliminating non-productive functions in the areas of administration, manufacturing supervision, and marketing and distribution. The net results of these efforts were the elimination of several managerial personnel and a reduction in Precision Co.'s losses from 44% of sales in 1984 to just over 3% of sales in 1985.

In 1986 the company moved from its California facility to the East, consolidated administration and distribution systems, and began to reposition themselves with respect to manufacturing, distribution and marketing. Petrush comments, "We called out our 'skilled' people, we wanted a critical mass of manufacturing business to come east with our company. We wanted career oriented people, not lightly skilled people".

Employees and Personnel Systems

Petrush commented that, in 1985, "we needed to re-train and work our skill-base, in terms of both its depth and breadth. We needed to change our bonus system to reward skill breadth and functional abilities . . . raises based on skills. We were/are looking for people with long-term interest in being in our workforce".

General Management and Leadership Style

Socially Responsive Management

The philosophy of investing substantially in the areas that count, while husbanding resources in other activities extended substantially beyond Precision Co.'s manufacturing and product development activities. The grinding of scissors and other high-quality surgical instruments produced substantial amounts of microscopic metal dust. Because of this and his concern for his talented, skilled workforce, Petrush installed an air filtration and ventilation system at a cost of nearly $100,000 that reduced the average particle concentrations in the air at the Precision Co. plant to levels from 25 to 1000 times lower than OSHA standards.

He explained that this investment would pay for itself many times over in the long run by reducing his medical insurance costs, by extending the working life of his skilled manufacturing employees, and by avoiding a union and the constraints in productivity that union regulations often bring. He further observed that even though he was positive about realizing such long-term returns, he was obliged to make the investment even without them because it was "the right thing to do".

Petrush also explained that he was able to afford such investments because he saved money in non-critical areas by "shopping around" and not being concerned with "show". Thus, he had recently saved Precision Co. several thousand dollars by buying the majority of the file cabinets used in the company's warehouse at an auction associated with the closing of a nearby firm.

In addition, Petrush spent substantial time giving plant tours and lectures to students in local colleges of business in order to try to communicate some of his concerns and beliefs about management's roles and responsibilities to them.

Functional area skills and management
Competencies
Information and evaluation systems

Information not available or collected in these categories for this time period.

EXPANDING MARKETS AND CAPABILITIES: 1989–1992

Organizational Goals and Objectives

The following information comes from interviews with Precision's general managers regarding goals and objectives of the company in the early 1990s:

> In terms of company objectives, we want to be number 1 or 2 in every market we participate in. You just know if you are, in terms of reputation, position, direct sales competitions . . . . but you don't know the other guy's numbers.

> We think that there is a six to nine month barrier to get into the minimum invasive surgical market, we'd like to secure a position as one of the top four players in this segment.

I want to be able to tell my sales reps that they have between 150 and 1000 non-discount instruments to sell . . . that they won't have to talk price . . . they will sell at premium.

We strive to achieve the highest levels of craftsmanship, combined with the most advanced metallurgical technologies, and overlaid with the most advanced designs (techniques) of surgery.

We are committed to maintaining our leading edge in this industry with ongoing research and development in the field of metallurgy and hard-metal technology. Our desire is to continue to provide the surgical community with the finest in instrumentation.

We are proud to continue the Precision Co. tradition of innovation and excellence—a tradition begun in the 1950s when one of our founders introduced to the surgical world the first tungsten carbide instrument. Since that time Precision Co. has developed the first tungsten carbide scissors, forceps, bone rasps, and serrated scissors. Continued research and development has led to the introduction of titanium instruments and other applications of hard-metal technology. We are committed to innovation!

We try to deliver any product within 30 days. The others sometimes have to wait six months.

I'm not going to wallow in the back-wash of mediocrity . . .

We are a quality driven company.

Corporate and Business Strategy

As of 1992 there was no corporate-level strategy. At the business level, according to Precision management:

So the bottom line is that we try to produce a good product and be extremely knowledgeable about it. Part of that has led us into where we are now, into what we call "procedural specific instruments". It is not that we can make something that can make everything, we make something that can do something very specialized, which will lead back into something that will do everything. I don't think we will ever be real big . . . we are not going after the major market share . . . most people want to buy the vanilla, not the real spices. The real spices is where you make the money.

Success in this business comes from pursuit of small niches.

Our strategy in the MIS (minimal invasive surgery) segment was initially defensive—then we decided we could grow in this segment, and be aggressive.

There is a lack of supply now in the endoscopic market . . . which we are trying to rapidly position ourselves in . . . . there are only small players in this game right now. We are going to "full market" in the plastic surgery and endoscopic markets. We think the endoscopic market will eventually cannibalize the general market, and eventually the others.

> The pace of the firm's growth is limited by manpower . . . in product management and sales . . . and the craftsmen in the back. We don't want to grow faster than we can support with qualified people to keep quality at its best.
>
> So, bringing it back to the definition of ourselves relative to most of the other people in the market . . . anybody can go out and get machine-shop work done very quickly, anybody can do machine work. What we find though, is, to react very quickly to the market, we can take our skilled instrument makers and, with some training and working together with us as market managers, with some of the surgeons that we bring in here . . . we give them a broader viewpoint, and working with our engineer, we can come out with a better design, faster than anyone else in the market.

Strategy Formulation

According to Petrush, one needs to formulate strategy to support the market needs driving force:

> Now the market is defined by market segment as the end-user defines it:
>
> General surgery
> Plastic/reconstructive surgery
> Cardiovascular surgery
> Endoscopic surgery
>
> You have to define the business by what the customer wants and needs, not by some internal examination of technology or manufacturing. This is a complex environment by segment. We are in "hard" instruments [non-disposable].
>
> You need to ask how product-specific is your market [down to the instrument, in this case].
>
> My competitors are moving into areas they don't have expertise in. Risk the real-time changes on their reputation.

Driving Force

The driving force since the turnaround has continued to be *markets served*, where the firm builds unique long-term relationships with a well-defined customer base and attempts to meet the various needs of that customer base through a broad and innovative product mix.

Organizational Resources and Resource Allocation

The managers at Precision identify the following as resources:

> Well, we've got our own manufacturing as one resource . . . we have my partner and my ability to oversee developmental projects, or manufacturing

projects, off-site . . . we have our engineer's ability to oversee, and my manufacturing director, to oversee the outside machine shop work. So, all of that takes time . . . we've pushed ourselves to the limit, then your resource gets to be your time . . . .

We're not limited by financial resources . . . we are limited by time . . . and I can pull in more management, but they'd not be competent, and it'd take more of my time to grow them . . . and right now I don't have that, I have a window in the market . . . and in three more months I'll have my own people that will have been seasoned in the field, that I can bring in, that will have a higher level of competency than somebody I could bring in from the outside.

Uhmm . . . it is two things . . . it is resources, which encompasses financial and my time . . . training and placing them . . . . One thing is, you can't grow too fast and not have enough leadership to train these folks . . . so . . . that is a limiting factor . . .

We're outsourcing . . . we need equipment to be competitive in the broad sense of the market, with the other players in our segment, so we've had to go outside and design, and have made for us, state-of-the-art equipment. We've now employed as a consultant, a very experienced, proven, electronics design engineer.

Functional Goals and Objectives and Strategy

Attention to functional level objectives and strategy at Precision is manifest in the following observations from management:

It is critical that businessmen know the difference between features and benefits and develop skills needed to support benefits. You need to understand the buying process of your customers.

Everyone is responsible for satisfying a customer request. I will interrupt whatever I am doing to accept a call from a surgeon.

You need to know the dynamics of the sale. Market share doesn't trade at retail price, and pricing is not a moral issue.

The life cycle concept is critical to running the business, in terms of the product and the skills of the people. You need to know the life cycle of each instrument and each person. Everyone at this company is on the growth or development stage of the life cycle.

In the endoscopic market there is a primary sale, which consists of the TV monitors, lights, blowers etc . . . . and the secondary sale, which includes the endoscopic instruments. We have been focusing only on the secondary sale, but have gotten into a joint venture with a firm to produce the primary market items for us (using the Precision name), so we can now offer a full-line product to the surgeon and the hospital.

When we got into the endoscopic area, we could make certain aspects of the instrumentation . . . and there were some things we could not do.

Although in the long term we want to increase our levels of skill. We also identified that we do not have the machine-shop capability, nor the equipment, so we've out-sourced nearly all the machining of the jaw-parts. So, we do that everyday . . . . From a micro standpoint, I can take you back in the back, and I can show you a new gal that has just come on, I am aware of her abilities on a daily basis . . . .

Functional Systems and Skills

But, prior to the first of the year [1992], actually prior to the endoscopy revolution, the system was Len and I would talk about it, if it was a good idea, then we would tell the plant manager what we were doing, then we would get together with our senior instrument craftsman, and work it out. Generally the ideas came from doctors, or we saw an idea in the marketplace and we thought we could do better. Now, with the endoscopy, we get together . . . we have so many projects pending, that we've had to do this, so we prioritize them, put a list, and we have several ways it can go. It can be done in our existing hand skills, what we call traditional instruments, if it's endoscopic, it is most definitely our owner and he works it through the engineer and then the factory, if it is something from the outside vendor, we have a person that handles that . . . . If it is a plastic surgery item that is not an instrument, the product manager will work with that. It all kind of coordinates through us . . .

What makes Precision what it is today is the people that allow us to produce those first two factors, relative to our competition, in a much more efficient and a much speedier way. So, in other words, every company may identify a new design, in fact in endoscopy they have. But they've lagged six to nine months behind us in being able to produce that instrument because of the people we have in the plant. And I think the key item here is the quality of the people gives us the quality of the product.

All of our surgical instruments are hand crafted. In these times of assembly lines and automated mass production, there will never be a substitute for the skilled instrument craftsman. Precise, durable, and finely adjusted surgical instruments must be made by hand. This, combined with superior grades of stainless steel and titanium and an extra-hard working surface, gives rise to an exceptional level of performance and longevity.

We have experienced guys in each market and in the manufacturing plant that can handle all these markets. A solid engineering and manufacturing staff.

In a niche or segment business [which this is] you need knowledgeable, experienced people working for you that know the details of the business. It is great to have a college education and be bright, but it is not enough in a niche business.

We are currently adding distribution capability in the plastics, cardiovascular and endoscopic markets.

We are in the specialty medical equipment and instruments business. Expertise in marketing: knowledge of what surgeons are doing.

Skills in the industry, in terms of craftsmen, are becoming less available, while the demand remains the same. Automated tool-making reduces the quality of the instrument.

Well, my perceptions will be both from a marketing and a skill base. We've gone from being stable to strong, meaning the skills we had were enough to make money and stay in business. But we weren't prepared to grow. We have strengthened, put in processes to get greater economies in manufacturing, standardize certain things, put together systems so that the old "Precision" runs by itself, the new "Precision", which is the vision, of a growing and developing business, is in place.

Functional Management

Our company is small, we don't have committees to look at things. We see an opportunity, get our engineers and marketing people together in a room, and tell them, "get this done" . . . and off they go.

We don't spend a lot of time in meetings around here. And that is one of the things that I am adamant about . . . is we don't do book reports and we don't do meetings, because those are done to reassure insecure people . . . they can be insecure in trusting the individual they've assigned to do the tasks . . . then they are going to want a very heavy reporting package back from them on an ongoing basis, because they don't trust them. So, we have a lot of trust in our group here.

A major error that marketing and MBA-types make is that they fail to understand the day to day factors in the market. They do not analyze the field situation. The field managers are left out of decision making, service levels are not assessed. You have to understand distribution and service time in the field. You must understand sales and not have a "sales" vs "marketing" situation.

All of Precision Co.'s manufacturing supervisors were working supervisors who were chosen on the basis of their skill and knowledge of the production steps involved. This means that most of their supervision is done indirectly through training and example, not by order and edict; and that they are viewed by those they supervise as both a source of help and a role model to emulate.

Employees

Well, one, before they go anywhere in the factory they have to have a demonstrated level of hand-skill. And not everybody possess those, some people are clumsy, some people can not use machines, some people are not able to acquire hand-skills. We give them a test, we give an instrument to grind . . . and another thing, they have to be patient people . . . they've

> got to be able to focus their attention closely to the task at hand, and not deviate from a pattern or a quality standard. So, that I think is the core requirement before they progress up the ladder of overlaying skills that we're going to teach them.
>
> We have then brought in the factor of flexibility . . . are they willing to learn new skills and be flexible, and still carry over this attention to detail, this focus on quality . . . and increasingly, bringing to bear new hand skills in their production capabilities. And I think that is the main thing we rely on. And another thing, I like to see a positive teamwork attitude out there . . . and we've got that, we've got a real good team out there.
>
> I can go out there and throw a new instrument out there and people get excited about it, not like "ahhhh, god, not another new pattern!". Hey, great, another new opportunity, something new! People get excited about that, and that's what we want.
>
> We always had an engineer . . . the one we have now came a year ago . . . we really never got the engineer involved in the past, until endoscopy. The engineer was really more for metal quality, he is a metallurgist . . . . With the endoscopy, it requires us to use more automation, more machine centers, a robot . . . you have to do a CAD of it . . . and drawings, source out parts . . . . We do the drawings, and we have bought up a machine shop's entire production . . . . we don't think it is a good investment at this time to buy a $300,000 machine when we need a new building, and are trying to build a new sales force . . . so we will take a higher unit cost. The drawings that we make feed right to his program and the machine makes them. That is mostly on endoscopic jaw parts . . .

Personnel Systems

The successful completion of the move east in the mid-1980s allowed Len Petrush to implement the third phase of his productivity plan in late 1986, namely, the re-training and re-motivating of the company's plant personnel in order to reduce scrap costs. Because of the high-quality nature of Precision Co.'s products and the labour-intensive adjustments needed to "finish" each instrument, Petrush did not seek to impose output-oriented production targets on his workforce. Instead, he felt that increased output was best achieved through better workforce motivation and training, an effort he increased substantially in early 1987. This effort had the added benefit that, with increased training, "scrappage rates", which contributed significantly to total production costs, dropped substantially.

A variety of techniques were used to strengthen those training efforts. One was a policy of 100% final inspection of all Precision Co. products. This not only guaranteed superior product quality, but also identified any potential training needs quite rapidly since each instrument was marked in ways that allowed the immediate identification of the persons who worked on it. In addition, whenever a surgeon visited Precision Co.'s plant in conjunction with the firm's new product development efforts, Len Petrush made sure that he or she met the employee(s) who had made the surgeon's instruments—a policy that he felt added substantially to employee product identification and motivation.

General Management and Leadership Style

> If anything like that [obsolescence] happens in the plant that is my fault. If I had not taken these people in the plant into endoscopy a year and half ago, then they would be facing obsolescence, because they would be working on instruments that are being used on a declining basis. So, that's the challenge . . . our challenge as managers is to insure that our company is on the leading edge of the market . . . and that drags everybody into that vortex . . . and they have to increase their levels of skill applying to new products. This obviates any obsolescence.
>
> I want my people involved in the business. I want them to "be the customer". I personally go on the shop floor and am quality assurance. We don't need "caretaker managers" in our distribution network.
>
> These people back here can do anything. It's a teamwork thing, leadership must start at the top. They (the workers) know I love this business, and that I respect them . . . . I bonus them . . . . I can't deny them the ability to achieve . . . they have an ownership position.
>
> Now, that's on the craftsman side of the business, the same situation applies at a management level. All of us in management can basically shift our attention, as needs arise . . . our knowledge base carries us across specialties. So, I think if I were to define Precision today, I would not define it by products, I would define it by the people that are allowing us to come to the market with better designed products, of a higher quality, faster than our competitors.
>
> It's a passion for what we do. If you look at the management team and most of the sales reps, we live and breath what we do . . . . All the people, it is really hard to verbalize . . . you can go back and talk to our senior craftsmen, they all care about what they do. The plant manager cares. When we interview people, it's kind of hard to verbalize, what I look for is someone that has been traumatized at some point in their life . . . be it death or, working their way through school . . . so I know that they are tough . . . the second thing I look for is loyalty . . .

Competences

> In any function, you should be able to identify and reward what I call your core of competency. Which we try to do. We know that there is a core of competency in that plant back there. In the guys that came with us . . . we treat them better than anybody out there . . . we reward them better at the end of the year. Up here our core of competency is, of course, everybody in this team, at the top, is in the core of competency area. And

> when I look at a distributor . . . I will look at the core of competency in the distributor. It may be in the one guy that is running the whole thing . . . and it depends on how he organizes, whether some of that competency can be dispersed out, and at what rate . . .

Evaluation Systems

> And we overlay the same quality, because those people are excited about their future, they understand that they need to do this. That they are competing not just on a quality issue here, but on a time issue with bringing a new design out. So we've got a great deal of flexibility built into the plant. The other thing is, we've not generally rewarded people for being with us on a time basis. Generally we will reward people for learning, like if somebody wants to move from six dollars an hour to eight dollars an hour, they are going to do that by learning different and new skills, and augmenting their abilities to perform different tasks in the factory on different types of instruments. Or different tasks on the same instrument. So, with a heavy amount of cross training, we build up our people skills.

CHAPTER SUMMARY

In this chapter, data that were collected at the Precision Instruments Company have been presented, using the initial data collection protocol for the Pilot Study as a presentation guide. In Chapter 5, a case history of Precision is told with respect to the development and management of its skill-base and resources and an idiographic-level theory of skill-based strategy and entrepreneurial leadership is developed.

itself many times over in the long run by reducing his medical insurance costs, by extending the working life of his skilled manufacturing employees, and by avoiding a union and the constraints in productivity that union regulations often bring. He further observed that even though he was positive of realizing such long-term returns, he was obliged to make the investment even without them because it was "the right thing to do".

HISTORY OF PRECISION INSTRUMENTS, SECTION IV: THE INNOVATION ENGINE COMES TO LIFE AND THE COMPANY LEARNS TO RESPOND TO MARKET CHANGES

When Petrush and Demaria purchased the company in 1984, they recognized that reducing costs through consolidation and a move to the eastern part of the USA would not be sufficient to ensure Precision Instruments a competitive position in the business. "Having a finance orientation, cutting costs and being efficient just won't cut it", according to Petrush. The company needed to move forward and begin to innovate again. He continues:

> In the mid-1980s the competition in our business was 60% German, and 40% U.S./U.K. We felt we needed to re-position ourselves versus Germany, and we pursued this re-positioning by: (1) Beginning to improve our technology and products; and, (2) in the long-range we wanted to broaden into specialty segments, and split production into domestic and international components.

So, while Petrush was beginning to rebuild the company in terms of productivity, he also set out to reshape its capabilities as an innovative designer and manufacturer of leading-edge surgical instruments. While some 30,000 different types of surgical instruments exist, Precision Instruments had, from the start, concentrated on a very small, high-quality segment of this market where it could offer instruments of superior design and quality. Unfortunately, the major product problem confronting Petrush was the fact that within the high-quality, hard instrument segments of the market, only a few new designs had been added to Precision Instrument's line in the ten years prior to the 1984 purchase of the company.

The primary objective of Petrush's innovation efforts was to restore Precision Instruments' position of leadership in the specialty surgical instrument market through a series of new product developments that could be sold on the basis of superior quality and design, rather than on the basis of "price and personality".

The strategy for competing in the market rested squarely on Petrush being able to rebuild, redefine, and leverage his skill-base, as reflected in this statement of business definition: "We strive to achieve the highest levels of craftsmanship, combined with the most advanced metallurgical technologies, and overlaid with the most advanced designs (techniques) of surgery".

What Petrush had set out to do was refocus his business from a technology-driven orientation to a customer-driven orientation. New instrument design, led by close relationships with leading surgeons, would drive the show, and metallurgical technology and manufacturing skill would keep pace. According to Petrush:

> Historically, the business was defined by type of metal and metal technology; now the market is defined by market segment as the end-user defines it . . . . You have to define the business by what the customer wants and needs, not by some internal examination of technology or manufacturing.

Petrush focused his attentions on learning about what surgeons truly needed, and would spend nearly 10% of his time actually observing new surgical procedures by leading surgeons in various specialties in order to develop ideas and leads for new products that Precision Instruments might develop. As Petrush puts it, "you need to get into the surgery room to stay in the hunt", and, "success in this business comes from pursuit of small niches". More specifically, Petrush wanted to have superior products for every critical task that such a surgeon faced during an important operation, but to avoid any products that offered no such advantages. "It is critical that businessmen know the difference between features and benefits; and develop skills needed to support benefits. You need to understand the buying process of your customers".

The new instruments Petrush wanted Precision Instruments to sell would contain the very latest in design and metallurgical technology, and would require the utmost of his craftsmen and engineers. If he wanted to offer the surgeon an instrument that offered benefits, and not bells and whistles, he had to focus on increasing the overall skill and capability of his plant. He felt that if he could do this he could simultaneously enhance his quality image and develop barriers to entry to other competitors.

The shift to a more customer-driven orientation had profound implications for the way management conceived of its markets and managed its human capital. In terms of markets, as Petrush noted earlier, "success in this business comes from pursuit of small niches". The successful pursuit of these niches required that the company's personnel become more willing to cope with changing markets and changing requirements for skills. In order to optimize the use of the skill-base, Petrush and his management team began to reorganize the business, in terms of what it did in-house, in an effort to better leverage his craftsmen, given changing market conditions:

> Prior to my arrival, Precision Instruments manufactured 98% of what we sold, but in order to compete today, and be full-market oriented, we are beginning to outsource non-critical products. We are keeping the products and processes that require the skilled craftsman we have . . . craftsmanship is essential . . . skills in the industry, in terms of craftsmen, are becoming less available, while the demand remains the same.

One of the big changes in the instrument market has been in the plastic surgery markets and in the minimal invasive surgery or endoscopy markets. Petrush comments:

> Our strategy in the MIS (minimal invasive surgery) segment was initially defensive—then we decided we could grow in this segment, and be aggressive . . . . There is a lack of supply now in the endoscopic market, which we are trying to rapidly position ourselves in; there are only small players in this game right now. We are going to "full market" in the plastic surgery and endoscopic markets. We think the endoscopic market will eventually cannibalize the general market, and eventually the others.
>
> In the endoscopic market there is a primary sale, which consists of the TV monitors, lights, blowers etc., and the secondary sale, which includes the endoscopic instruments. We have been focusing only on the secondary sale, but have gotten into a joint venture with a firm to produce the primary market items for us (using the Precision Instruments name). So, we can now offer a full-line product to the surgeon and the hospital.

Entry into the growing and challenging MIS and cardiovascular markets put pressure on the company to improve its current capabilities and develop new ones. In order to compete in these segments, the company had to learn to design and build an entirely new type of product. The demands on all areas of the company's skill-base were increased. The endoscopic instruments require the use of CAD/CAM engineering, the latest metallurgical applications, and an intensive craft ability to assemble and finish the instrument.

As Precision Instruments began to enter the plastic surgery and minimal invasive surgery market segments, it also had to carefully protect its positions in the traditional general surgery and care products segments, and also in the cardiovascular market, which it had entered just prior to the major take-off of the MIS segment. Petrush pushed new technologies into these more mature and traditional areas, and worked on revitalizing the skill-base in these areas, so that by 1990 the Company had made progress in both gaining market share in the general surgery area and in dominating the care product area.

This change in product development strategy required that the entire company be involved in its implementation. He firmly believed that the company could not succeed to the degree he wanted it to without the cooperation of all its people. It was not a straightforward step to move from a technology base that was 40 years old into new technologies, even if the product was still surgical instruments. The new product developments demanded new skills of his people, and there was initial resistance, as William Zesinger, current marketing manager (1992) remembers:

> When we came on board [in 1984], the company was so hung up in focusing on that business . . . the traditional Precision Instruments, old, stale products . . . and putting all their efforts into that, without looking outward for new things. The challenge was, mainly on Len's part, was to get control

> of their attention, and say, "look, this is old, and this is new, you put together some system for maintaining the [traditional products], and let's go develop some new hand skills, let's develop some outside sources that can do it".
>
> As the new products started being brought in, and these were at a greater profit, everyone started getting on board because they saw that they could put in the same amount of effort and get twice as much money. We started bringing in more plastic surgery items. Then everyone was receptive, the plant manager, the senior craftsmen . . . and when the endoscopic hit, there wasn't a [negative] attitude . . . there was a "let's try it" attitude.

The company put in processes to get greater economies in manufacturing, standardized certain things, and put together systems so that the *Old Precision* could run by itself. The *New Precision*, which is part of the vision of a growing and developing business, is in place and everyone is part of that vision. The basic thrust of all these efforts was to focus Precision Instruments' unique but limited resources on those specific manufacturing tasks where it could generate the greatest possible degree of customer satisfaction and/or competitive advantage.

Petrush was able to compete in both the newly emerging segments, as well as the more mature segments, because he had incorporated the life cycle concept into his overall strategy. Interestingly, he considered it critical to understand the effects and the interactions among the life cycle effects of each market segment, and each product, on the "skill life cycle" of his workforce:

> The life cycle concept is critical to running the business in terms of the product *and the skills of the people*. You need to know the life cycle of each instrument and each person. Everyone at this company is on the growth or development stage of the life cycle. [our emphasis]

Figures 5.1 and 5.2 tell the story of Precision Instruments in terms of the life cycle of the various market segments. Figure 5.1 illustrates where each segment was in its product life cycle, as of 1992. The figure indicates the percentage of the units that are produced in-house, using the skill-base directly (60% for cardiovascular, for example); it also shows the percentage of sale and profits for each segment, as well as indicating the general buying cycle in the business.

At the early stages of the buying cycle the surgeons and key scrub nurses are responsible for instrument selection and purchase, and make their decisions based on benefits and not price. Figure 5.2 shows how the life cycle for various segments has changed for Precision Instruments over just two years, and how many competitors there were in each segment in 1987.

An example of how the New Precision, which focuses on providing the surgeon with leading edge instrument designs, operates is illustrated in the company's development of an all-new superlight and hard titanium

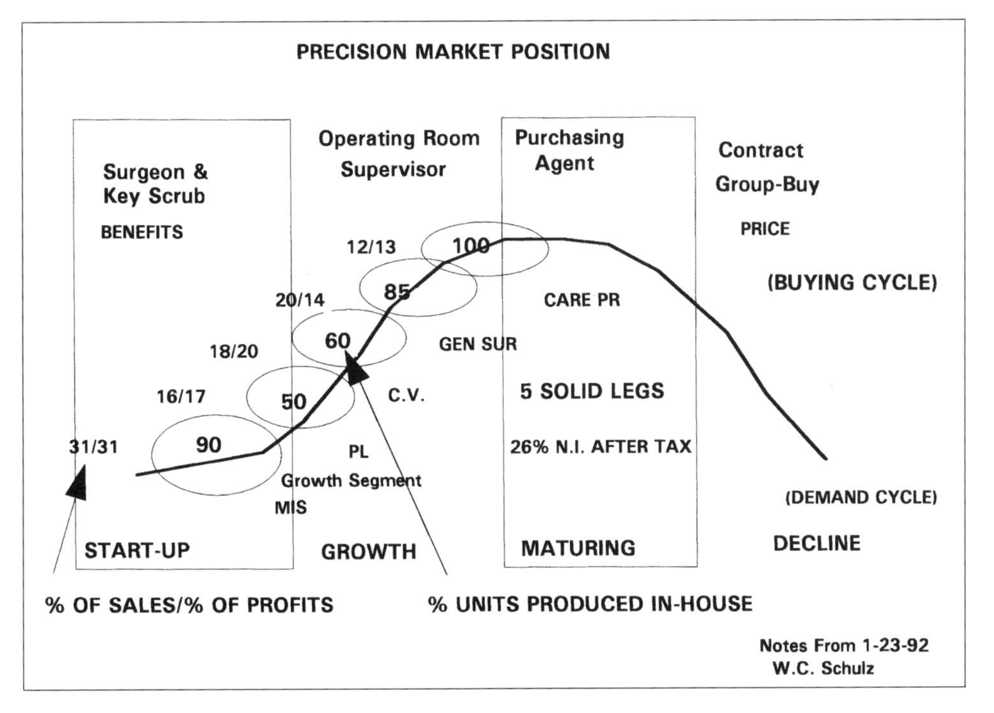

Fig. 5.1. The market position life cycle of Precision Instruments.

forceps that can be used in an innovative new heart bypass surgery procedure. As a consequence, Precision Instruments, in slightly less than a year had become the major supplier to this market, while adding to the list of its products that are on the leading edge of current surgical procedures.

With the attention Precision Instruments gave to developing innovative new products, and with its focus on providing the leading-edge customers with high-quality, high-benefit products, Precision Instruments had expanded its product line by over 70% from 1986 to 1989. By 1989 it had nearly triple the sales of its next leading competitor in carbide instruments. In fact, by 1989 Precision Instruments' product line of specialty needle holders, scissors, forceps, and rasps exceeded the combined total of its top three competitors by more than 40%. Equally important, these innovations allowed Precision Instruments to increase the average price of its instruments by almost 25% during since 1986, and to claim, without fear of contradiction, that it was the world's product and quality leader in the market segments in which it competed.

Precision Instruments has been equal to the challenge of entering new market segments and bringing the most advanced new instrument designs to market. In terms of company objectives, "we want to be number one or two in every market we participate in", according to Petrush, and, "I'm not going to wallow in the back-wash of mediocrity. My firm is going to move

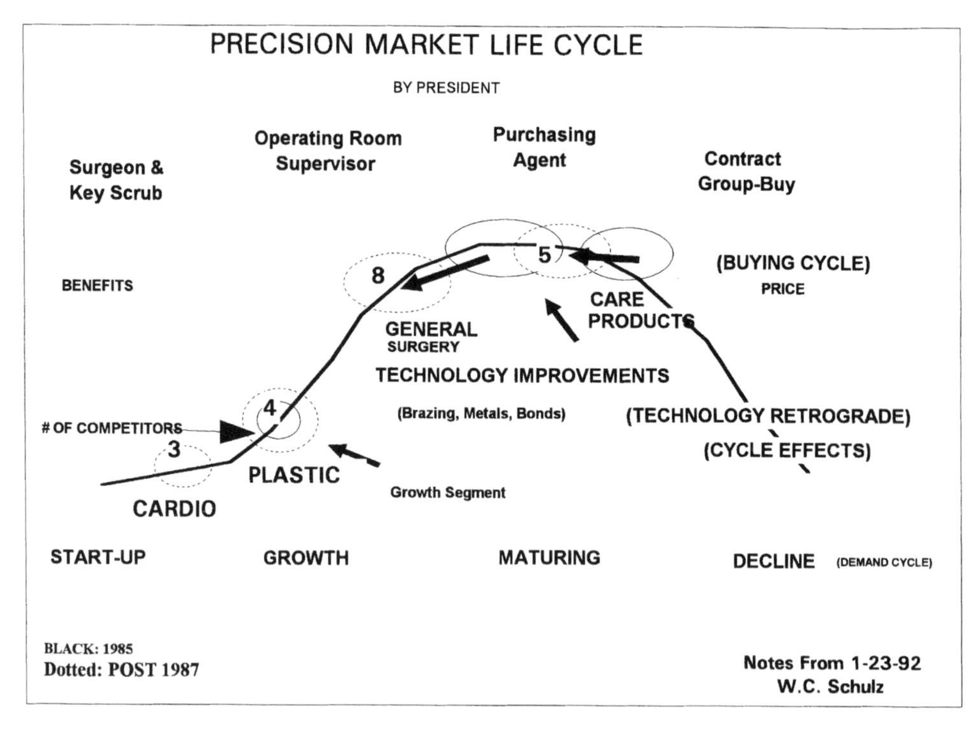

Fig. 5.2. The market and technology life cycle at Precision Instruments.

ahead". Petrush credits much of his company's recent progress to the effort and spirit of all his employees:

> I think the key item here is the quality of the people gives us the quality of the product . . . . So, I think if I were to define Precision today, I would not define it by products, I would define it by the people that are allowing us to come to the market with better designed products, of a higher quality, faster than our competitors. And anybody who looks at this business is going to understand that.
>
> They [the workers] know I love this business, and that I respect them. I bonus them, I can't deny them the ability to achieve. They have an ownership position.
>
> These people, to the person, have allowed us to take this business where we've taken it. From the office staff . . . everybody in this organization is willing to go the extra mile, and put out the extra effort to support us.

As the Precision Instruments Company moves into the middle 1990s it is prepared to move aggressively into promising segments that will challenge the skills of its managers, staff personnel and craftsmen, and that will offer opportunities to be the world's best.

SUMMARY OF PRIOR CONSTRUCTS RELEVANT TO SECTION IV

The following theoretical constructs are relevant to Section IV of Precision Instruments' history:

Absorptive capacity (Cohen and Levinthal, 1991)
Benefits strategy (Chrisman *et al.*, 1988)
Boundary spanning (Thompson, 1967)
Core competence (Prahalad and Hamel, 1990)
Dynamic capability (Teece *et al.*, 1992, 1997; Sanchez *et al.*, 1996; Gorman and Thomas, 1997)
Economies of scope (Teece, 1980)
Flexibility (Nonaka and Johansson, 1985; Rothwell and Whiston, 1990)
Hollowing the corporation (Quinn *et al.*, 1990)
Human capital (Becker, 1964; Schultz, 1961, 1990)
Implementation (Galbraith and Kazanjian, 1986)
Institutionalization (Barnard, 1938; Selznick, 1957)
Knowledge link (Badarocco, 1990)
Learning by doing (Dutton and Thomas, 1985)
Leadership (Eastlack and McDonald, 1970)
Markets served (Abell, 1980; Day, 1981; Dutton and Freedman, 1985)
Outsourcing (Prahalad and Hamel, 1990)
Product-market life cycle (Fox, 1973; Hofer, 1975)
Skill-based competition (Naugle and Davies, 1987; Klein *et al.*, 1991)
Technology push and need pull (Burgleman and Sayles, 1986)
Value and value creation (Peters, 1984)

See applications section for details.

APPLYING EXTANT THEORY AND DEVELOPING FIELD-BASED THEORY FROM SECTION IV

Extant Theory Application

The shift in Precision Instruments' strategy from a technology-push (Burgleman and Sayles, 1986) orientation to a more customer-driven, need-pull perspective was a fundamental and profound one. Perhaps the most challenging aspect of the shift was getting Precision Instruments out "into the hunt" and in touch with leading-edge customers again. The company had lost its ability to understand the benefit needs (Peters, 1984; Chrisman *et al.*, 1988) of its customers and to respond to them. It had lost its absorptive capacity (Cohen and Levinthal, 1990).

Petrush, who has a very strong inclination for being in the field and getting to know the customer's needs intimately, began to provide the company with more absorptive capacity, which is an organizational counterpart to entrepreneurial ability (Kirzner, 1979). The absorptive capacity of a firm reflects the sum of the abilities of individuals (boundary-spanning individuals in particular (Thompson, 1967)), such as Petrush, to perceive opportunities and then make links to the productive capacity of the firm.

The additional capacity of the firm to exploit external knowledge of market needs and processes can only be of benefit to the firm if it can capitalize on the knowledge through the exercise of its skills and capabilities. Petrush recognized the need to be able to compete on skills (Naugle and Davies, 1987; Klein *et al.*, 1991) and began to aggressively reshape the skill-base of the company through a combination of training, recruitment, and careful hollowing and outsourcing (Prahalad and Hamel, 1990; Quinn *et al.*, 1990). Petrush was careful to protect what he thought was the core of his firm's skill-base, for as Prahalad and Hamel (1990, p. 84) note, "outsourcing can provide a shortcut to a more competitive product, but it typically contributes little to building the people-embodied skills that are needed to sustain product leadership".

The ability of Precision Instruments to build a dynamic capability, or ability to develop new competences (Teece *et al.*, 1992), depended heavily on Petrush being able to get the cooperation and belief of everyone in the company that it was in their best interest to learn how to design and build a variety of new products (Galbraith and Kazanjian, 1986). At this point Petrush's previous commitments to a better working environment and profit sharing served him well. He had begun to build an institution that had a core of shared values (Barnard, 1938; Selznick, 1957).

His workforce did respond and began to learn new design skills, new hand skills and new skills at dealing with customers. The company learned by doing (Dutton and Thomas, 1985) and by taking Petrush up on the challenge to build new instruments. Petrush had tapped into the latent economies of scope (and skill) of his human capital (Teece, 1980). He recognized that his company's ability to innovate was integrally linked to the talents of his people (Amendola and Bruno, 1990). The reward at Precision Instruments for the investment in human capital was much more flexibility (Rothwell and Whiston, 1990). The flexibility of Precision Instruments, which is fundamentally a skill-based strategy issue, was a critical component of its strategy to serve niche markets (Day, 1981).

Substantive Grounded Theory Development on Skill-Based Strategy and Entrepreneurial Leadership

(1) Skill-based strategy enhances the effectiveness of entrepreneurial actions.
(2) Economies of scope provide an effective mechanism for moving on emergent opportunity
(3) Core development often requires entrepreneurial competence.

(4) Understanding and managing the skill and product life cycle of a company is a critical aspect of effective skill-based strategy.
(5) There is a strong interaction between skill-base development and the development of institutional character

Skill-Based Strategy Enhances the Effectiveness of Entrepreneurial Actions

In crafting the skill-based strategy of Precision Instruments, Petrush recognized that by focusing on the human resources he had at his disposal and growing them, he could affect positive change in both the operating capacity of his firm and its strategic flexibility. Skill development at Precision Instruments is both an operational and strategic factor to the business:

> You can't separate the two of them, if you don't have skills and competency, you're . . . my scrap rate would be on the moon. It would kill me operationally, financially. Now strategically, we couldn't do what we are doing, we couldn't be moving into new markets . . . so one of the things we are, we are actually planning right now based on the level of competency in our plant.
>
> We are having meetings now about what we can do, how much we can expand, and how fast. Because right now we are like standing in the middle of a gold mine, and it's how big a shovel we have to knock nuggets down and haul them out the door.

The switch to a more customer-driven orientation made the need for a strong skill-base at Precision Company, and strategic attention to the skill-base, more important than ever. The company had to have the broad capability to respond to discrete and specific requests from "the leading edge customers [early adopters] that others look to for buying behavior". This orientation put new demands on the company's skill-base in terms of instrument design and craftsmanship capability and metallurgical knowledge.

The strategic effects of the investment and development of Precision Instrument's skill base are apparent in the explosion of new products and markets the company has entered the past three years. Petrush has been able to blend two very important skill-based resources into a powerful engine of value creation: (1) his and his marketing manager's skills in searching and perceiving market opportunities, and (2) his firm's overall skill level and capability in designing and manufacturing the highest quality, most functional instruments that can meet the challenge of the opportunity.

Petrush, whose scanning of his customer environment is really quite intense and ethnographic in nature—he goes into the field, listens, talks, watches, tries to understand the actual world of his customer—is a classic example of the entrepreneur that Kirzner (1979) envisions. He is a seeker of opportunity and disequilibrium in the field of medical instruments. He is also the classic Schumpeterian entrepreneur, who listens to and watches his habitat closely, seeking to leverage his skills in new ways and achieve protected niches of high value production. Petrush seems to recognize, as Dutton and Thomas (1985) observe:

> No niche is permanent, and all advantages are temporary. Optimization is local in space and temporary in time. Nevertheless, by combining state-of-the-art knowledge [embodied skill] in any applied field [technique] with effort [persistence] and luck, major temporary advantages can be discovered and with skill can be converted into substantial economic and social gains (p. 51).

The entrepreneurial process at Precision Instruments clearly shows that opportunity is perceived as external to the firm and that it is customer driven. The internal entrepreneurial skill that is required is to be able to know which opportunities to explore. The experiences at Precision Instruments suggest that the ability to make effective entrepreneurial choices, those that consciously match customers with skills where value can be created and sustained, is greatly enhanced when the entrepreneur makes explicit reference to the skill and technology base of the company or resources under his or her influence.

Economies of Scope Provide an Effective Mechanism for Moving on Emergent Opportunity

Petrush has stated that his company is defined by the relationship between its ability to: (1) recognize a design opportunity for a new instrument, (2) come up with the latest instrument design that incorporates the most advanced metallurgy and engineering, and then (3) utilize the highest craft skills available to manufacture to the highest quality standards, providing the customer with maximum benefit.

> So, bringing it back to the definition of ourselves relative to most of the other people in the market . . . anybody can go out and get machine-shop work done very quickly, anybody can do machine work. What we find though, is to react very quickly to the market, we can take our skilled instrument makers and, with some training and working together with us as market managers, and with some of the surgeons that we bring in here . . . and working with our engineer, we can come out with a better design, faster than anyone else in the market.

This process of opportunity recognition and capitalization provides a solid and general description of the core competence of this firm. It has developed an ability to blend various technologies and assemble an array of resources to pursue opportunity in a nearly *ad hoc* manner. That is, the strategy is to have a broad and deep set of skills with which the firm can generate economies of scope. These economies let it flexibly move into un-anticipated areas of opportunities as they are told directly to Precision Instruments by surgeons, or are discovered by Petrush and others in the hunt. The firm's skill-base really drives its ability to exploit emergent processes quickly. The need for speed, without a loss of quality, puts direct pressure on the management to make sure that the skill-base of the company is continuously improving, and it also puts an entrepreneurial responsibility on the leadership to find the right markets and new markets to achieve dominance in.

Core Development Often Requires Entrepreneurial Competence

During the eight years that Demaria and Petrush have managed Precision Instruments they have led the effort to reshape the skills and the focus of those skills in the company. They have, in essence, come to know their firm's skill-base intimately, and have developed and then segmented their skill-base so that the diverse needs of both the general surgery market and the minimal evasive surgery market can be well served by Precision Instruments. Petrush has been able to bring the quality products to market faster than his rivals through a mixture of outsourcing and core development. He has, as Quinn *et al.* (1990) discuss, hollowed out his company around the core skill elements, and has expanded the core as necessary to better serve the customer.

When he has been unable to bring core skills to the company because the underlying technology and skills are so new, he outsources for them and participates in joint ventures:

> We knew when we got into the endoscopic area that we could make certain aspects of the instrumentation . . . and there were some things we could not do. Although in the long term we want to increase our levels of skill . . . . We're outsourcing. We need equipment to be competitive in the broad sense of the market, with the other players in our segment, so we've had to go outside and design, and have made for us, state-of-the-art [video and electronic] equipment. We've now employed as a consultant, a very experienced, proven, electronics design engineer.

This ability to find the proper skill resources externally that complement the internally controlled skills is an important element of competence as we have discussed the term. The concept of competence, of the blending of resources and assembly of skills, goes beyond owning the resources: it is really about coordinating them. *Entrepreneurial competence*, which is a new term, addresses the ability of the firm's boundary spanners (leaders, entrepreneurial managers and mavericks) to link all these externally controlled (but influenced) and internally controlled resources to market opportunities that: (1) they have recognized, and (2) they feel will afford the company a chance to create new value production activities. Petrush believes that it is his responsibility, as an entrepreneur, to ensure that the skills of his workforce are always developing and improving—that the skills do not become obsolete—by seeking new markets and applications for those skills:

> If anything like that happens in the plant it is my fault. If I had not taken these people in the plant into endoscopy a year and half ago, then they would be facing obsolescence, because they would be working on instruments that are being used on a declining basis.
>
> So, that's the challenge: our challenge as managers is to insure that our company is on the leading edge of the market; and that drags everybody into that vortex, and they have to increase their levels of skill applying to new products. This obviates any obsolescence.

Understanding and Managing the Skill and Product Life Cycle of a Company is a Critical Aspect of Effective Skill-Based Strategy

An interesting tool that Petrush uses to assess his skill position is the product-market life cycle. He explicitly links the product life cycle effects, down to the single instrument, to the skill life cycle of his personnel. This involves an understanding of the various hand-skills that are required to produce different instruments, and of the skills of his engineers, and even his marketing and staff personnel. Keeping everyone on the growth and development part of the skill cycle is an investment in human capital and is a direct challenge to the entrepreneurial abilities of the top management team. For as Petrush acknowledges, it is their responsibility to make sure that the skill base does not erode. This puts tremendous pressure to find new and successful more skill-intensive applications for his workforce. A potentially large side-payment, if you will, in this dynamic is that the company develops capabilities to create new opportunities and disrupt the market-place as it attempts to find new applications for its new skills. The skill-based firm and its leadership become more like the Schumpeterian innovator and less like the Kirznerian entrepreneur as it presses to sustain itself.

There Is a Strong Interaction Between Skill-Base Development and the Development of Institutional Character

The final element that has enabled Petrush to turn the company around so successfully has been his commitment to building a strong institutional character to his company, one that values the skills and creativity of its workforce. The nature of the skill-base, overlaid with the institutional character of the company, enables it to bring products to the market fast and with quality. There is a real craft mentality in all areas of the firm and a passion for quality, service and innovation.

The institutional character (Selznick, 1959) that exists now at Precision Instruments is profoundly different from the organizational character

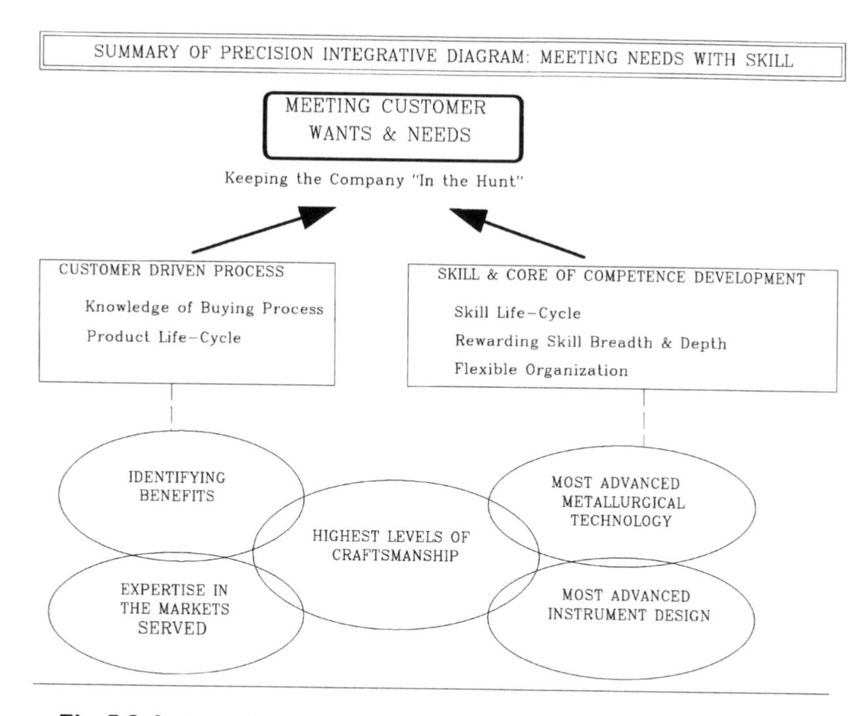

Fig. 5.3. Integrative diagram summary of Precision Instruments

(Selznick, 1957) of Precision Instruments under the stewardship of the professional managers of the big medical firm that owned it from 1971 until 1984. As the skills of the workforce have gained explicit attention, and have gained more commitment from top management as a vital strategic resource, the character of the company has begun to mature and provide a base from which more aggressive actions, to pursue more difficult opportunities, are being taken.

INTEGRATIVE DIAGRAM SUMMARY OF PRECISION INSTRUMENTS

Figure 5.3 on the next page summarizes the practitioner-based, or idiographic-level, theory that has been identified at the Precision Instruments location. This summary will be used for cross-case analysis in Chapter nine.

IDIOGRAPHIC-LEVEL THEORY SUMMARY: PRECISION INSTRUMENTS

Firm Growth and Pace of Development

(1) Building a craft-oriented skill-base takes time and commitment.
(2) Even a short-term harvest can severely damage the skill-base.

Development of Systems to Protect Skills

(3) Building skills requires consistent implementation of skill-based systems.

Entrepreneurial Leadership and Skill Development

(4) Skill-based strategy enhances the effectiveness of entrepreneurial actions.
(5) Economies of scope provide an effective mechanism for moving on emergent opportunity.
(6) Core development often requires entrepreneurial competence.

Life Cycle Effects on Skills

(7) Understanding and managing the skill and product life cycle of a company is a critical aspect of effective skill-based strategy.

Institutional Competence and Skills

(8) There is a strong interaction between skill-base development and the development of institutional character.

NOTE

1. Available information about the early history of Precision Instruments is extremely limited, as it was a private company and there are no surviving members still with the company.

6

ENTREPRENEURIAL LEADERSHIP AND SKILL-BASED STRATEGY IN A TECHNOLOGICALLY SIMPLE, CRAFT-ORIENTED, STABLE BUT CYCLICAL MANUFACTURING ENVIRONMENT: THE CASE OF CUSTOM CABINETS COMPANY

INTRODUCTION

Custom Cabinets Company (CCC) is a manufacturer of custom-made wood cabinets designed for relatively expensive homes. The company competes primarily in the Southern USA and has one of the most competitive turnaround times for custom design, manufacture and installation of cabinets in the nation. The average turnaround time in the industry for an entire room of cabinets, from initial design conversations with the customer to final installation, is nearly six weeks. Custom Cabinets Company's turnaround is less than 10 days.

Custom Cabinets Company has been chosen for inclusion in this book for a variety of reasons. One is that the company relies heavily on its craftsmanship skills and its human capital, in a relatively low-technology intensive setting, to design and manufacture its high-quality products as quickly as possible. CCC thus provides an excellent source for examining skill-based strategy concepts, especially as they relate to the development of organizational systems and communication. Second, the 1992 leadership of Custom Cabinets Company engineered an impressive turnaround of the company after it went into Chapter 11 Bankruptcy in 1985. Thus the case provides an interesting context for exploring organizational learning phenomena, especially in terms of how the founding entrepreneur learned to adapt his strategy in order to better exploit the skill-base of his company. Third, Custom Cabinets Company is also one of the leading practitioners of the "Theory of Constraints" (TOC), or "Synchronous Manufacturing" and provides a rich source of information on how implementation of TOC links operations and strategy in a process of continuous analysis and improvement.

The Custom Cabinets Company case differs most from the others in this study in that it exercises its craft skills in the face of a simple but highly cyclical market that is often affected by external elements, such as the weather. It also operates, unlike the other three cases, in a relatively low-technology setting, although it can be readily argued that this is changing fast.

CCC'S HISTORY, SECTION I: THE FOUNDING OF CUSTOM CABINETS COMPANY AND TRIM

In 1969 Mike Viera arrived in Town, South, to begin study in secondary education at South University, and to prepare for a career teaching biology and coaching football. In his junior year he was married and began to work part-time. Despite the fact that he had gone to South University for three years and had studied in a craft, teaching, his perception once he got married was that, "the only thing I knew anything about was carpenter work. And I went to work [part-time] for a carpenter and subcontractor . . . " [our emphasis].

Due to the cyclical nature of the building industry, Mike was laid off after six months of work, but he capitalized on an opportunity to do all the "finish work", such as building the cabinets, baseboards and panelling, for two speculation houses that an architecture professor from South University was building. "I borrowed the equipment, and during the holidays I did all the finish work on those two houses . . . the first full-sets of cabinets I'd ever built".

Mike had a natural aptitude for the carpentry and cabinet business, and enjoyed the work and the direct link between the work and its rewards. Although he had graduated from the University's secondary education program, he pursued the cabinet and finishing business.

> I liked what I was doing, I liked it particularly because if *I was smart and did a better job*, I'd make more money . . . . Within the field of school teaching, whether you did a real good job or a lousy job you got paid the same. *It wasn't a very good system.* [our emphasis]

Mike felt particularly strong in his abilities "for observation and paying attention to details, and . . . to think through how, and what would the steps be that you need to make something, and put it together". After carefully analyzing how much time he had put into the job at the speculation houses, and how much money he had made, Mike decided to commit to expanding his efforts in the cabinet and finishing area.

> The next month I went to a hardware store and bought a table saw on credit, thirty day terms. My wife cried when I told her, and I bid on some other jobs and got those . . . . so that was how the business began . . . . I started with a motorcycle, a skill-saw and an extension-cord—and no money.

SUMMARY OF PRIOR CONSTRUCTS RELEVANT TO SECTION I

The following theoretical constructs and theoretical propositions are relevant to Section I of CCC's history:

Deep or embedded knowledge (Winter, 1987; Mintzberg and Westley, 1989; Waddock, 1989; Badarocco, 1991; Spender, 1996; Grant, 1996)
Entrepreneurial skills (Herron, 1990)
Learning by doing (Dutton and Thomas, 1985)
Path dependency and serendipity (Barney, 1986b)

See applications section for details.

APPLYING EXTANT THEORY AND DEVELOPING FIELD-BASED SUBSTANTIVE THEORY FROM SECTION I

Extant Theory Application

There is an important link between the skills of entrepreneurs, their deep or embedded knowledge (Winter, 1987; Mintzberg and Westley, 1989; Waddock, 1989; Badarocco, 1990) of an activity or business, and the potential success of new ventures that they lead (Herron, 1990). In the case of Mike Viera, a pressing need to support his family, combined with his skills as a wood craftsman—along with a set of serendipitous events (Barney, 1986b)—led him to form a new company despite never having planned to do so.

Once in business, Mike learned by doing (Dutton and Thomas, 1985), and found satisfaction in that as he learned to do better, to leverage his technical and administrative skills (Herron, 1990), he was rewarded for his efforts.

SUBSTANTIVE GROUNDED THEORY DEVELOPMENT ON SKILL-BASED STRATEGY AND ENTREPRENEURIAL LEADERSHIP

(1) Businesses emerge from what their founders know and enjoy—leveraging key skills.
   (a) Skills and opportunity
   (b) Experience as knowledge
   (c) Craft and knowledge skills
   (d) Leveraging skills in making entrepreneurial choices
   (e) Rewards: the positive effects of leveraging key skills

Businesses Emerge From What Their Founders Know and Enjoy—Leveraging Key Skills

Skills and Opportunity

Every organization that exists has, at its genesis, a founder or group of founders that literally forged the organization out of the raw materials of knowledge, money and machinery. These founders often provide the substantive base, or roots, upon which the firm grows and develops in terms of energy, vision, skill and values. More often than not one might suspect that the founders of a successful firm began with a planned, rational, grand vision of their organization and what they wanted to achieve with it. However, as was the case with Custom Cabinets Company and Trim (now Custom Cabinets Company, Inc.), another archetype of the founding and development of firms is that they emerge over time out of unique and serendipitous mixes of: (1) personal characteristics and needs, such as the skills of the founder, and his or her abilities to recognize an opportunity, and then to learn and innovate, and (2) environmental circumstance and opportunity.

Experience as Knowledge

In 1969 Mike Viera had little idea that he would be at the helm of a small custom cabinet manufacturing company five years later. However, there are some important factors, which are not necessarily unique to Mike, that bear on the fact that he eventually came to establish a company.

First, when confronted with the need for income, Mike fell back on what he felt he knew, and what he felt he knew was what he had experience with. Mike seemed to perceive that experience was a critical form of knowledge, and that knowledge was the key resource to finding work that would pay adequately. Mike's knowledge-base was a key guiding and limiting resource in his pursuit of opportunity. This suggests that skill-based resources, at the least, are a prerequisite to any opportunity-seeking behaviour by entrepreneurs, and more generally, organizations.

Craft and Knowledge Skills

Mike's knowledge-base was rooted in two distinct types of general skills which Mike himself identified as "craft" skills and "knowledge" skills. (1) Craft skills are those that involve the hands-on application of specific techniques and expertise with regards to the products or services in question, in this case woodworking; (2) knowledge skills are general skills that are related to identifying, understanding and reasoning through problems and opportunities.

Leveraging Skills in making Entrepreneurial Choices

In pursuing the cabinet-making business Mike had made an entrepreneurial choice in terms of how he could best utilize his skills and knowledge to create value. He had leveraged the skill-based resources he had at his disposal, his natural and practised craft skills and aptitude at working with wood, and his analytic knowledge skills in "putting things together" to create the core of his business. He had few other resources, in terms of money and equipment, but ultimately built his business from the craft and knowledge skills he had developed through all the experiences of his life. Interestingly, it appears that little of Mike's substantive and important skills were picked up formally in school. It also appears that Mike went out of his way, in terms of what his formal training had prepared him for, to respond to an external opportunity that would allow him to leverage what he thought were his key skills. He made a resource-based decision to exploit an opportunity.

Rewards: the Positive Effects of Leveraging Key Skills

The ability of Mike to have learned and improved his craft leads to a second major factor that addresses why Mike would form his own company: he both "enjoyed the work" and enjoyed the fact that as his experience, skill and knowledge-base grew, and his work subsequently got better, he was rewarded for improvement. The intrinsic rewards of the craft, combined with the positive external feedback (making more money) he received as he worked smarter hooked Mike, in a sense, into the business. He could work smarter in an area he had aptitude and experience in, he enjoyed the work in part because he could work smarter in it, and could get commensurate rewards for his improved efforts. This, combined with the fact that Mike desired more control over the process of making cabinets, and possessed a strong set of analytic and knowledge skills, suggests that the logical outlet was for him to start his own company. The substantive relationships between some of the characteristics of Mike and the founding of his company are illustrated in Fig. 6.1. Note that both Mike's knowledge and craft skills were integral to his ability to learn and then apply that

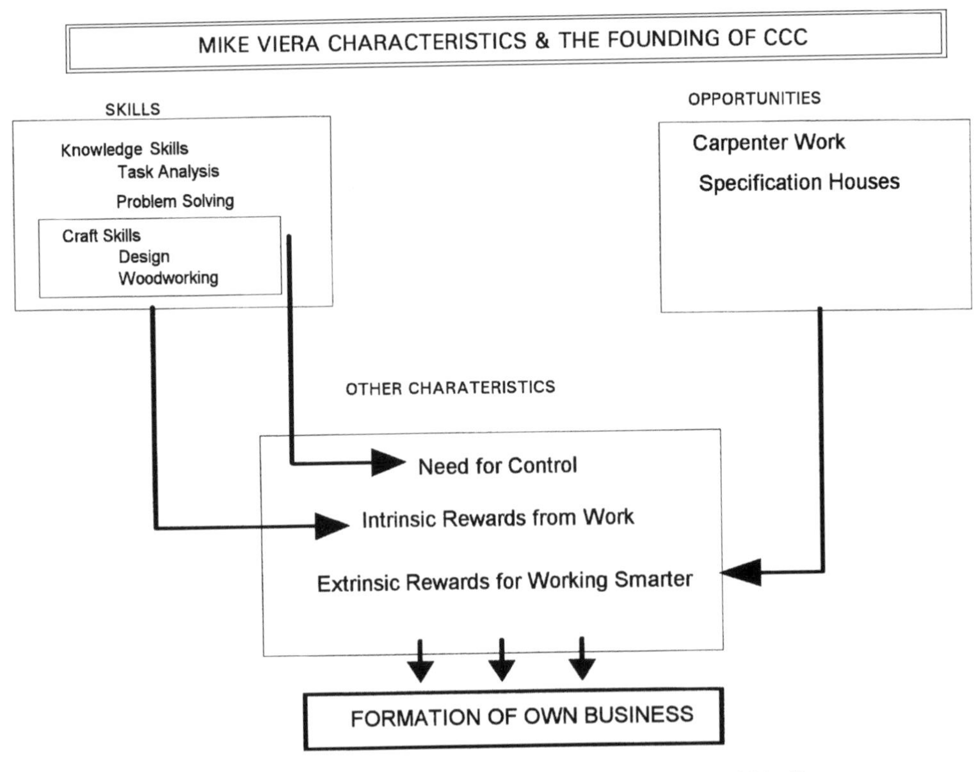

Fig. 6.1. Mike Viera: characteristics and founding of his firm.

learning in the form of "working smarter". This figure suggests that the formation of Mike's business occurred because, in part, it provided a means for him to leverage all his skills at the time, and it also provided him a means to exercise control and achieve proportional external rewards for his efforts.

SECTION II: CUSTOM CABINETS COMPANY, INC. (CCC) EARLY FIRM DEVELOPMENT

In 1972, Mike Viera was a one-man company. However, by 1975 he had hired 10 employees, including Joe Banks as a cabinetmaker and draftsman (Joe is CCC's current Vice-President of Development) and Mike had built a 1500 square foot building that he needed to build certain door styles for cabinets. Most of the firm's cabinet-building at this point was prepared "on-site" and to exact customer specifications. However, Mike wanted to build more of the "custom" cabinets in the shop so he could control the size and quality better. Because of the expectations by the customers that he would build to tight tolerance specifications, and because he had to build the doors in the shop, Mike felt he had to define a system that would help him build an entire cabinet set in the shop, and that would still meet precise customer needs.

> So I had to develop a system that could build-in a lot of variety, and be able to change the sizes in the shop . . . . In '74–75, it became apparent that I was going to have to get a lot better at figuring all these numbers for parts lists . . . to find some other way.

At this point Mike himself had become a constraint to his firm's growth:

> As we started doing more work in the cabinet shop . . . the routine was that I'd measure a job, and I'd sit down at night and draw up everything and give them all the dimensions . . . it became apparent that I needed to come up with some other kind of system or I was going to be trapped into doing this the rest of my life. There was more work available . . . I didn't set a goal, a certain number of dollars in sales, but I wanted to begin to continue to grow . . . grow based on what we could handle.

In order to tackle the problem, Mike enlisted the help of Todd Weaver, a brilliant computer software pioneer and friend, and Joe Banks. Mike and Todd worked one night a week for two years until they had developed a sophisticated, powerful tool (CABSYS) to automatically generate, on a personal computer, the manufacturing instructions for any "custom" job (an entire kitchen, room etc.) that had been designed by Mike. This was no mean task. Back in the late 1970s, when Mike and Todd, "embarked on a project of taking what was in my [Mike's] head, the formulas and all . . . making the calculations and putting them in the computer", there were few if any such automated design/manufacturing translation programs or expert systems available. "There was certainly nothing available commercially", according to Mike. Despite the inevitable hardware problems that Mike and Todd Weaver faced, Mike saw the new program as, "being such a powerful tool for us that I was willing to put up with all the mechanical problems we had . . . but it gave us a big advantage because we could generate this cutting list quickly". In order to get the computer to actually produce a cutting list that the shop floor could use, Mike worked with Joe Banks and others to debug the program. Joe had been a trim carpenter and cabinetmaker with CCC since 1975 and was trained to do the drafting that Mike had done earlier. This put him in an ideal position to be able to help with the implementation of CABSYS. He understood both the needs of the cabinetmakers and the shop-floor personnel, and he knew how the drafting function fit into the CABSYS process. Working together, Mike, Todd, Joe and the company were able to develop and implement a tool that would prove to have a significant impact on both how Custom Cabinets Company would function and compete in the business.

The CABSYS innovation was so powerful that, according to Mike, "even now, [1992] there's like huge [commercial] programs, but none of them are nearly as sophisticated as our program. Most of them will do a cabinet at a time, ours does an entire job at a time". Mike continues: "Now we have a full-time programmer on staff, we have moved from a CP/M environment to a Prime mini-computer . . . there are thousands of lines of codes . . . we are way beyond where Todd Weaver had started".

The conversion process from initial field measurement of a job to the generation of actual parts lists for assembly had become streamlined and the innovation had both freed Mike to work on other areas of his business and provided a strong foundation from which to structure the entire company.

> It allowed me now, to go work on other things—we started training drafting people that would take the field measurements, and they would draw up the job, and they would put in the designations for the different cabinet boxes, and then an entry person would produce a list . . . and that is essentially the way we are today.

With Mike able to leave the complicated measuring–drawing–cutting conversion process to Joe and others, including Stasi Bara, who was an installation crew foreman (and future Vice-President of Operations), he was free to pursue more business opportunities for the company. The fruits of this early freedom are illustrated in Fig. 6.2. This shows the monthly sales from November 1979 until September 1982, and shows the exponential trend of that series.

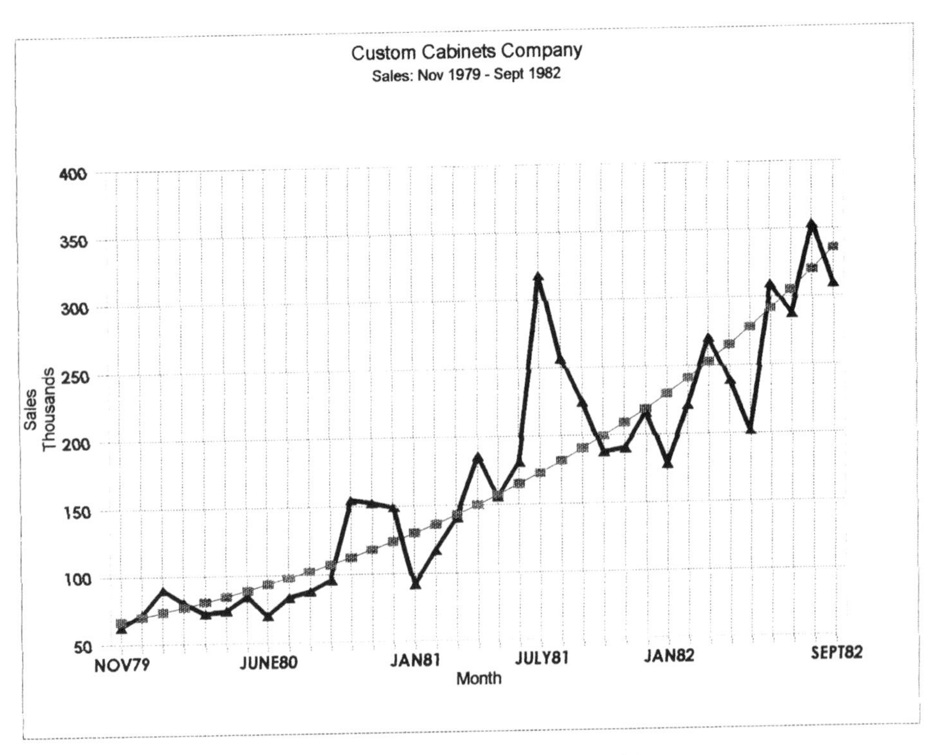

Fig. 6.2. CCC sales, 1979–1981.

As can be seen in Fig. 6.2, Custom Cabinets Company did begin to grow rapidly in terms of sales after the development of CABSYS. Between December 1979 and December of 1981 sales more than doubled, going from just less than $75,000 per month to nearly $180,000 per month in the two years. Despite the cyclical nature of the business, CCC did not post any significant losses in 1980, as measured by a rolling three-month average. Sales growth through 1980 and into the summer of 1981 was brisk. By the end of 1981 annual sales had doubled those of 1980 and reached $2 million.

However, in late 1981 CCC experienced its first real fall-off in sales, and also lost money, in terms of a quarterly average, for the first time. In order to cope with the strains of growth and to prepare for the future, Mike hired Scott Smith as Chief Financial Officer in 1982. Smith joined a company where the seeds for long-term success appeared to have been sown. By 1982 Mike had hired all but one member of what would evolve into his 1992 top-management team, and his company was improving its technical capabilities each year. Yet, despite Mike's technical success in building the CABSYS computer and organizational system, CCC lacked the basic financial and administrative systems necessary to sustain its growth and enable it to respond to rapid changes in the market-place. Smith emphasizes the point: "when I got here they didn't know how much money they had in the bank or anything".

The losses in late 1981 foreshadowed a much larger problem that caught Mike and the company by surprise two years later. Sales had rebounded in early 1982 but went flat in late 1982. They remained flat in 1983 until the summer, normally the peak time in the business, when they suddenly fell nearly 70%, as illustrated in Fig. 6.3, which shows sales from the September 1982 period until November 1984—a striking contrast to the earlier sales figure.

To put this in some perspective, the US national economy in the late 1970s was in very poor health, but the South economy, and especially Big-city, CCC's largest market, were somewhat insulated from the shocks that had hit the automobile manufacturing sector and the north-east. By late 1983, however, the difficult economic times had finally caught up with South and Custom Cabinets Company, Inc. It would prove to be a defining moment for Mike and his company.

SUMMARY OF PRIOR CONSTRUCTS RELEVANT TO SECTION II

The following theoretical constructs and theoretical propositions are relevant to Section II of CCC's history:

Absorptive capacity (Cohen and Levinthal, 1991)
Deep or embedded knowledge (Winter, 1987; Mintzberg and Westley, 1989; Waddock, 1989; Badarocco, 1991)

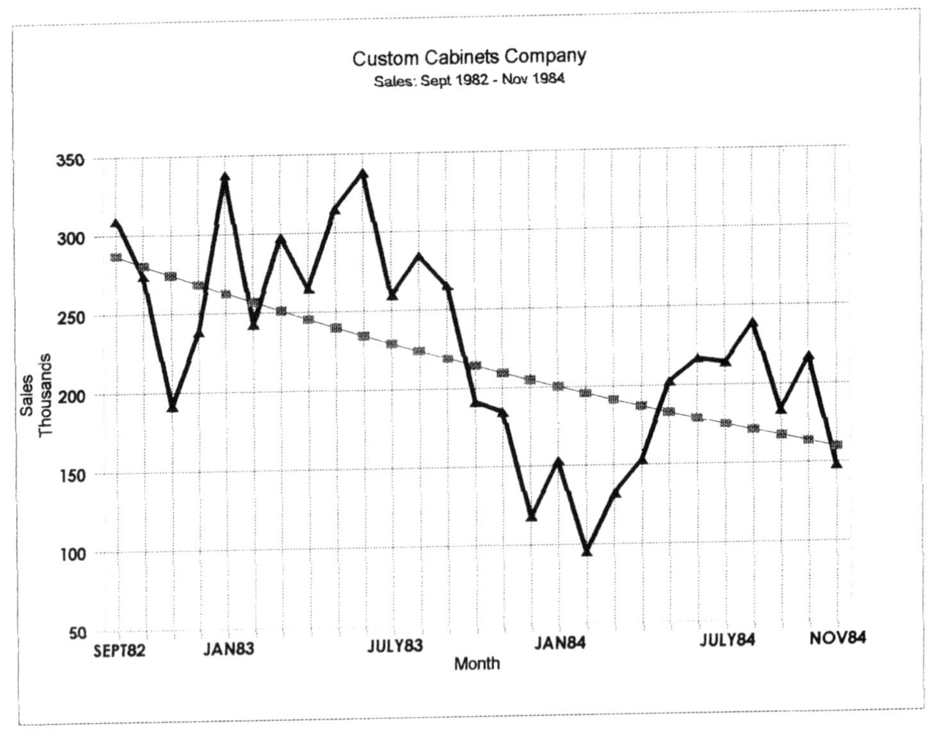

Fig. 6.3. CCC sales, 1982–1984.

Implementation (Galbraith and Kazanjian, 1986)
Innovation (Drucker, 1985)
Organizational routines (Nelson and Winter, 1982)
Schumpeterian innovation (Schumpeter, 1934; Nelson and Winter, 1982)
Systems (Ansoff, 1965; Goldratt, 1990)
Transilience (Abernathy and Clark, 1985)

See applications section for details.

APPLYING EXTANT THEORY AND DEVELOPING FIELD-BASED SUBSTANTIVE THEORY FROM SECTION II

Extant Theory Application

During the early years of CCC, Mike was a constraint on the growth of the company because much of the deep knowledge (Winter, 1987; Mintzberg and Westley, 1989; Waddock, 1989; Badarocco, 1991) he had about the business was in his head, and he had to do the work. The ability of Mike to go into the market and learn from his customers was limited by his need to be in the shop; thus the absorptive capacity (Cohen and Levinthal, 1990) was very low, and his ability to recognize new opportunities (Kirzner, 1979) was also limited.

Mike looked to a friend and outside expert to help him develop a truly architectural innovation (Abernathy and Clark, 1985), based on computer technology and software development, that would free him to become an entrepreneurial leader. Mike's move into the computerization and routinization of much of his expert knowledge was an example of a Schumpeterian innovation (Nelson and Winter, 1982) because it provided the base from which CCC could enter a market and destabilize it via strategic change. The strategic change, in this case, was that CCC could build quality custom cabinets in a much shorter time, and with more scale economies than many of its competitors. This innovation was the first step to CCC's eventual dominance of the Big-city market.

The development of CABSYS, and its subsequent impact on the structure of the firm, shows how process innovations, through careful implementation (Galbraith and Kazanjian, 1986), can lead to new organizational routines (Nelson and Winter, 1982) and structures.

SUBSTANTIVE GROUNDED THEORY DEVELOPMENT ON SKILL-BASED STRATEGY AND ENTREPRENEURIAL LEADERSHIP

Organizational systems as innovations and structural foundations for competence.

(1) Following up on entrepreneurial choices: innovations to work smarter
  (a) Breaking barriers and serving the customer
  (b) Developing new skills and competences

(2) Systems, skill and competence evolution
  (a) Organizational competence develops with systems
  (b) Skills drive systems and systems are the bridge between individual skills and organizational competencis.

Following up on Entrepreneurial Choices: Innovations to Work Smarter

Breaking Barriers and Serving the Customer

Once Mike's company began to grow, the challenges of meeting customer expectations, combined with the finite capabilities of Mike to do all the technical drafting work, put pressure on him to work smarter and break through barriers to growth, which we saw earlier is something that is central to his thinking. His ability to define and analyze tasks led him to conclude that he needed to develop a "system" that would, to some degree, allow others to do routinely what only he was capable of doing at the time.

Mike had made an entrepreneurial choice to build truly custom cabinets, off the construction site, and needed to further leverage his own skills in order to have the company grow. He followed up by linking a technical achievement, CABSYS, which is an automated translation program for measuring, drawing and cutting cabinets, with an organizational innovation: the creation of the "Custom Cabinets Company Way", which is how various individuals would interact and use CABSYS to build quality cabinets more quickly than had been possible before.

The evolution in technical, and later, organizational, systems marked "a real breakthrough" for Custom Cabinets Company, in that they provided a means for the company to structure itself to better meet customer needs in terms of both quality and custom variety. It also freed Mike to pursue other important issues, breaking a potentially dangerous barrier to growth for both Mike and his new company.

Developing New Skills and Competences

In the process of developing the cabinet manufacturing system, which was truly an important innovation for the firm, Mike and others in the firm acquired new skills and thus expanded the firm's skill-base. The company also began to develop new competences, in terms of skill and resource-assembly and blending, that would ultimately be the foundations of their long-term competitive advantage in building high-quality custom cabinets. For example, prior to Mike's working with Todd Weaver, a local computer expert, Mike and the individuals in his company had virtually no skills in computer software development or hardware use. Likewise, other than Mike, few people in the company were adept at developing and drafting jobs so that they could be used in an accurate and timely manner by the manufacturing people. Once the computerized process had been introduced, it enabled Mike to offload many of the now routinized skills and train others to do the tasks that were preventing him from looking for more business and improving the company. Indeed, the following statement from a 1992 CCC information package illustrates the "Custom Cabinets Company Way", and drives the point home that CABSYS is at the heart of CCC's organizational competencies.

> Producing the custom-made product the builder demands requires the cooperation of CCC's many departments. After the builder awards a bid, his CCC sales representative takes measurements at the construction site and sends in the cabinet order to CCC's home office via fax. Each job's cabinet layout is drafted and entered into the computer, which calculates parts and generates paperwork. These cutting instructions and identification labels follow the parts through the shop as they are cut, assembled, and finished. Most customers also take advantage of CCC's delivery and installation services.

Systems and Competence Evolution

Organizational Competence Develops with Systems

In terms of organizational competence evolution, the systems innovation that Mike introduced was the starting point for the development of important competences that would become critical to CCC's competitiveness. Prior to the innovation, however, the *company*, as a collective of individuals, really had no organizational competence, in the way we have talked about the concept in this book: Mike, by himself, was responsible for: (1) the measuring of a job, (2) the drawing of the job, and (3) manually converting the job drawing into dimensions that manufacturing could understand and cut. Finally, (4) Mike was (and still is, to a much lesser extent) also a vital resource on the manufacturing floor, as Scott Smith, CCC's Chief Financial Officer comments, "as far as somebody able to solve problems, especially out there [on the shop floor], you know, he's what I call close to a genius . . . I mean he can walk through there and see things other people never see".

Clearly Mike had, and has, skills, but since he seemed to be doing almost all the coordinating and assembly, or blending, of tasks, skills and resources (time being the most pressing) by himself in the early stages of the company's start-up, the company could not be said to have developed any organizational or core competence—all the relevant skills resided in one resource, Mike. Only after the computerized process had been implemented in terms of organizational systems and structure, and the skills and tasks had been disseminated to others, could there be a chance for the company to develop competence.

After the technical and organizational innovations: (1) new people were hired and trained to draft jobs that could be entered into the "expert system", based on the measurements provided from others in the field; (2) people were trained to use the computer and enter the data from drafting; (3) Mike and others learned how to enhance and modify the computer code based on feedback from both drafting and manufacturing personnel; and (4) manufacturing learned and gained the confidence to cut jobs based on the printed work-orders that the new, multi-person and multi-area "system" had generated. The programming code was more than a technical way to reduce time in processing an order; it was also a commitment to a structure and a system: a way of doing business that built on CCC's organizational competence in assembling, coordinating and leveraging the unique skills and general material resources at its disposal. In fact, CCC's Prime system keeps track of more than just cabinets. It handles the accounting functions such as accounts payable, accounts receivable, invoicing, and keeping the general ledger. It also maintains payroll and personnel data and records materials inventory and prices.[1]

However, where CCC developed no effective systems, there was also no organizational competence, or ability to assemble and utilize resources to a particular end. Where no resident skills resided, the organizational systems needed to enable the firm to grow and succeed could not be developed. This is illustrated by the fact that while impressive organizational systems were developed at CCC in the drafting, manufacturing and delivery activities in the firm, where Mike had strong skills and had concentrated on developing the more technical systems, there were no such skills or systems in marketing and administrative activities. This indicates that the company was not being managed as a whole system. The financial systems were weak, and the subsequent strategy or tactics of the firm reflected no real direction, in terms of the effective utilization of the firm as a complete system.

Skills Drive Systems and Systems are the Bridge Between Individual Skills and Organizational Competences

The information above provides a base to suggest that skills, and the routinization of key skills, are at the heart of the development and maintenance of organizational systems. Further, because organizational systems involve multiple people and "skill-sets", and because systems provide the context within which "basic resources", including individual skills, are coordinated and blended, they are also the bridge between individual skills and organizational competencies. The "organizational routines" that Nelson and Winter (1982) suggest are the firm's equivalent of individual skills are the working-out of, and day-to-day evolution of, the organization's systems, and hence the basis for the organization's competences.

In essence this means that competence, in the way we have defined it, cannot exist without the existence of organizational systems, which develop in response to the need for the organization to embody key skills and which provide the processes that link these skills in ways that consistently, and predictably, produce valuable services for the firm. The organizational systems that were developed at Custom Cabinets Company appear to be "skill utilization technologies" and are vital aspects of the firm's memory and learning capabilities. The link between resources, systems and competence is made indirectly by Mike in the following statement:

> I think that . . . when I look at our company, and at any company, you've got a building, a lot of equipment, and you've got a lot of people . . . and raw materials . . . by themselves none of those things really accomplish anything. *What makes all that work is a system.* [our emphasis]

Here Mike is implying that systems are the mechanisms by which his company integrates and blends the firm's resources, which have little value in and of themselves until they are brought together in pursuit of a common opportunity. Figure 6.4 illustrates the relationships between the skills Mike had, the opportunities he recognized in the market-place, and the subsequent need for systems development and organizational innovations. Note that the same set of skills and characteristics as were presented earlier, along with Mike's need for development and growth and his ability to learn, provide a context for his growing the business as well as starting it.

SECTION III: THE FIRM NEARLY DIES

By the winter of 1983 sales for CCC were back to their 1980 levels and profits were consistently negative. As Fig. 6.5 shows, profit as a percentage of sales plummeted throughout 1983, 1984 and 1985. The economic slowdown in the South in the mid-1980s, combined with CCC's lack of adequate administrative and financial systems, and a poor strategic choice in terms of target markets served, proved to be too much for the company, "we went from having a significant retained earnings, to a negative retained earnings and having to file for Chapter 11 bankruptcy [protection] in 1985". Mike reflects, "So, we went through a long period of time that was very difficult, and did a lot of unpleasant things . . . we went from about 150 people, to 35 or 40".

A contributing factor to the need for reorganization was a continuing lack of adequate financial and administrative systems at CCC in the early 1980s. Scott Smith's commentary speaks to this point:

> You know . . . all of a sudden they started losing money and couldn't figure it out. I told them, well, looks like to me that you've been selling to these

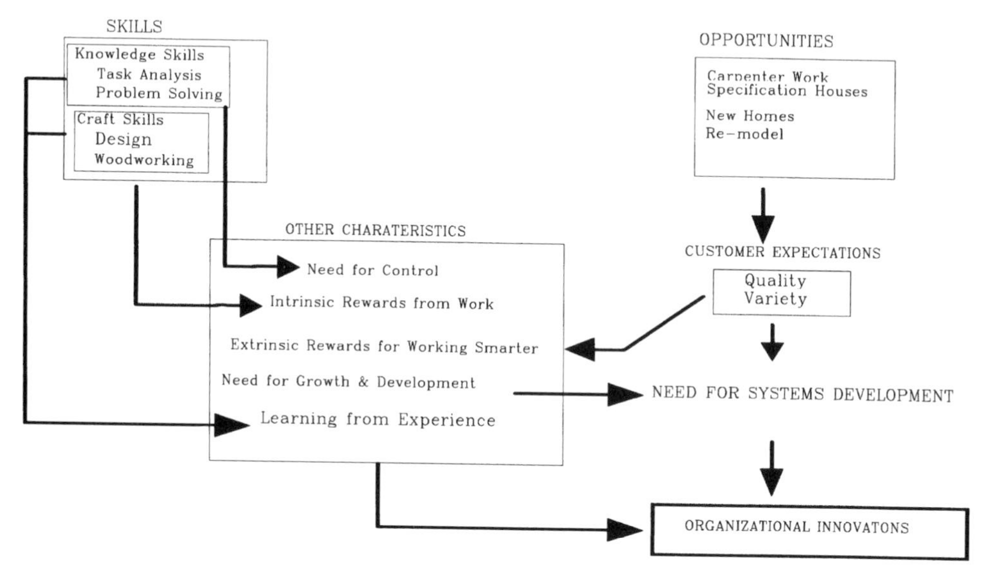

Fig. 6.4. CCC and the development of the initial system.

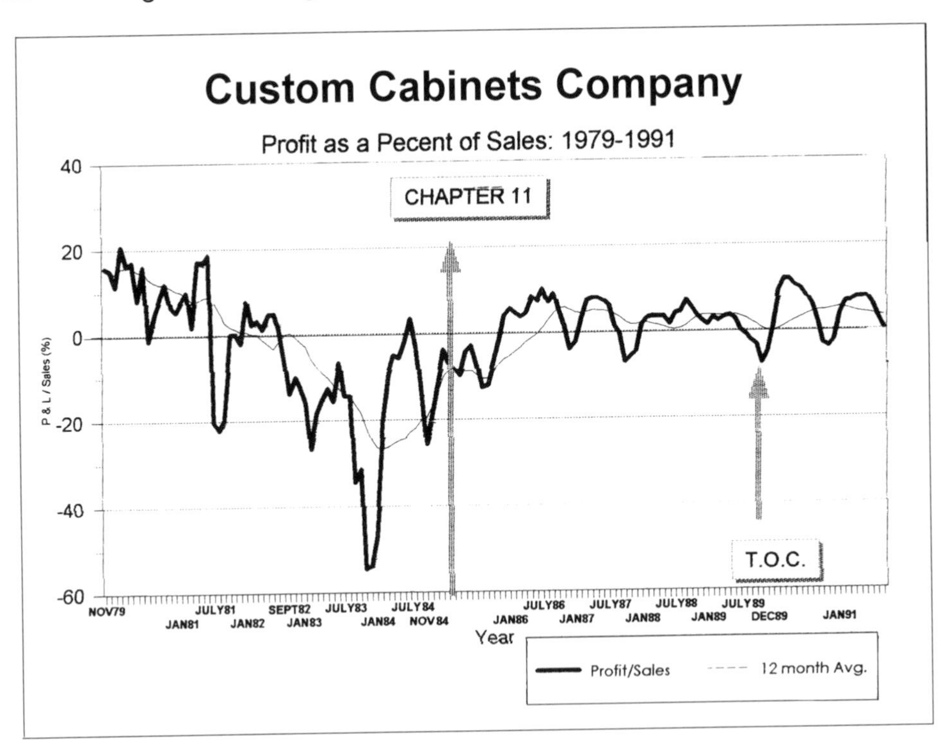

Fig. 6.5. CCC profits as a percentage of sales, 1979–1991.

> companies [in the Big-city, South market] below cost . . . and as long as you [have] these local [Town, South] customers, it makes up for it and you make profit.

Part of the problem, according to Smith, was that CCC had made a push to serve the high volume low-priced jobs in the Big-city area in an attempt to grow. "But then this local market went down for us", according to Smith, "and we lost a lot of customers . . . the low end, non-profit business in Big-city was dominating our schedule . . . so these people, the local high end, got mad and quit . . . and the economy went down at the same time".

The fact that Mike had let the low-price, high-volume jobs dominate his delivery schedule and permitted the company to slip in its overall performance delivery is a significant point, given Mike's recognition of the importance of being "reliable", which is measured by the ability to deliver a job to the site on time:

> This has probably, for me in the long run, been the most important thing, to be known as being reliable, whether I made or lost money on an individual job wasn't as important . . . as was my reputation for reliability . . . and that's what we guard against now. We don't want to take on more business than we can be reliable with.

Since the reorganization Mike has been careful to pace the growth of the company according to its capabilities to meet customer expectations. Despite the difficult times, all was not bad for the company. Custom Cabinets Company kept a strong cohort of skilled and loyal employees, continued to improve and refine its CABSYS program and cabinet-making and delivery systems, and the core of the CCC's future management team had begun to form. Mike himself made a strong commitment to turning the company around. As one company document states, "his personal efforts during CCC's 1983–85 financial crisis show his dedication: he sold his home, sold his boat, and cashed in his life insurance policies to try and keep the company afloat".[2]

SUMMARY OF PRIOR CONSTRUCTS RELEVANT TO SECTION III

The following theoretical constructs and theoretical propositions are relevant to Section III of CCC's history:

Capacity utilization (Ansoff, 1978; Goldratt, 1990)
Core competence (Prahalad and Hamel, 1990)
Strategy as fit, building on strength (Hofer and Schendel, 1978)

See applications section for details.

APPLYING EXTANT THEORY AND DEVELOPING FIELD-BASED SUBSTANTIVE THEORY FROM SECTION III

Extant Theory Application

When Custom Cabinets Company began to grow and exercise its new systems and structure it found that it had an abundance of capacity early on. The company had developed a core competence (Prahalad and Hamel, 1990) but had not been aware of its limitations and strategic implications. Mike utilized his firm's capacity (Ansoff, 1978; Goldratt, 1990) so that the good money, from high-end local work, was driven out by the bad money from low-end work in Big-city.

The firm did not pace its growth on the careful and strategic utilization of its capacity and constraints, and failed to craft a strategy that would build on its strengths (Hofer and Schendel, 1978). Ultimately this mismatch between firm capacity and skill and their utilization was a major factor that led to bankruptcy in 1985.

SUBSTANTIVE GROUNDED THEORY DEVELOPMENT ON SKILL-BASED STRATEGY AND ENTREPRENEURIAL LEADERSHIP

(1) The failure to understand the boundary of core skills can be fatal.
(2) The need for a skill-based strategy that is rooted in competent systems.

The Failure to Understand the Boundary of Core Skills can be Fatal

The middle 1980s were trying and nearly fatal years for the cabinet-making company in South. Due to a combination of poor economic conditions, and poor strategic judgement by management, the company lost over 100 people and the confidence of many of its most important customers.

Perhaps the most telling mistake management made was its failure to understand the boundaries, or "contour guidelines", of its core skills and competence, especially in terms of appropriate strategic actions for the firm. For while skills and systems may provide many avenues for action, there are inherent limits to their effective deployment; and it is the role of the strategist to understand these limits and to either work within them or seek other fitting ways to use them.

The lack of sufficient understanding of CCC's core competence and its link to competitive advantage led to an overloading of the system's capacity, especially its capacity to deliver the service it was designed to provide—custom-crafted cabinets. There was a subsequent loss of quality and reliability, which was one of Mike's guiding principles. Obviously CCC lost sight of who they were and got hurt when the market went bad. It appears that in the early 1980s Mike did not appropriately utilize the skills and systems that CCC had crafted since the mid 1970s. He failed to recognize the strategic implications of the systems he had in place in terms of how he could best serve customers. He elaborates:

> In the 1980 to 1985 time period . . . we, locally, if one of our builder customers was going to build a $300,000 house, we'd build cabinets for him. If he was going to build a $50,000 duplex, we'd put cabinets in for that duplex . . . and we did. *But our whole system was designed to build custom things*. Where we had the advantage; where we were most important to the customer was in the custom house [segment], where quality and variety and flexibility and good service were the most important things. More important than just what the base price was.

If one reviews the description of the CCC Way, and then looks at the specific skills and competence needed to blend those skills together, it is apparent that the system is tightly defined to provide a high-service, custom-crafted product. Mike describes some of these skills:

> Well . . . one element in that [the skills in the company], is our sales staff . . . they are really the link between what the customer wants and our manufacturing capability. They've got to describe to manufacturing what the customer wants as quickly and accurately as possible. And that's a big undertaking for us because we have such a wide set of options available. So our sales position requires a pretty good level of expertise between knowing what we can do manufacturing-wise and trying to make that match for what the customer wants. Drafting sits right in between sales

> and manufacturing . . . they've got to take that information and convert it into language that manufacturing can produce a product with.
>
> So I see that we have . . . we should have people in sales that are very good at communicating and capturing details . . . in drafting we should have . . . and obviously we do . . . not to the level I would like to have, but we have very talented people that now take that information from one format and turn it into another format that the computer can understand, that then produces the list.

This entire passage reflects the essence of the core skills CCC has built, and which have provided the base for their primary core competence: the ability to quickly communicate and translate specific customer needs into a generic manufacturing blueprint which can then be custom assembled both quickly and with high quality. In this case the stock resources which CCC competes on—quality, flexibility, speed, productivity and capacity—derive from CCC's capabilities to gather information that is used for scheduling and manufacture, and then to assemble the work. These capabilities derive from the effective blending of more specific organizational routines and material resources, which themselves can be traced back to core skills and skill-sets in individuals and groups. Again the "systems" are driven by the need for particular skills at the area and individual level, and the coordination of these elements is dictated by the system. The "system" in Mike's words, is really the key to the skill-assembly process, or the building of competences.

The Need for a Skill-Based Strategy that is Rooted in Competent Systems

After the bankruptcy, Mike saw the connection between his company's competence and the needs of the customer much more clearly. He began to restructure the firm's strategic thrust to ensure that there was a "fit", that began with a knowledge of the core competence, between the ability of his firm to offer services and value, and the needs of the customer. He began to realize that the "whole system" needs to be grasped by the strategist, and that the firm needs to be seen as a system of value creation that leverages the important skilled resources of the firm.

This resource-based view of the firm has implications for how one should manage capacity, which is regarded more intimately and directly as a function of skill. The critical resources must be protected. In this case drafting and the CABSYS activities must not be overwhelmed by either non-essential sales jobs (low-quality, high-volume), or there must be more of the skills available. As we shall see, by 1989 Mike had fully realized this "constraint" concept and had taken dramatic steps to improve his firm's ability to obtain throughput and maintain quality. He had begun to see the firm as a complete integration of systems, and his firm's evolution in competence, in terms of integration, began to give him the leverage he needed to solve the problems of both his customers and employees.

SECTION IV: CCC'S REBIRTH: THE DEVELOPMENT OF A COMMITTED ORGANIZATION AND A "NEW" FOCUS FOR THE COMPANY

The 1985 reorganization of Custom Cabinets Company, Inc., was painful for all involved, but as a result of the severe cutbacks, and despite relatively flat sales through 1986, CCC was able to post positive profits through 1986 and most of 1987 and was able to get a bankruptcy court to approve its reorganization plan in 1987. Under the plan, the company agreed to pay 100% of its debts to its creditors within 10 years. Mike also hired Ron Rostow in 1987 as Vice-President of Marketing. Rostow had previously founded several companies of his own and brought valuable experience to the management team.[3]

The experience of the bankruptcy, and the subsequent "burning desire to make it work" on Mike's part, tempered the members that remained and created a loyal cohort of employees from which to rebuild the company. Scott Smith comments that, "back in 1985 when we filed for Chapter 11, I think that loyalty was what kept everybody together going . . . with Mike". Mike acknowledges that since the reorganization, "we have a lot of determination, and understand what adversity is and how to handle it . . . and that if you're diligent maybe there is hope".

This determination and diligence is a part of Mike's expectations for his managers and employees, as Stasi Bara, Vice-President of Operations emphasizes, "we like to knock obstacles down . . . that is one thing that Mike has really stressed". A critical part of this "knocking down of obstacles" involves the attitudes of Mike and his employees towards change. Bara continues, "Mike has always been a person of change . . . he's always been a visionary that could look down the road . . . I always thought we'd get to a plateau and stop . . . [but we haven't]".

In part because of the needs of the company to re-examine the way it was doing business in the early to mid-1980s, and in part because of Mike's desire to "build something that I can look back on and be proud of", *change* became an integral part of the "CCC way". The primary agent for change revolved around the company's passionate commitment to the reorganization of itself into a quality and service organization, and to the idea that "quality is a process of ongoing improvement". Mike expands on this theme:

> When we started to come out of it in 1985, we had made what Tom Peters describes as a "passionate commitment to quality and service", and I think that was a very important strategy for us . . . we were determined to be the best at listening to customers, and doing what customers said they wanted us to do.

Figure 6.6 shows how effective CCC has been since its commitment to quality service after the reorganization. Figure 6.6 is an "Interrupted Time Series Analysis Chart",[4] which tracks quality in terms of both the percentage of cabinets that are rejected by the internal CCC inspectors (dashed line) and the percentage of "no-charge work orders (NCWO)" (solid line), which measures how many work orders CCC had to cut to repair cabinets in the field. Notice the large and permanent change in level to no-charge work orders after December 1987. Prior to December, NCWO's were averaging over 10% of sales, and by June of 1988 CCC had cut that number to 3%, where it has stayed. The implications of this commitment to quality seemed clear to the management team in terms of what customer segments CCC would seek to serve in the future: the "upper-medium" to "high-end house" segments.

These segments would "allow CCC to focus on customers who require quality, options, 'hand holding', and extra services."[5]

This "new" focus also forced changes in the ways CCC personnel did business in the latter 1980s, as David Hone, a draftsman with the company since 1987, remarks:

> When the company changed its market strategy from a plain Jane cabinet maker to a custom cabinet maker, it killed us all the time. We were constantly changing the rules . . . we needed to invent ways to do it . . .

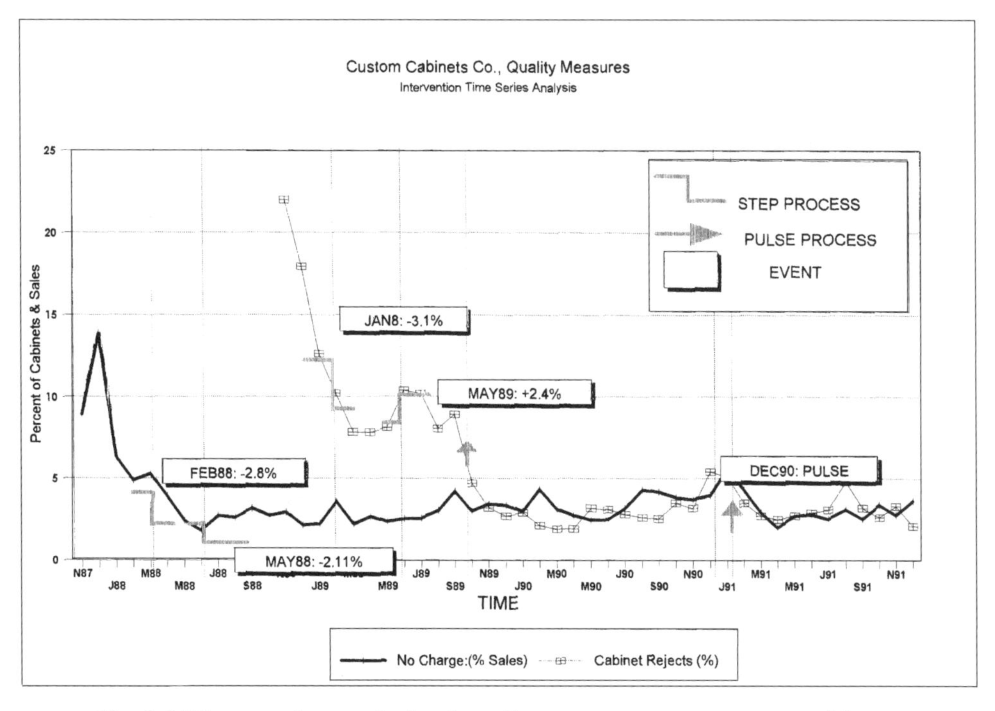

Fig. 6.6. Time series analysis of quality measures at custom cabinets.

By 1987 the company was reinventing itself, and Mike and his management team were poised to meet the challenges of new growth and sustained profitability. From the first quarter of 1986 until late 1989 CCC averaged a 3% profit on sales, and sales had begun to grow exponentially again.

SUMMARY OF PRIOR CONSTRUCTS RELEVANT TO SECTION IV

The following theoretical constructs and theoretical propositions are relevant to Section IV of CCC's history:

Commitment (Ghemawhat, 1991)
Institutionalization process (Barnard, 1938; Selznick, 1957)
Markets served (Abell, 1980; Day, 1981; Dutton and Freedman, 1985)
Quality (Peters and Waterman, 1982)
Strategy as fit, building on strength (Hofer and Schendel, 1978)
Transilience (Abernathy and Clark, 1985)

See applications section for details.

APPLYING EXTANT THEORY AND DEVELOPING FIELD-BASED SUBSTANTIVE THEORY FROM SECTION IV

Extant Theory Application

When CCC went into bankruptcy, Mike made many personal sacrifices to ensure that the firm would survive. This commitment provided the base from which the company was able to build a loyal and talented cohort of people to rebuild the company. Mike was able to instil a sense of purpose into the organization (Barnard, 1938; Selznick, 1957) that was focused on building on the company's core competence (Prahalad and Hamel, 1990).

Mike changed both the nature of the workplace and the markets that it served (Abell, 1980; Day, 1981; Dutton and Freedman, 1985). He targeted the high-end custom markets and aimed to provide that segment with high-quality products and services. He paced the growth of the company on his ability to deliver his products within the time-frame he had promised, and thus had made a strategic commitment (Ghemawhat, 1990) to reliability as a measure of service quality as well as to product quality.

SUBSTANTIVE GROUNDED THEORY DEVELOPMENT ON SKILL-BASED STRATEGY AND ENTREPRENEURIAL LEADERSHIP

Commitment and The Roots of Competence

(1) Commitment is the baseline for institutional competence.
 (a) Commitment to solving customer's problems: linking the core.
 (b) Commitment to employees: protecting the core.

Commitment is the Baseline for Institutional Competence

Commitment to Solving Customer's Problems: Linking the Core

By 1986 the hard-learned lessons of the bankruptcy and reorganization had put a temper on the emerging management team of CCC that had not been there before. Mike had begun to shore up his company's weakness in marketing by hiring Ron Rostow, and more importantly, was now focused on getting the most out of his firm's core competence and linking it to the customer's value needs. His focus was manifest through a personal and company-wide *commitment* to providing the best quality products possible, and to listening to the customer.

Mike began to see his company as the solution to his customer's problems:

> Well . . . I think any company that is successful is basically solving a problem for somebody, and what I see our company doing is . . . what I think helps distinguish us is that we try to . . . uhh . . . solve the problem. We basically take on the problem that the superintendent or the builder has and provide him something that is appealing to the market.

This perspective naturally led Mike to seeing that the relationship with customers is far more of a team effort than he might have thought before.

> We've basically taken the attitude that we are a partner in this, we want to be accommodating to the customer. We want to provide what they tell us they need. We try to differentiate our product by making it more of a service than just a commodity.

This, of course, builds directly on the company's strengths and core competence: the ability to quickly communicate and translate specific customer needs into a generic manufacturing blueprint which can then be custom assembled both quickly and with high quality.

CCC is viewed by the managers as a service company that has a manufacturing component. The service is described in terms of problem-solving for the customer, which is an extremely broad way to view the business. This problem-solving perspective implies a higher order of thinking about the skills that are or may be required to run the company, in that the firm must be highly flexible and capable of exploiting economies of scope with regard to customer needs.

Mike talks about the management team's commitment to the service the company offers, and that commitment is unique and important to the process of offering a custom product. The commitment is also important in keeping communications open between the firm and the customers.

Commitment to Employees: Protecting the Core

Mike's commitment to the customer was matched by his, and his management team's, commitment to including the employees of the firm, who had loyally weathered the storm, in a new profit-sharing plan (called the "bottom-line incentive plan"). This reflected Mike's desire to build a business where his employees could also grow and prosper:

> My objective is, in all honesty I feel my calling, what God intends for me to do, is to build a business. For people to have a job that they like, that pays them a wage that they can support themselves and a family with, is a very high calling for someone to have.

The bottom-line incentive plan helps the firm, as an institution, build shared values and goals. It also provides a mechanism whereby individuals can push their training and education to help themselves, then the firm, then themselves again. This is a positive feedback loop that is directly related to skill and competence evolution in the company.

The bottom-line incentive plan, along with other training and continuing education programs, is building upon the trust that Mike showed in his employees, and in which they shared for Mike after the reorganization. The building of trust is an outgrowth of the development of institutional competence at CCC, of its transformation from an organization to an institution (Selznick, 1957). This evolving institutional competence is related to the firm's culture and leadership and is definitely a "stock asset" with which the firm can compete and grow.

SECTION V: THE BLOSSOMING TREE. CUSTOM CABINETS COMPANY IN THE 1990s

From early 1988 until late 1989, CCC had managed to accommodate a large growth in sales and a gain in market share without losing money. Custom Cabinets Company celebrated its banner year of 1989 by moving down the highway from its plant in Town, South, to a much larger renovated 89,000 square foot manufacturing and office facility in Mauraville, South.

With the larger facility the company was potentially poised to handle new growth opportunities. Unfortunately, the US economy had weakened again, and the winter "trough" of 1990 was another difficult time for the company, as sales declined from a record peak of $600,000 in November 1989 to half that in February 1990. During the same period the company suffered its first sustained losses since 1987.

Mike, who as Stasi Bara stated earlier, "has always been a person of change", decided that he had to find a better way to manage his company's resources, especially during the off-peak winter months.

While he had plenty of room to grow, in terms of physical space, Mike realized that he could not afford to continually re-size his workforce, his human assets, to manage the bottom line. He knew that his company could not accommodate the customer in the peak months if he lost good people in the winter. Mike continues:

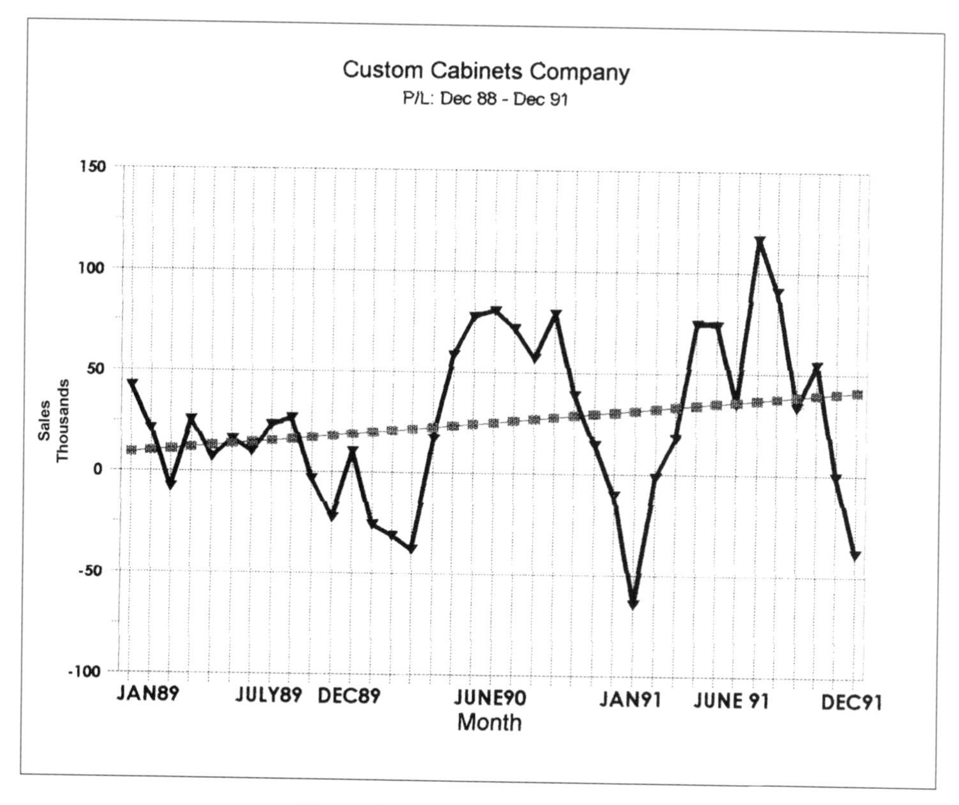

Fig. 6.7. CCC sales, 1989–1991.

> we haven't laid people off in years . . . but, until last year we had reduced hours. We decided last year (1990) that everyone would work 40 hours . . . you see, when the hours are cut . . . the good ones, that have initiative, and want to get their bills paid, they're going to get another job . . . and the ones that really don't care, and the ones that I'm not really interested in having in the first place, are the ones that stay. So we want to protect a significant asset, and we see our people as being a significant asset.

In late 1989, Mike had been sent materials on "synchronous manufacturing", or "theory of constraints" from the Center for Manufacturing Excellence at Another University in Big-city. He was interested by what he read, and in January of 1990 he attended a two-week residential seminar in New Haven, Connecticut. This "Jonah" course, offered by the Avraham Y. Goldratt Institute, had a profound impact on how Mike viewed and managed his business afterward:

> When I returned to South I was convinced that we needed to make some drastic changes. Although no one from the Goldratt Institute ever told me what do, I was sure of my intuition and Implementation Plan. My first step was to convince everybody else that I had a good plan. To accomplish this, I taught seven two-day Functional Education Workshops (FEWs) using the Institute's computer simulations.[6]

Mike's goal was to convince his people that the theory of constraints (TOC) was different from those ideas his company had tried and had been following in the past, and that it would be the best way to ensure that the company started on the important road to continuous improvement. The cornerstone to Mike's plan was based on educating his management team, then other key personnel, and eventually everyone in the plant, about the theory's approach to decision making, or "*the thinking process*". He built a state-of-the art classroom, with personal computers and multimedia, at CCC in order to do this, and created a new position in his organization called "educator". The thinking process, or effect–cause–effect method of analysis, is a critical aspect of the theory of constraints, and is described by Mike:

- Use step-by-step logic to build a *decision tree* that pinpoints what you consider to be undesirable effects.
- Link the undesirable effects to a minimum number of "root causes". The root causes are at the lowest level in the tree.
- By solving or changing the root causes, you should be able to reverse the chain reaction of events and evaporate the undesirable effects that appeared higher up in the tree.

In essence, what Mike was committing his company to was an entirely new way to communicate and make decisions at both the operating and strategic levels. The "decision trees" Mike discussed are actually maps which teams of managers and employees develop which list undesirable attributes in the organization, and which are used to attempt to logically map and identify the root causes of those effects. As of February 1992, all the members of Mike's top-management team have attended at least one Goldratt Institute "Jonah" seminar, and over 60 others in the company have been trained to do effect–cause–effect analysis. "We've taken all our salesmen, team leaders, foremen, and draftsmen through our class", says Viera. "In the slower winter months we'll have the line people going through the class. We're trying to push the concept down to the line workers; give them some thinking tools to solve the problems on the floor instead of the solutions coming from the top."

The ongoing implementation of TOC has had an impact on operating performance within the company, and is opening up new avenues for CCC to utilize its assets strategically. From an operations standpoint, "before we got started in the theory of constraints, we were assembling cabinets 40 hours after we released the paperwork," says Viera. "Now we assemble eight hours after we release the paperwork. We can build our hardest doors in 12 hours, the easy doors in four hours." The company has gone from producing a peak of 200 quality cabinets/day to nearly 375 quality cabinets/day with no net increase in the number of production employees. This has helped ensure that the company continues to make good on its promise to make 10-day delivery on a totally custom-designed product of high quality.

From a strategic management standpoint, TOC and the thinking process have given CCC's managers powerful new tools for understanding their business and the opportunities they might be best suited to exploit. It also gives them an effective means for working together as a team in developing strategy and tactics. Mike explains how the management team evaluates and discusses business strategy now:

> We meet quarterly, and each man will develop an [effect–cause–effect] tree . . . and their implementation plans . . . each guy critically analyzes their own work. I guide them, I think that as a group we have a much better understanding of our company, and where we need to go, and what we need to do to get there.

He continues:

> Probably, the biggest impact [of TOC], is the understanding and willingness to challenge our assumptions . . . we are willing to challenge our marketing assumptions . . . who we sell to, what is it we will build, how much do we sell it for, where do we sell it . . .

John Zayac, Vice-President of Production, agrees with Mike's assessment and elaborates on how the quarterly "internal Jonah" meetings allow him to better understand how all the elements of the company need to work together in an effort to execute the "company implementation tree", which represents actions that each area must take if the "root problems" or opportunities are to be dealt with successfully. "The biggest value of TOC is when we get together at our [quarterly] two or three day internal Jonah meeting", Zayac said, " . . . and we have a company tree . . . and I see where I am on that tree . . . and how it affects Ron Rostow in marketing. We are able to see what impacts the decisions I make have somewhere else". Mike adds:

> Plus, we are all forced to be accountable to each other. If there is a very undesirable effect that is having serious negative impacts in several areas of the company, we are all very aware of it and come to a consensus and commitment to our solution. Some of us may have to make sacrifices so that a solution can be implemented in another area. But we all understand the importance, and together we have assigned the priorities that will have the greatest impact on our constraints.[7]

The ability of CCC to coordinate its resources, based on the "company trees" that are the result of many different people's direct input in the thinking process of TOC, has given it a tremendous increase in productivity and capacity in its throughput potential. The company "tree" is literally beginning to blossom, both in terms of its overall capabilities, and with respect to its commitments to its customers and employees.

The commitment to TOC, and the thinking process, has led Mike to re-evaluate what he needs and is looking for in employees, for example. "So, we've started building internally and getting more skilled people, training them, making sure we can meet demands", according to Scott Smith, "we've plowed back a lot . . . in continuing education. We've brought new people on, and each time the emphasis is more and more on education". The company, in the last two years, has begun to put much more of an emphasis on building the skill level of the company as part of its business strategy. Mike explains:

> One thing, if you do E–C–E, you come down to 2 or 3 root causes generally, and a lot of the trees you'll find in there is a lack of training or . . . something missing in the skill level. When you've come through that analysis and you see that as a root cause . . . well, it's something that you really can't ignore . . . and it . . . then dictates . . . if you've come to that conclusion, then you start to make an implementation plan. Then you've got to make a commitment to the training.

"Everybody needs to be developed up", as Zayac, the Vice-President of Production, notes: "the demand for the employees to make more decisions, and to be responsible for the product will require somebody that does have [more education], and is capable of being more educated . . . and developing more skills". In return for the increased productivity CCC has instituted a *bottom-line incentive plan*, where, after the accounting team figures the Profit and Loss Statement at the end of a month, 10% of the profit is set aside for the bottom-line incentive plan. The amount is divided among all eligible employees, based on the proportional share of that employee's base pay rate.

Mike puts the emphasis on development into some historical perspective:

> I think that we've also come to the conclusicn that not very often can we go hire the skills that we need . . . . In the past, I guess, the first ten years we were in business, a lot of the emphasis was on breaking down the work so that this worker only knows how to run this machine . . . and this is all they have to know . . . and in an hour everything they need to know we can teach them. Our attitude now is very much different, in that we want that person to be very versatile . . . we want them to know this machine . . . and this one and this one . . .

What Mike and his management team are hoping to build is an environment where individual and team learning are a vital and continuous part of every employee's job. The increase in overall skill levels and the better coordination of resources provides a resource base from which to improve quality and service to the customer.

The customer is still the force that is driving the improvements at the company. Mike, today more than ever, is striving to have his company offer the highest quality product, with the most custom service, while being reliable in delivery time. "We've basically taken the attitude in this that we are a partner in this", Mike said, "we want to be accommodating to the customer . . . we want to provide what they tell us they need." As one CCC document stated, "being committed to this ideal, CCC remains flexible and sensitive to customers' needs. While others in the industry just "sell boxes", CCC strives to provide an atmosphere of teamwork and partnership.

To see how the company's commitment to internal development and continuous improvement are tied directly to satisfying the customer, the following mission statement from a Custom Cabinet Company document is telling. Four things will ensure CCC's continued success, according to the statement:

(1) Stay ahead of competitors by offering the highest-quality products, unsurpassed service, and shortest lead time.
(2) Keep up with technological advancements in the woodworking industry to keep production efficient.
(3) Regard the company's people as its most valuable investment. Assemble a highly-trained work force by improving and motivating current employees and recruiting skillful new ones.
(4) Create an environment of ongoing improvement, where employees enhance and streamline their individual tasks by continually seeking out better methods and processes.

It remains an ongoing challenge at Custom Cabinets Company to provide the best quality cabinets to their customers in a timely and reliable fashion. CCC is learning to improve itself on a continuous basis and is learning to adapt challenges both in the market-place and from elements beyond its control, such as the weather.

SUMMARY OF PRIOR CONSTRUCTS RELEVANT TO SECTION V

The following theoretical constructs and theoretical propositions are relevant to Section V of CCC's history:

Dynamic capability (Teece *et al.*, 1992; 1997)
Human capital (Becker, 1964; Schultz, 1961, 1990)
Implementation (Galbraith and Kazanjian, 1986)
Integration (Newman, 1992; Whiston & Rothwel, 1992)
Invisible assets (Itami & Roehl, 1987)
Learning by doing (Dutton and Thomas, 1985)
Organizational learning (Hedberg and Starbuck, 1981; Dutton and Freedman, 1985)
Skill-based competition (Naugle and Davies, 1987; Klein *et al.*, 1991)
Theory of constraints (Goldratt, 1990; Dettmer, 1997)

See applications section for details.

Extant Theory Application

By the late 1980s Mike Viera had seen his company rebound from a near fatal position to one of relative strength and health. He realized, however, that he was letting the cyclical nature of his business affect his decisions with regard to the human capital he had at his disposal in a negative manner. When the company was in the winter trough, he was reducing hours and letting talented people go. After being introduced to the "Theory of Constraints" (Goldratt, 1990), he recognized that once again he was not utilizing or protecting his capacity, which was largely a function of his experienced human capital base (Becker, 1964; Schultz, 1961, 1990).

He also realized that his management team's capacity to integrate and utilize the firm's resources could be improved, and that the theory of constraints could help. He took aggressive steps to implement a company-wide plan to learn how to learn (Hedberg & Starbuck, 1981), using TOC.

The continuing implementation of the theory of constraints has had a remarkable impact on the way the company does business, especially in terms of its strategic decision making and implementation processes (Hofer and Schendel, 1978). The Theory of Constraints is a theory of dynamic capability (Teece *et al.*, 1992), and is a powerful integration (Newman, 1992; Whiston & Rothwel, 1992) tool that is especially effective in providing a means for a company to craft skill-based strategies (Naugle and Davies, 1987; Klein *et al.*, 1991).

SUBSTANTIVE GROUNDED THEORY DEVELOPMENT ON SKILL-BASED STRATEGY AND ENTREPRENEURIAL LEADERSHIP

Organizational Learning, Competence Evolution, Strategy and Performance

(1) The theory of constraints as a resource-based process theory.
(2) TOC is also a theory of organizational competence.
  (a) The "thinking process" generates future competence maps.
(3) The firm is a net generator of resources: this affects factor values.
  (a) Using slack as an investment.
(4) The thinking organization: TOC and human resource evolution.
  (a) The development of institutional competence.
(5) Competitive advantage from commitment and competence: summary chart.

The Theory of Constraints as a Resource-Based Process Theory

We saw earlier that Mike and the company had made a passionate commitment of quality and service to the customer, and that solving the customer's unique problem was what the company was all about. Given this commitment, Mike realized that the company must continually improve its capabilities; but he was faced with a highly cyclical business in which it was hard to maintain profitability in the winter months. He was continually struggling both to hang on to good people in the winter and then to utilize those people well. If the company succumbed to the pressures to maintain the bottom line during the winter, it would not be prepared for the rapid growth in the spring and summer months in terms of both capacity and quality, and overall profitability might be hurt even more severely.

Intuitively, Mike knew that he had to find a way to coordinate and manage his resources more effectively so that he could meet and exceed the expectations of his customers. When he reviewed the materials sent to him on the Theory of Constraints, or synchronous manufacturing, he thought that this indeed might be a way of thinking about business that would let him manage his resources much better. Interestingly enough, Mike's commitment to introducing TOC to his company was the beginning of a bold and systematic attempt to reinvent his company yet again, this time from a resource perspective. With TOC, the focus would be on improving the capabilities of the firm as a whole without ever losing sight of the need to solve the customer's problems and creating value for them.

As was argued earlier, the Theory of Constraints is a process-oriented resource-based theory of the firm. In order to seize opportunities and receive value, the firm must: (1) have some *a priori* capabilities to create value and (2) then be able to learn to evolve and transform its capabilities as customers change, and as the needs of customers and the capabilities of competitors change. The ability to transform capability, as a process built into the strategic perspective of the firm, can only come from a resource-based process theory such as the Theory of Constraints. The fact that CCC has been able to continually improve both its operating performance and its market penetration since beginning to learn about TOC indicates that the focus on capability, then opportunity, is indeed effective.

TOC is also a Theory of Organizational Competence

The adoption of TOC had profound strategic and operations effects on CCC. Since the very nature of the thinking process, or Effect–Cause–Effect (E–C–E) analysis in TOC, is integrative, the management team began to learn how to coordinate and subordinate their own area's resources in the effort to improve the throughput of the entire firm. This, of course, is at the heart of our earlier discussion on the concept of competence, which reflects the firm's ability to assemble and blend its resources in the pursuit of opportunity. In this sense, TOC, as a resource-based process theory of the firm, is also a theory of organizational competence. As Mike and John Zayac suggested, the most important impact TOC has had on the company is: (1) how it has taught them to systematically review and understand how all the parts in the organization fit together to produce value and (2) how it has taught them to challenge time-honoured assumptions about the value of their resources, the activities of the market-place, and what they, as a management team, stand for.

The "Thinking Process" Generates Future Competence Maps

An important point is that the E–C–E process is inherently interactive and integrative, which is a necessary condition for competence-building. This process is both a formulation and implementation process, and involves the development of consensus and team-building, two critical parameters for competence development and sustainability. The decision trees that are part of the outcome of the thinking process in TOC are *impact diagrams* that are based on a total system's perspective. The *future tree* is really an intended competence growth map which shows what capabilities are needed and what tasks need to be done to get to the desired resource allocation position in terms of both behaviour and physical resources.

The management team has learned quite a bit about itself from the reorganization effort, and has, through the use of TOC, begun to develop the strategic competence required to grow and thrive. The ability of CCC to grow without undue hardship and chaos stems from an evolution in its competence. If competence is about the adroit blending of skills and other resources to pursue opportunities, then contained growth is an elementary aspect of it. A firm that grows beyond its capabilities in the effort to grab a market opportunity will in all likelihood not be able to sustain itself and in fact may lose key human resources in the retraction phase which will follow opportunity-based growth.

The Firm is a Net Generator of Resources: This Affects Factor Values

If one looks at the firm from a resource-based process perspective, like the Theory of Constraints, then it becomes apparent that the firm is really a *throughput generation process* phenomenon. The firm is a net generator of resources, so the primary challenge for management is to build and expand throughput, not cut or limit resources in an effort to reduce cost. This implies that the resource-based approach to the strategic thinking process be centred first on how a resource can contribute to the entire system's ability to generate money through sales, rather than on how much the resource costs. Traditional approaches to valuing factor market resources (inputs to the firm, such as labour and equipment), which are based on the economics of scarcity, are not useful in the resource-based process setting.

In the case of CCC, a real challenge is in managing the interaction between the development of both specific and generalized capability, as reflected in the skills of its workforce (a factor input) and measured in operating expense (OE), and the subsequent expansion of the system's throughput capacity. The capability–capacity cycles are linked, and throughput is negatively affected at the peak if OE is artificially reduced to maintain the bottom line. The excess capacity that Mike builds into his firm is really a stock asset that he can draw upon; it is slack which he can use to ensure that he has an adequate reliability buffer.

Using Slack as an Investment

Slack is also an investment for the company. During the sluggish times in the sales season, Mike has begun to buy slack through an investment in training and continuing education—he has begun to build his human resource base. Since his human resources are being invested in during a time when there are no internal physical constraints to restrict throughput, the cost of worker time is nominal.

While such investment is an operating expense flow, the throughput effects that may be gained later, and the overall capability of the organization to learn and adapt to new demands should always offset these expenses. Indeed, the very heart of the process of continuing improvement is that investments in skill, thinking capability and communication, trust, and then complementary equipment, will lead to a positive multiple increase in the throughput of the entire system. The key is to have a solid understanding of the resources of the firm, which are constraints, and how they can be assembled and disassembled, and how they can be improved. This is especially true in the case of the investment in human resources, as they are the ultimate source of ideas for improvement and contain the most potential for synergy, flexibility, growth and long-term returns, as Teece *et al.* (1997) suggest.

The Thinking Organization: TOC and Human Resource Evolution

In order to improve the capabilities of the firm and achieve improvements in throughput, Mike understood that he must make a deep and concerted effort to improve the firm at all levels, and that the only way to do that was to commit to involving and educating all his managers, and then supervisors, and eventually all employees, in the disciplined thinking process of the theory of constraints.

Further, he understood that a general commitment to education and the continuous improvement of each employee's skills was vital to the overall continuing improvement process that TOC would introduce into the firm. As Mike and the management team learned how to think using the effect–cause–effect analysis trees, they saw that the overall skill level of the firm needed to be raised and that it was their job to provide the tools for their employees to better themselves.

The investment in human capability affords CCC strategic options that they have not had in the past. With an increasing overall skill-base in the company they are now able to purchase more sophisticated equipment, much of it computer-based, that allows them to capitalize on economies of scope, which is a vital component of their custom-shop strategy. Likewise, the improved workforce has been able to increase its quality and reduce its lead time in producing an ever-increasing variety of custom products. As Scott Smith comments, "so we've really got a competitive advantage with the fact that we've got a custom product that is high quality, we have a short lead time, and we just cover them up with service". CCC has developed a strategic competence in that its strategy leverages its core skills and competences in a customer arena that recognizes the value of the CCC product/service mix and is willing to pay for it.

The Development of Institutional Competence

As the company continues to make investments in people, both in terms of training and continuing education, what CCC is really doing is evolving into a "thinking organization". It is, to use Selznick's concept as we discussed it earlier, developing a unique character and is building "institutional competence". The character of the firm is rooted firmly on being a partner with customers in solving their problems. It does this through the exercise of its skills and its commitment to teamwork, human resource development, technology enhancement and systems development. The entire firm, as a system, is coordinated through the use of the theory of constraints, which, as we stated earlier, is an ideal resource-based process model for strategic decision-making. The theory of constraints is entirely driven by human thinking, collaboration and action, and is about the development of competence in an organization.

Competitive Advantage from Commitment and Competence: Summary Chart

Figure 6.8 is a theoretical schematic which summarizes many of the issues discussed during the analysis of this case. It depicts how Custom Cabinets Company attempts to meet and exceed its customers' expectations in terms of quality, service, variety and reliability, by committing to being a custom-shop and, more importantly, by committing to solving customers' problems by using the theory of constraints as a way of defining the company's character and competence. Through the use of TOC, the company focuses its human resource development investments (in terms of both time and money), its systems and technology development investments, and its teams, on solving the customer's problem. To the extent that CCC

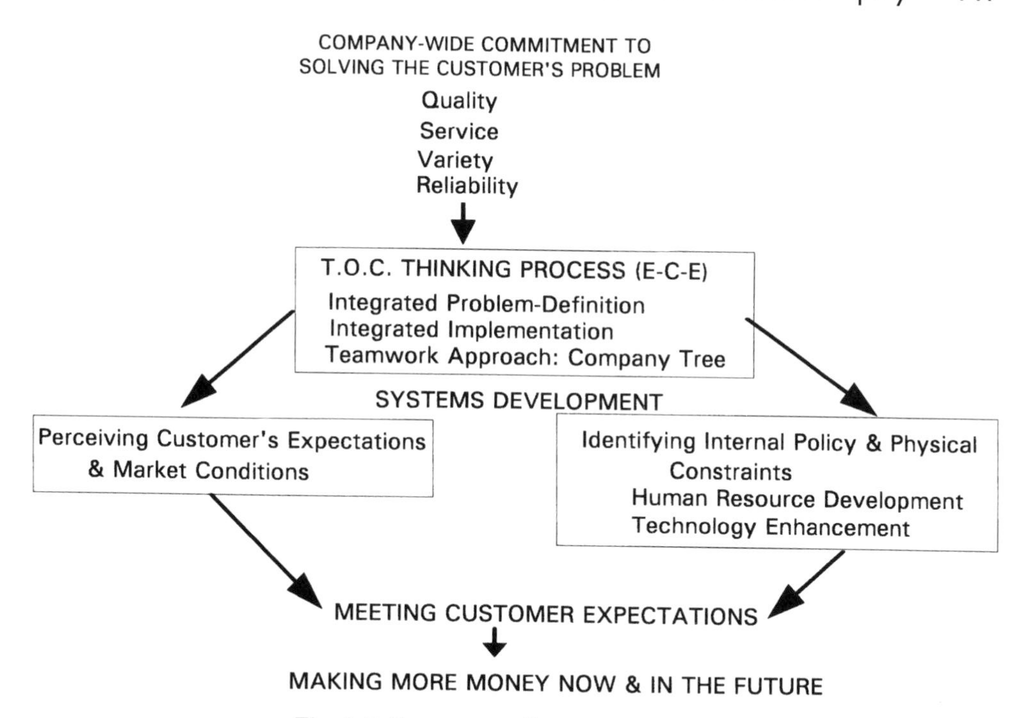

Fig. 6.8. Summary diagram of CCC.

is able to solve the customer's problem better than other companies, they achieve a competitive advantage in the markets they compete in.

IDIOGRAPHIC-LEVEL THEORY SUMMARY: CUSTOM CABINETS COMPANY

Deep Knowledge and Skill Types

(1) Businesses emerge from what their founders know and enjoy—leveraging key skills.
  (a) Skills and opportunity.
  (b) Experience as knowledge.
  (c) Craft and knowledge skills.
  (d) Leveraging skills in making entrepreneurial choices.
  (e) Rewards: the positive effects of leveraging key skills.

Organizational Growth and Skills

(2) Following up on entrepreneurial choices: innovations to work smarter.
  (a) Breaking barriers and serving the customer.
  (b) Developing new skills and competences.
(3) The failure to understand the boundary of core skills can be fatal.
(4) The need for a skill-based strategy that is rooted in competent systems.

Organizational Systems and their Relationship to Skills

(5) Systems and skill.
- (a) Organizational competence develops with systems.
- (b) Skills drive systems and systems are the bridge between individual skills and organizational competencies.

Institutional competence: commitment

(6) Commitment is the baseline for institutional competence.
- (a) Commitment to solving customer's problems: linking the core.
- (b) Commitment to employees: protecting the core.

(7) The thinking organization: TOC and human resource evolution.
- (a) The development of institutional competence.

Theory of Constraints as a Resource-Based Theory of Management

(8) The Theory of Constraints as a resource-based process theory.

(9) TOC is also a theory of organizational competence.
- (a) The "thinking process" generates future competence maps.

(10) The firm is a net generator of resources: this affects factor values.
- (a) Using slack as an investment.

(11) Competitive advantage from commitment and competence: summary chart.

NOTES

1. CCC Internal Document, 1992.
2. CCC Staff Document, February, 1992.
3. CCC Staff Document, February, 1992.
4. This analysis chart displays all the statistically significant permanent changes in quality levels, and all significant temporary "pulse" changes in the time series. Dated events represent statistically significant, and permanent decreases in defects, while "E" represents a temporary upward pulse shock in January of 1991.

   The underlying analysis is based on statistics generated by the "AutoBox" software program, which uses artificial intelligence routines to diagnose and identify significant interruptions in the level, rate of change and variance of a time series. See Cryer (1986), McCleary and Hay (1980) and McDowall and Meidinger (1980).
5. CCC Staff Document, February, 1992.
6. Viera, Mike, "TOC in Action: A CCC Case Study", working company document, 1992.
7. Viera, Mike, "TOC in Action: A CCC Case Study", working company document, 1992.

7

ENTREPRENEURIAL LEADERSHIP AND SKILL-BASED STRATEGY IN A TECHNOLOGICALLY AND SKILL-INTENSIVE, "MARKET-LESS" PROFESSIONAL SERVICE ENVIRONMENT: THE CASE OF HAUSER CHEMICAL RESEARCH, INC.

INTRODUCTION

Hauser Chemical Research, Inc. (HCR) is one of the world's leading manufacturers of extracted natural products, and as of 1992 was the world's only commercial scale manufacturer of the anti-cancer agent taxol. The company offers a wide range of chemical and engineering services and offers its talents internationally. The company is known for its fast turnaround times and quality service. Hauser Chemical research has been chosen to be included in this book for a variety of reasons. One is that the company relies heavily on its base of highly trained scientists and technicians, its special kind of human capital, as a means for solving the particular problems of its customers. Hauser Chemical Research thus provides an excellent source for examining skill-based strategy concepts. Second, the company has no marketing capacity *per se* and operates in a "market-less" environment. That is, the company is primarily in the business of solving a vast array of problems, each of which has its own market context. It is a professional service company. Third, despite the "market-less" nature of its current business, the company has begun to learn how to tap into the value it creates for a particular customer and generalize its solutions. It is beginning to learn how to develop its own application-oriented businesses, which is a departure from past practice. The case thus provides an interesting context from which to explore the issues of organizational learning and entrepreneurial leadership and development.

The Hauser Chemical Research case differs most from the others in this study in that it exercises its scientific or knowledge skills in a service environment, without operating in a few distinct "markets".

HCR'S HISTORY, SECTION I: THE SLOW GROWTH OF HAUSER LABS, 1961–1983

In the winter and early spring of 1961, Dr Ray Hauser was a materials engineer who had developed a specialization in the application of polymers as plastics, adhesives, coatings and elastomers.[1] Ray had worked in materials engineering with a large aerospace firm in Colorado until the day in March 1961 that he was assigned to be a member of a "Materials Technical Staff" and was asked to punch a time clock.

Not wishing to participate any further in a work environment that was fraught with "excessive internal politics" and which no longer seemed to value the autonomy of professionals, Ray Hauser entered the wilderness of entrepreneurial choice and leadership:

> Nineteen-sixty-one was my year of venture into uncharted waters, or the year to say a prayer and take a leap of faith. At the time, I was well aware that the Denver area could not support a materials consulting/research firm . . . . I was on my own to sink or swim financially.

On April Fool's Day 1961, the new firm of Hauser Research and Engineering Company began operating out of the basement of Ray and his wife Connie's house in Littleton, Colorado. The first projects were very small and Ray found that getting the doors to government research contracts open was nearly impossible.

> My perception was that having come out of aerospace I would have the opportunity of getting contracts on a national scale. I was wrong. If you are a sheepherder from Colorado, you don't get their attention, even if you've done comparable work, is what it boiled down to . . . . I thought that we were going to survive on government contract work, and it took eight years before we got our first government contract . . . so there were some difficult local limitations.

Despite the severe limitations that remaining in Colorado meant in terms of market access, Ray and Connie, who had joined the company as a partner, moved from Littleton to Boulder, Colorado, in early 1962. "Gross income increased 50% per year from 1961 through 1966, with repeated contracts from satisfied commercial clients. A petroleum testing lab was acquired in early 1966, financed by a second major bank loan. That acquisition was seldom profitable, but it broadened the client base and it became the launching pad for a future spinoff".[2]

By the late 1960s the company had begun to develop a broader range of experience in materials, and had become active in petroleum testing (through the 1966 acquisition), medical materials and instrumentation, and forensics. In 1973 the company moved into a new 27,000 sq. ft building, which it shared with another company and which was financed by an emergency loan from the Small Business Administration.

During the decade of the 1970s the company grew at a rate of about 20% per year. The growth was limited by capital constraints, and according to Ray, "has been driven, I'd say, with technological opportunities. The business aspect of things came along where they may". Through the 1970s Hauser Labs also continued to grow in terms of key personnel, albeit slowly (average of one person per year). Hauser proceeded to add to its skill base through the hiring of talented and self-motivated professionals whenever he could identify a person he thought would both contribute to the growth of the company, and fit into the company, where "[my] intent was, providing a workplace where other employees could do their best. In general we gave people the opportunity to pick up as much responsibility as they wanted". Ray continues:

> The senior people were generally hired because of the individual, even if there wasn't an opening. Like Tim Ziebarth, who came on as chief chemist in 1974. Up to that time we had a petroleum lab, we did have a chemist, but we didn't have a chemical group. We were flat broke when we hired Tim Ziebarth. It represented more of a risk on his part than he was aware of . . . but, nonetheless, a fair bit of faith on our part that here was an individual who could make things happen, who could build a chemical group, and he did.

Ray cites another example:

> Dick, our chief engineer, we didn't have the financial resources to bring him on, but I knew that he was going to be a heck of a lot better chief engineer than any other that I had come across . . . and uhh, we brought him on. In general, some of the people have been hired with the idea of, "is this a person who can make things happen, who can make a certain market segment grow?".

The importance of having talented and self-motivated professionals was critical to Hauser Laboratories, which, Ray says, "has always worked on the basis of a good technical capability, and on the willingness to take on new and sometimes impossible projects . . . on the excitement of new challenges".

Indeed, beginning in the mid-1970s Hauser Labs began to diversify and expand into new fields of technology and markets, and began to work on the design and testing of thermal insulation, medical electrodes, and transcutaneous electrical nerve stimulation (TENS). In the late 1970s, Dr Tim Ziebarth, Hauser's chief chemist, left the company to co-found another venture and Dr Dean Stull, who had come aboard Hauser Chemicals in 1975, stepped into the chief chemist spot. Stull, who Ray Hauser had thought of as the heir-apparent of his company, helped the company continue to expand its experience and skill base in petroleum testing and other engineering services. By 1983 the company, and its leadership, was poised for a major structural change—the spin-off of Hauser Laboratories Chemical Research, Incorporated.

SUMMARY OF PRIOR CONSTRUCTS RELEVANT TO SECTION I

The following theoretical constructs are relevant to Section I of HCR's history:

Boundary spanning (Thompson, 1967)
Culture (Ansoff and Baker, 1986)
Deep or embedded knowledge (Winter, 1987; Mintzberg and Westley, 1989; Waddock, 1989; Badarocco, 1990)
Dynamic capability (Teece *et al.*, 1992, 1997)
Entrepreneurial skills (Herron, 1990)
Knowledge–technique base (Lenz, 1980)
Learning by doing (Dutton and Thomas, 1985)
Path dependency (Barney, 1986b)
Technology push and need pull (Burgleman and Sayles, 1986)

See applications section for details.

APPLYING EXTANT THEORY AND DEVELOPING FIELD-BASED SUBSTANTIVE THEORY FROM SECTION I

Extant Theory Application

Put off by an excessively controlled and bureaucratic organizational culture (Ansoff and Baker, 1986), Ray Hauser left his job in 1961 as a materials engineer with a large firm and set off to start his own company. Hauser had little entrepreneurial experience, but felt he could rely on his deep knowledge (Winter, 1987; Mintzberg and Westley, 1989; Waddock, 1989; Badarocco, 1990) of materials engineering and his experience in working with the government to bring contracts to his new company. However, Hauser did not understand the procurement process, from a non-technical viewpoint, and the contracts did not come through.

Hauser built the company very slowly, since local demand for his expertise was thin, and exploited his knowledge of materials technologies whenever he could find an opportunity (Burgleman and Sayles, 1986). The business was essentially technology driven according to Hauser, and the expansion of the business depended largely on his ability to add to the firm's knowledge–technique base (Lenz, 1980) and to learn by doing (Dutton and Thomas, 1985).

Ray chose to expand his firm through selective acquisition of other small technical service companies, and through recruitment of talented people. In particular, Ray would hire, at senior levels, whenever he found someone that was both technically adept, and who possessed entrepreneurial skills (Herron, 1990). The people he brought on were given the chance, and expected, to grow business in their area of expertise.

This provided a breadth to the firm's knowledge–technique base (skill-base) from which it could leverage general problem-solving capabilities—creating a dynamic capability (Teece, *et al.*, 1992, 1997) at Hauser Labs that attracted competent and entrepreneurial scientists.

SUBSTANTIVE GROUNDED THEORY DEVELOPMENT ON SKILL-BASED STRATEGY AND ENTREPRENEURIAL LEADERSHIP

(1) Pure skill-based growth is slow.
(2) Investments in technical and entrepreneurial capital are a means to quality growth.

Pure Skill-Based Growth is Slow

In 1961 Ray Hauser entered the strange and seemingly uncharted world of entrepreneurship when he and his wife Connie founded Hauser Laboratories. Ray's background was that of a scientist, and while he had developed a strong technical background in chemical and material engineering, he had little experience as the owner of a business. He had a foundation to build his business, however, in the knowledge and skills that he and his wife possessed, and in his desire to create an atmosphere that was independent of corporate politics and policies, and which valued professionalism, credibility and autonomy.

Ray had counted on the fact that he could win government contracts to build his business, and when this did not occur he had to struggle to keep the company alive and to grow it as he could. The ability of Hauser Labs to grow was severely limited by market access and financial capital constraints. Technological opportunities, and Hauser's skill-base to exploit them, really determined where, when and how the company could grow. Given Ray's personal characteristics and the need to be seen as credible by his clients, only when the skills in the company could truly support an effort to exploit a new technology would the firm offer a new service. Since growth in the skill-base of individuals takes so long, Ray had to look outside the company if he wanted to expand the capabilities of his firm.

Given the other operating constraints, Ray was forced to adopt a slow and deliberate skill-based growth.

Investments in Technical and Entrepreneurial Capital are a Means to Quality Growth

This growth dynamic led Ray to look for outstanding people that he could hire in the hopes that they could help the firm grow. Specifically, if Ray found a person with outstanding technical skills and an entrepreneurial, "build a business" outlook, he would do all he could, despite the financial constraints, to bring that person aboard and give them the chance to grow a new area of expertise for the Labs. In essence Ray was investing in technical and entrepreneurial capital, and hoping that he could find the mix of the two in one talented individual.

Ray's strong technical expertise, his training as a scientist, and his values and professionalism provided the base from which he could assess both external opportunities and the skills of his company. As Ray saw opportunities in the factor markets (labour), he could act on them and build the capability of his company to credibly serve current markets and grow new ones. Hauser Labs' growth, for nearly 25 years, was paced by the skills it could acquire, given both capital and market access constraints.

SECTION II: THE SPIN-OFF AND DEVELOPMENT OF HAUSER CHEMICAL RESEARCH

By 1983 Dean Stull, who had been running the Chemistry Department of Hauser Laboratories as his own business in many respects, had worked at Hauser Labs for about seven years. Dean comments that, "I had set a goal for myself earlier that if I stayed there seven years, that was long enough. My idea was that there were lots of opportunities and that I should take advantage of those, rather than stay with [the Labs]". Dean wanted to build his own company, and began to look for opportunities to do so. Interestingly, the spin-off started without a specific market or product focus. Dean explains:

> In 1983, Randy Daughenbaugh, who is my good friend, and I were chatting and his company, which had its research center in Boulder, was leaving to go back to Houston, and he wasn't interested in moving back to Houston.
>
> So we were just brainstorming, and it occurred to me anyway, that we had an interesting set of skills as problem solvers. We were people who enjoyed taking other people's problems and finding solutions. In particular, chemical solutions to chemical problems.
>
> So we looked around to find a way to start a business that would do that for people, and in the process of solving their problems create business opportunities downstream. So we really sort of did have a business, but

> we didn't know what it was yet. But problem-solving was the orientation we took, and we were going to look for the actual business we would do after we got into problem-solving.

Daughenbaugh, who had managed an extraction and distillation laboratory in Boulder for the Chemical Exchange company, brought unique skills to the team. Ray Hauser, who recognized the value of Dean and Randy working together, and who, as a manager, had always encouraged his employees to take as much responsibility and initiative on the job as possible, supported the spin-off effort. Ray states, "so, when in July of 1983, Dean [Chief Chemist at the Labs in 1983] said he'd like to spin-off the petroleum work of Hauser Laboratories into a new venture . . . and when he said he would put some cash into the venture, I appreciated this self-motivated employee".[3]

Dean and Ray were able to work out a deal where Dean would work at Hauser Labs 2/3 time, and with the spin-off company, Hauser Laboratories Chemical Research, Inc. (HLCRI) 1/3 time. The new company acquired the petroleum section of Hauser Labs, and began to look for new opportunities. According to Kurt Ammon, one of the founding employees at HLCRI, "the petroleum business was basically a pay-check to keep things going for those other investigative things that we were trying to do, to find out what kind of a niche we wanted to get into". Regardless of the specific markets the new company began to serve, Ray Hauser observes that, "the synergism of the two businesses was apparent immediately. The energy of the new executive chemists and the 22 year history of Hauser Laboratories brought new work quickly".[4]

The new company faced nearly all the challenges of a new venture startup. At times during the rocky first year cash was a problem. As is traditional in an under-financed venture such as theirs, Stull and Daughenbaugh often took no salary from the company. However, they never missed a payroll for their three employees, the money frequently coming from their personal bank accounts.[5]

The new spin-off also started with very few material resources and had to "bootstrap" much of its equipment and facilities, as Greg Huckabee, another founding employee in HLCRI, notes:

> So the resources we brought together were Dean and Randy's ability to see things and to do things without . . . really on a shoestring. Randy and I would work from 7:00 a.m. until 6:00 p.m. I would go over to his house, his wife would fix us dinner, and then we would go out to his wood shop and we would build whatever we needed . . . . The real, down-to-earth resources is what started the whole company . . . the ability to work with your hands and to think beyond the science of what we were doing . . .

Despite the lack of capital, the company, renamed Hauser Chemical Research (HCR), continued to expand its capabilities and services, and continued to look for opportunities that could allow it to exploit its skills as a problem-solving team.

SUMMARY OF PRIOR CONSTRUCTS RELEVANT TO SECTION II

The following theoretical constructs are relevant to Section II of HCR's history:

Absorptive capacity (Cohen and Levinthal, 1991)
Entrepreneurial skills (Herron, 1990)
Human capital (Becker, 1964; Schultz, 1961, 1990)
Serendipity (Barney, 1986b)
Synergy (Ansoff, 1965; Hofer and Schendel, 1978; Clarke and Brennan, 1990)
Value and value creation (Peters, 1984)

See applications section for details.

APPLYING EXTANT THEORY AND DEVELOPING FIELD-BASED SUBSTANTIVE THEORY FROM SECTION II

Extant Theory Application

The entrepreneurial culture (Baker and Ansoff, 1986) that Hauser had instilled at the labs had the positive effect of attracting a high-quality human capital base (Becker, 1964; Schultz, 1961, 1990), but it also provided the seeds for its own destruction, if not handled properly. The managers that were growing along with Hauser Labs were ambitious and capable of running their own companies.

Rather than lose potential synergies (Ansoff, 1965; Hofer and Schendel, 1978; Clarke and Brennan, 1990) and opportunities to a rival firm, Ray Hauser let the entrepreneurs he had seeded create an affiliated yet independent company in Hauser Chemical Research.

The spin-off, lead by Dean Stull, was formed in part serendipitously (Barney, 1986b) because Randy Daughenbaugh was looking for some opportunity to stay in Boulder, and just happened to be good friends with Stull, and happened to have a complementary set of skills and knowledge. The combination of Stull's entrepreneurial drive and expertise and Daughenbaugh's technical skills provided a base from which they believed they could create value for customers and themselves.

In this case, the formation of HCR is an example of a start-up that is more Kirznerian than Schumpeterian in nature. That is, while no concrete opportunity had been identified prior to start-up, there was great potential to exploit the knowledge-base of the small firm in a general problem-solving context. That is, the firm possessed great absorptive capacity (Cohen and Levinthal, 1990). There was no great process or innovation, in the beginning, that had been developed that could change markets and provide growth for a new firm.

SUBSTANTIVE GROUNDED THEORY DEVELOPMENT ON SKILL-BASED STRATEGY AND ENTREPRENEURIAL LEADERSHIP

An Incubating Firm Must be Able to Accept Spin-offs and Redevelop its Skill-Base

The entrepreneurial climate that Hauser Labs had developed over the years, and which was at the core of its ability to recruit talented executives like Tim Ziebarth and Dean Stull, was a key ingredient in the formation of a new, but highly interdependent company led by Stull. Recognizing the potential of Stull and Daughenbaugh to create a new set of opportunities, and wanting to maintain a close affiliation with them, Ray Hauser negotiated a deal to spin-off part of his company.

The close tie between the two companies and the amicable nature of the spin-off illustrates an important point in the management of firms that are characterized by a skill-intensive and entrepreneurial workforce: the ultimate price (reward) in fostering technical and entrepreneurial growth in a company is the formation of new businesses. If the leadership of the parent company is going to benefit from such spin-offs, it must be able to accept its role as incubator and then work both to stay close to the new company and to develop new areas of expertise and competence by restocking the technical and entrepreneurial talent that has gone. Ray Hauser was able to do both, and subsequent events led to a reuniting of the two firms – ultimately creating a firm with a broad and solid set of skills and competencies.

HISTORY OF HCR, SECTION III: THE PROBLEM-SOLVING EQUATION AT HAUSER CHEMICAL RESEARCH

The new spin-off, like the original Hauser Labs, was essentially a service business. In this case, the services came in the form of technical and general problem solving in many areas of chemistry. The petroleum lab business that HLCRI acquired from Hauser Labs was the nucleus of the company in the first year, and according to Kurt Ammon, an early employee with HLCRI, the petroleum area grew, "by offering service . . . offering the same tests [as others], but offering it at midnight if somebody needed it then . . . a real service issue".

At the heart of this service orientation of the spin-off was the idea that the company's *raison d'être* was really to solve customers' problems. Tim Ziebarth notes:

> Most of our clients are unable to solve their own problems because their views are too narrow: they don't really see the problem, they don't have a *suite of knowledge* that allows them to apply very many technologies to solving a problem. Our people are very broad-based technologists, who are good problem-solvers. They know how to ferret out the true problem, and draw in a lot more than one narrow discipline to solve these problems. [our emphasis].

Dean Stull elaborates further, noting that his company is a central part of the need fulfilment process of his customer, "generally speaking our customer has a need that a problem is preventing him from fulfilling, so we solve the problem so he can fulfill the need. So we are part of his need fulfilment process".

This perspective demanded (and still does) that the employees at HCR be generalists as much as possible, that they are on the leading edge of their technologies, and that they be able to communicate well among themselves. These are central issues to the top management team, as Stull notes:

> We are worried about protecting our competitive advantage from our proprietary technology viewpoint; but not nearly as concerned as we are about keeping our people abreast of technology changes and working as generalists. Which I think is a key differentiation we have from a lot of other companies.

The ability to respond, as a team, to a wide variety of client problems is what the company was, and is, all about. It was no surprise then, in March 1984, that the company was able to hit its first "home run" in an area it had little formal expertise in, natural products extraction. The company literally stumbled into an opportunity to extract sanguinaria, an anti-plaque compound, from the bloodroot plant. Dean Stull explains:

> we were sitting in our office one day in 1984 trying to solve a problem someone brought to us and we had this idea [for an extraction process]. We sketched it on the board, then tested it in the lab. We asked ourselves, "Why doesn't everybody do this?" It's so obvious. We went and looked at the literature and nobody used it or knew about it.[6]

The extraction process innovation that the company had developed, called "Continuous Dynamic Liquid/Solid Extraction™", produced more concentrated extracts at higher yields, often resulting in higher quality and lower raw material production costs, than other extraction technologies.[7] This competitive advantage won HCR a contract to extract sanguinaria for Vipont Laboratories of Ft. Collins, Colorado.

Ray Hauser comments that, "as Vipont doubled or tripled its business each year, so did HCR". In fact, the work for Vipont accounted for nearly 85% of the company's revenue and HCR soon became an authority in extractions from natural products. The dependence on Vipont was alarming to HCR's leaders, however, and they began to pursue aggressively other opportunities that would allow them to leverage their skills in natural product extraction, particularly as they related to high value-added applications. Stull elaborates on this strategy and how it builds on economies of scope that the company has developed:

> We have actually tried to intentionally stay as broad-based as possible, kind of the opposite of economies of scale. As problem solvers we really don't care what the market is that the problem is solved in. For example, in the natural product area, which is really our biggest area, we are in the food business, the pharmaceutical business, the cosmetic business, the flavor business and the medicinal business, which are all in specific and different markets. We don't know anything about those specific markets, but we solve problems for particular people in each one of those markets. And the technology we have happens to be horizontal, it doesn't care if you are doing pharmaceutical or cosmetic [applications].

This philosophy of maintaining an open and broad net, in terms of application of the company's skills and resources, is in part driven by the fact that the scientists and technicians at Hauser are "general technologists" and not "research scientists" *per se*. Randy Daughenbaugh, the Chief Technology Officer at HCR, explains how the technical and problem-solving orientation and culture at HCR interacts with the technology to create the flexibility that is reflected in the company's strategy:

> The point I was going to make about the technology side, with the problem solvers . . . we don't focus on really being inventors . . . we focus really on technology transfer. Maybe there is something we've learned in the plastics or paint area that we transfer over to some different project that doesn't have anything to do with plastics or coatings or paint . . . .
>
> And so we try to learn, we try to apply what we've learned in the different areas to the things we are wrestling with at the moment. It is more of an application of a technology rather than an invention. But still, there are inventions that come out of that effort too . . . .

Tim Ziebarth, the current Senior Vice-President of New Ventures at HCR, states that, "we're technologists, we are not fundamental discovers . . . . We try to know as much technology as we can so, when problems or opportunities come up, we can draw on a bunch of resources to come up with a high-value solution that we can then do something with".

A critical component in HCR's effort to expand its expertise in various technologies and to be effective problem solvers has been its ability to communicate internally. The company has developed an electronic mail system that is accessible through one of 130 stations on a personal computer network, and a voicemail system for practically all employees. The electronic mail system is a particularly powerful tool that the company uses to leverage its entire skill-base. One observation or note from a scientist or technician about a particular problem can become a "hot-button, or focal point for ten or twelve responses" from all over the company, according to Stull.

Through the use of these communications networks, and through the use of "hallway meetings" and other informal *ad hoc* "whiteboard meetings", the various members of the organization can draw upon each other's talents and insights to solve customer problems as they are identified, and the organization can learn. As Stull comments, "we are fully team oriented . . . people just sort of gravitate to the teams . . . [and] more decisions are made in the hallway than anyplace else".

SUMMARY OF PRIOR CONSTRUCTS RELEVANT TO SECTION III

The following theoretical constructs are relevant to Section III of HCR's history:

Absorptive capacity (Cohen and Levinthal, 1991)
Convertibility of capacity (Ansoff, 1978, 1979)
Customer needs and functions (Abell, 1980)
Economies of scope (Teece, 1980; Pansar and Willig, 1981; Rothwell and Whiston, 1990)
Knowledge–technique base (Lenz, 1980)
Integration (Newman, 1992; Whiston & Rothwel, 1992)
Interpretation systems (Daft and Weick, 1984)
Organizational learning (Hedberg and Starbuck, 1981; Dutton and Freedman, 1985)
Schumpeterian innovation (Schumpeter, 1934; Nelson and Winter, 1982)
Technology push and need pull (Burgleman and Sayles, 1986)
Value and value creation (Peters, 1984)

See applications section for details.

APPLYING EXTANT THEORY AND DEVELOPING FIELD-BASED SUBSTANTIVE THEORY FROM SECTION III

Extant Theory Application

A key ingredient in the success of the new spin-off was the general problem-solving orientation the company took. The focus on packaging the firm's skills and knowledge–technique base (Lenz, 1980) to meet a broad range of customer needs and functions (Abell, 1980) enabled the company to develop dynamic capability (Teece *et al.*, 1992, 1997) and begin to benefit from economies of scope (Teece, 1980; Pansar and Willig, 1981; Rothwell and Whiston, 1990) and to build competence in the convertibility of its capacity (Ansoff, 1979).

All of these attributes contributed to the development of general organizational learning (Hedberg and Starbuck, 1981; Dutton and Freedman, 1985) and interpretation (Daft and Weick, 1984) systems within the company. The firm became an integrated (Newman, 1992; Whiston, 1992) problem-solving unit that could assemble and reassemble its skills and resources in a myriad of configurations—each targeted at fulfilling a need for its client, and then learning to generalize on its solution.

This problem-solving approach led to the extraction innovation that would ultimately change the business, and which was an architectural innovation Albernathy and Clark, 1985) in the Schumpeterian tradition (Schumpeter, 1934; Nelson and Winter, 1982).

SUBSTANTIVE GROUNDED THEORY DEVELOPMENT ON SKILL-BASED STRATEGY AND ENTREPRENEURIAL LEADERSHIP

(1) Conceptualizing the firm as a problem-solving entity was a development strategy.
  (a) It is a value creation and appropriation strategy.
  (b) It is a basis for systems and structures development.
(2) Problems are addressed by applications of technology, which are skill-driven.
(3) Skill-based strategy exploits economies of scope, which derive from economies of skill.
  (a) Technology transfer is a competence issue.
  (b) Systems and structure do matter.
  (c) The team formation process is at the heart of generating economies of scope, which derive from economies of skill.
(4) Summary figures

Conceptualizing the Firm as a Problem-solving Entity was a Development Strategy

A central part of Stull and Daughenbaugh's development strategy for HCR was to conceptualize the company as a problem-solving concern that: (1) would focus on fulfilling the needs of customers in terms of specific problems and then (2) would attempt to capitalize on those aspects of a specific problem that could be applied more generally.

It is a Value Creation and Appropriation Strategy

Solving the customers' problems is really a form of adding value and a way to think about the process. What HCR has done is to expand the problem-solving perspective, which is customer-focused, into an internal engine designed to propel the firm towards new opportunities which can allow it to keep some of the long-term residual value from its "generalizable" experience.

It is a Basis for Systems and Structures Development

The company's problem-solving orientation has been the basis for how it recruits its skill-base, operates and builds its internal systems, and expands its activities. In order to build and support a broad skill-base, the company hired individuals that were "generalists" by nature: that liked to tinker and play with different ideas and technologies. The company, as a whole, developed a broad "suite of knowledge" as Ziebarth calls it, which contained a variety of specific technical skills and capabilities (see Fig. 7.1 for a partial listing of some of the technical areas served), as well as a base of general knowledge that was at the heart of the firm's capacity to cross over and apply the specific technologies to a wide array of applications.

Problems Are Addressed by Applications of Technology, Which Are Skill-Driven

There appear to be two elements to the value creation and appropriation engine that HCR has built: (1) a broad knowledge and skill-base in its people and (2) the strategic application of that skill-base within a technologist's domain; that is, HCR limits its activities to the development, application and transfer of various technologies. HCR does *not engage* in systematic attempts to discover new knowledge or contribute to basic research. Invention and discovery are part of research; while the application of technologies to markets are aspects of innovation and entrepreneurship.

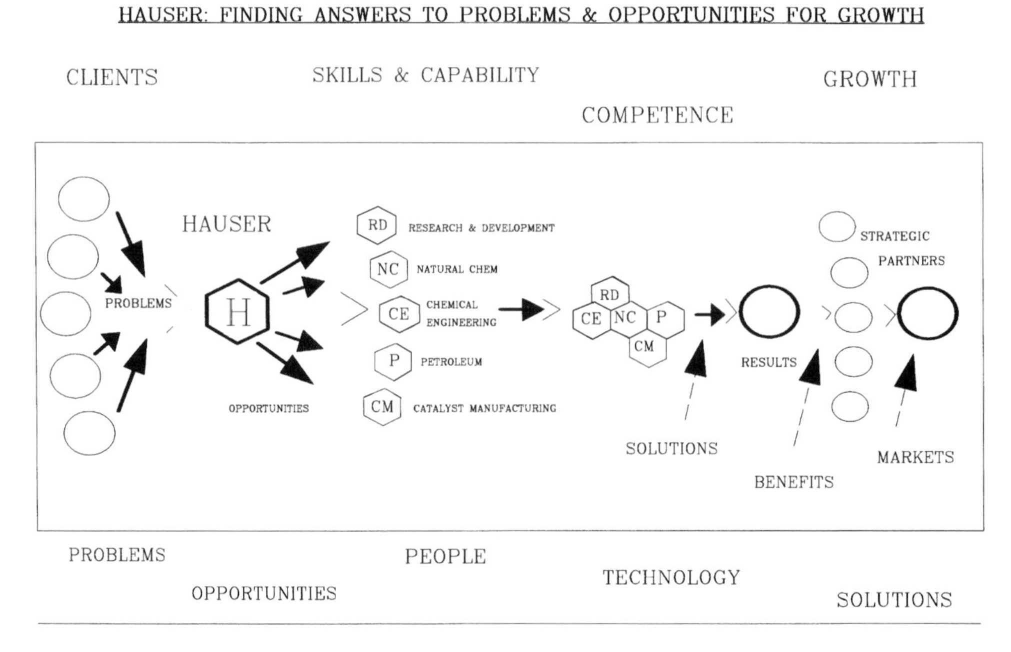

Fig. 7.1. The problem-solving approach at Hauser, from 1990 Annual Report.

The focus of HCR on technology application and transfer puts HCR squarely in a developmental, innovational and entrepreneurial mode of business where it must focus on being on the leading edge of technology delivery. To the extent that HCR is technology driven, it is really skill driven; for technology, broadly conceived, is embodied skills, it is a system of techniques which, when appropriately applied, consistently produces predictable outcomes. The development and appropriate application of techniques within the technology is wholly determinant on the skills of the people that use and extend the technology. There is evidence that HCR's leadership recognizes the strategic importance of its skill-base and the growth and development of that base in terms of its capability to apply technologies effectively. The need to stay abreast of the latest technologies, in terms of the firm's skill-base, and to remain generalist in nature is considered of more strategic importance to HCR than is the protection of the proprietary technologies they have developed in the course of doing business.

Skill-Based Strategy Exploits Economies of Scope, Which Derive from Economies of Skill

The horizontal nature of the technologies that HCR attempts to master is an important aspect of its ability to exploit economies of scope. This ability, in part, derives from HCR's competence, as we have used the term, in blending and combining its resources. In this case the resources are the knowledge of a technology, and the experience-base to judge how effectively the technology may be blended into a new application.

Technology Transfer is a Competence Issue

This perspective on technology transfer is interesting in that it really gets to the heart of the competence idea, and shows how central organizational learning is to the health of a technology and hence, to a skill-based company. Organizational learning is critical because all the elements of a technology transfer are not known by one person, there must be a blending of ideas, insights and skills of many people in order to transfer a technology across applications. In order to have such a meeting of the minds, there must be mechanisms in place that foster exchanges of ideas and organizational learning through shared experience. The ability of the organization to learn is critical to its survival in such a technology and skill-driven environment, as Dr Daughenbaugh recognizes, "if you aren't learning new things, new technologies, you are going to be obsolete very quickly. So the growth of the knowledge base is very important . . . ".

The success of the company, in terms of being able to solve problems and benefit from them, is in large part due to the broad-based scope and skills of the people in the firm and their ability to learn from their experience.

This is an attribute built into the company by the founders, who are somewhat general in their skill-base. The focus is not on technology *per se* but on the capability of the people to understand the new technologies and to adapt and change the technologies as needed, given customer problems.

Systems and Structure do Matter

Part of the way HCR attempts to ensure that organizational learning occurs is in the way it structures itself and communicates. In order for the company's broad skill-based strategy to work, the organizational structure required is inherently open, flexible, informal and inverted. By inverted I mean that the role of management is to support the needs of the individual scientists and technicians in terms of equipment, training, decision-making autonomy etc. This organic form of organization, which is rooted in flexibility, is a time-saving device and is a structural root for skill-based strategy success. Time-based strategy is a direct function of the skill-base of the company, especially in its breadth.

The "systems" that are developing at Hauser are built upon the "soft skills" of the people. Leveraging their science training with a broad perspective and drawing on resources they attack problems from a generalist view, which can yield unique and new solutions that they can potentially turn into a sustainable business. The organic structure also encourages cross-pollination of ideas and a more shared knowledge of the overall skill-base of the company, which is something each scientist and technician must know about, or be able to learn about, once he or she is confronted with a new problem from a customer. Access to information on what skill resources the firm has at its disposal is critical, since individual employees, who hold the specific knowledge, are encouraged to take responsibility and to form *ad hoc* teams, composed of whatever skill-sets of people are necessary to tackle a problem.

The Team Formation Process is at the Heart of Generating Economies of Scope, Which Derive from Economies of Skill

Given this *ad hoc* approach to team formation it is important that boundaries in the organization are fluid. At HCR the "hallway meetings" and electronic mail systems are mechanisms that attempt to ensure fluidity in team formation and decision-making. For a given problem, one person may become a focal point in the company and may become responsible for drawing in a good many others in attempting to find a solution. The decision to draw in others is essentially made on skill and knowledge-based criteria and supports the contention that the company's strategic processes are skill-based and skill and capability driven.

Teams at HCR seem to form much as snowflakes form, and the assembly rules for team formation seem to follow the general *process of crystal development in natural science*, that is: a person, or team of people (a skill-set) inside the firm provides the intellectual material from which potential "crystal habits" can form, based around the precipitating mass of a real customer problem. Given a particular problem, if the critical skill and capability mass in the firm exists, a unique "habit" will develop around the problem and new "crystals" of knowledge and action will form in the company.

In particular, the process is like the formation of snowflake crystals, which form from the assembly of very simple basic elements (frozen H2O molecules), but which constitute complex crystalline entities and variations. Based on the unique conditions that pattern the distribution of molecules in the vicinity of a precipitant, a singular snowflake "habit" will form that shares two basic characteristics with other snowflakes: it is composed of frozen water molecules, and it has six "legs" that radiate from its centre.

Like the snowflake habit, the same basic elements in the firm (chemists, engineers, technicians) give rise to a myriad and broad scope of activities and capabilities, each depending on the unique conditions of the problem that precipitated the formation of the problem-solving team, or "knowledge habit". This process, where a few central elements can be recombined in a myriad of ways to produce a much larger set of outputs, is at the heart of developing economies of scope, and perhaps more appropriately, economies of skill. The fact that there is an emphasis on maintaining a "generalist" perspective is what enables HCR to assemble their resources in so many unique ways. If one specializes too early, or fails to have some fail-safe check on the specialized approach to a business, it seems that a sort of "paradigm lock-in" can take place and inhibit a company from recognizing valuable opportunities to leverage its skills. This occurs if the conception of the skill-base is too narrow, which is not a problem at Hauser.

Summary Figures

The overall problem-solving approach, which guides how teams are formed, decisions are made and systems developed, is part of a larger process of fulfilling the needs of the customers and appropriating a share of the value in the process. The process of satisfying customer needs at HCR is driven by a problem-solving orientation, where HCR applies its suite of knowledge and technologies both to specific technical problems and to the development of more general solutions.

Figure 7.1 illustrates how Hauser Chemical Research itself presents how its problem-solving approach guides its business. This figure, which is adapted from the 1990 HCR Annual Report, suggests that Hauser, as a whole, is a catalyst between the problem-sets that clients and customers have and the discrete skills and capabilities of the scientists, technicians and the teams they form. The skills of the people, in terms of the knowledge and practice of various discipline-based technologies and their competence in combining these skills in novel ways, lead to solutions to customer problems and potential downstream benefits that Hauser can tap into on its own or with strategic partners.

SECTION IV: HAUSER MEETS HAUSER AND LEARNS TO TAP INTO DOWNSTREAM VALUE

In early 1986 Vipont's demand for sanguinaria extract prompted HCR to seek more adequate and larger production facilities. A New Jersey-based ventures group came up with the financing to move the company to better quarters, and the company went public in the process. At the same time, Dr Stull was aggressively pursuing other opportunities. Dean came across an announcement from the National Cancer Institute in the *Commerce Business Daily*. The NCI was funding a worldwide search of natural products that might have some benefit for the treatment of cancer and was asking for proposals from companies who could extract and purify from the raw materials.[8]

After failing to win an initial contract bid with NCI, Hauser Chemical Research tried again, and was awarded an umbrella contract for extraction which eventually led to HCR's exclusive position as the world's only commercial-scale manufacturer of taxol, a promising anti-cancer compound. HCR also began to develop flavour extracts for food companies and, by the summer of 1989, HCR was larger than Hauser Labs in terms of personnel, and had begun to encroach upon the parent company's environmental chemistry business.

"This became the first area of competition between the two friendly companies sharing the same surname and partial ownership", according to Ray Hauser. "Rather than butting heads, Dean, Tim Ziebarth [who had since rejoined Hauser Labs] and Ray discussed the advantages of a merger that could:

(1) Avoid competition between the two companies.
(2) Provide succession opportunity for Ray and the employees of Hauser Labs.
(3) Provide sufficient capital basis for HCR to qualify for NASDAQ over-the-counter trading.
(4) Cost very little cash.

Thus, on 1 January 1990, the private Hauser Laboratories Inc. became a division of Hauser Chemical Research, Inc.".[9] The merger had a large impact on the ability of the firm to offer a variety of effective services to a client. Kurt Ammon remembers that,

> the biggest part of when we merged the labs together . . . all of a sudden, with what they [HCR] could do, and what we [Hauser Labs] could do, the company could basically turn every phone call into a job. How both businesses complement themselves is real important, and that is why we

> have such a wide diversity of abilities, and it has definitely expanded now from where it was five or six years ago.

Tom Scales, HCR's current Executive Vice-President of Operations, believes that, "strategically, I think that there is a synergy that makes the company successful". Tom Scales comments that "we are technically sophisticated beyond the implications of our size".

As important as the breadth of the skill-base in the newly merged company was, perhaps even more important was the way that HCR now began to approach its problem-solving orientation. Tim Ziebarth expands on this point:

> Historically, the labs have always solved or addressed the particular technology problems with a particular client. They've been paid their service fee, then the client and Hauser went on their merry way. Hauser never really looked specifically for ways to gain the true value for those solutions that we provided . . . .

Ziebarth continues:

> The lab services business has always been a "window on opportunity" . . . [yet] all we got was fees for the time and materials spent. It was often the case, where we may have made a technological improvement or solved a problem that had a very wide generic benefit, or a very high value benefit to the client, that we never aimed on capitalizing on that by somehow participating in that bigger value, other than just the immediate need the client had.
>
> We feel that now, because of having the lab business inside HCR, while we are working on problems for clients in the service business, that exposes us to or gives us exposure to a lot of new potential opportunities in the general field of technology: to new technologies, to new ways of using technology, and to problems with existing technology that need general solutions, not just to specific solutions for a client.

With the merger, HCR began to systematize its efforts at capturing as much potential downstream value from its specific problem solutions as possible. Whenever the company could see that its specific solution may be generalizable to other applications, it began to think of ways to develop the general solution into a business.

In fact, Tim Ziebarth became the executive responsible for an entirely new function in the expanded HCR—New Venture Development. He was charged to "look at those opportunities which are outside the bounds of our current operating capabilities". His function's role "is to try and identify, to look for in the labs, and from outside people contacting us and so forth, new businesses that are based on technologies that we don't currently have in our operating units". One of HCR's strategic objectives, according to Ziebarth, is to learn how to tap into the value stream of both the generalizable technologies that the lab may develop, and also into "other sources of new value-added opportunities out there that have a real technology base . . . . We take known technology, as close to the forefront of technology that we can, and try to find useful, high-value applications. And we find the strategic partner necessary to do that in packaging and marketing". In fact, Ziebarth in particular acknowledges that he is the focal point in the organization for thoroughly understanding the skills and capabilities of HCR, and for comprehending the strategic implications and impact that the firm's skill-base poses for the firm.

The New Ventures function, along with the Business Development function, whose mission it is to "expand the use of existing technology to do more of the kinds of business we are already doing", and to expand existing businesses, were organizational innovations designed to ensure that HCR could continue to grow and have as broad a playing field of opportunity and choice as possible.

The merging of Hauser Labs and Hauser Chemical Research was also an important opportunity to build a working environment and culture that could lead to a sustained growth and development of both the company and its employees. For, while, "the corporate culture changed very little for all involved", according to Ray Hauser, there was much to be gained by bringing the two different teams under one roof. Tim Ziebarth comments that, "we wanted the research group of HCR to pick-up some of the operating style and so forth, that Hauser Labs had; and we wanted Hauser lab chemists to sort of broaden their horizons and understand some of the new things we were doing on the HCR side". This reasoning extends to the relationship between the Labs and the manufacturing personnel, again Ziebarth observes, "one of the things that the labs people need more of than anything is knowing production techniques, ideas and the manufacturing culture . . . [on the other hand] if we could have everyone in production, from the day laborer to the managers, thinking like the labs and problem-solving, we would have much smoother production".

The corporate culture at HCR is still very supportive of individual employee initiative and it encourages employees to take as much responsibility as they can and to learn as much as they can. Given the dynamic nature of the problem-solving environment, there is a tolerance for experimentation and failure, so long as individuals learn. As employees learn and do things better, they help contribute to the performance of the company, which comes back to help the employee through the company's employee stock option program.

There is also, given the *ad hoc* nature of many of the teams that form, an inherent bottom-up character to the company's culture. The employees are able to take responsibility, to empower themselves, based on their own initiative. The role of top management is to protect the employees and provide them with the raw materials and training they need to do the job better. Management is there, according to Ziebarth, "to just nurture that attitude, and we have to monitor to make sure that they are still thinking generally, solving the clients problems . . . giving them the support they need to be able to do their jobs . . . listening to them when they say I need some help getting that resolved".

This approach has led to time-based advantages at HCR, where both quality and speed of service have become competitive advantages. This point is illustrated by how fast the company was able to bring its taxol manufacturing capacity on-line:

> Our production facility here for taxol . . . we did not have anything in this building in the first of September . . . which was 10 months ago. Today this plant produces product [that meets the rigorous government standards]. It would have taken a normal pharmaceutical company 18–24 months to reproduce that process. We did that because all the people in the plant that were helping to design, build and install it had full authority to make all the decisions necessary to get to the end. And they made a lot of mistakes. But we told them all along that "if you make a mistake, I'll take the blame for it, just keep making the decisions to get the plant running", and they did . . .

This attitude towards employees is an important aspect of Hauser's ability to keep its current skilled employees and to recruit new ones. Sue Maynard, who was hired as the company's Human Resources Manager in early 1992, comments, "I was very influenced by the people-attitude that Dean Stull and Randy have towards their staff, in realizing that their human resources is a tremendous, valuable asset, that they not only want to take care of, *but that they want to nurture and grow* . . . " [our emphasis].

SUMMARY OF PRIOR CONSTRUCTS RELEVANT TO SECTION IV

The following theoretical constructs are relevant to Section IV of HCR's history:

Absorptive capacity (Cohen and Levinthal, 1991)
Capacity utilization (Ansoff, 1979; Goldratt, 1990)
Corporate entrepreneurship (Burgleman and Sayles, 1986; Hofer & Bygrave, 1993)
Empowerment (Barnard, 1938)
Flexibility (Nonaka and Johansson, 1985; Rothwell and Whiston, 1990)
Knowledge link (Badarocco, 1990)
Organizational learning (Hedberg and Starbuck, 1981; Dutton and Freedman, 1985)
Synergy (Ansoff, 1965; Hofer and Schendel, 1978; Clarke and Brennan, 1990)
Technology push and need pull (Burgleman and Sayles, 1986)
Time-based competition (Stalk and Hout, 1990)
Value and value creation (Peters, 1984)

See applications section for details

APPLYING EXTANT THEORY AND DEVELOPING FIELD-BASED SUBSTANTIVE THEORY FROM SECTION IV

Extant Theory Application

The growth of HCR, once it had developed the innovative natural extraction process, was largely paced by its ability to find new applications of this technology in an example of technology-push innovation and entrepreneurship (Burgleman and Sayles, 1986). The company had learned to manufacture bulk quantities of extract, which was a new capability for the company, and this led to relatively rapid growth for the firm—relative to the slow skill-based growth of the parent company.

The spin-off had begun to compete for business with Hauser Labs, and the merger of the two firms accomplished a number of positive things for both companies. The most important of these was that the combined company could leverage its complementary skills and achieve service economies of scope (Teece, 1980) through the synergies that existed in the combination (Ansoff, 1965; Hofer and Schendel, 1978; Clarke and Brennan, 1990).

The new company was flexible (Nonaka and Johansson, 1985; Rothwell and Whiston, 1990), fast (Stalk and Hout, 1990), focused on the creation of value for its customers through its ability to solve problems, and in a position to learn how to tap into the value it created for specific clients and generalize many solutions.

The new firm also began to institutionalize itself (Barnard, 1938; Selznick, 1957) as a "learning organization" (Hedberg and Starbuck, 1981; Dutton and Freedman, 1985). Through organizational innovations the leadership of HCR set up a New Venture Development process and team within the company, as well as a formal Product Development team. Both of these efforts greatly expanded the firm's absorptive capacity (Cohen and Levinthal, 1990) and provided mechanisms to ensure that the highly skilled human capital of the firm was utilized to its capacity (Ansoff, 1979; Goldratt, 1990).

SUBSTANTIVE GROUNDED THEORY DEVELOPMENT ON SKILL-BASED STRATEGY AND ENTREPRENEURIAL LEADERSHIP

(1) The merging of complementary skills both broadens and deepens the skill-base.
(2) Systems and structural developments must relate and feed the core skills.
(3) Entrepreneurial leadership is first about skills, then recognizing opportunity and capitalizing on it.

skills and capabilities and with its entrepreneurial outlook (Andrews, 1971, 1987; Peters, 1984; Hofer & Bygrave, 1993; Teece *et al.*, 1992, 1997)

SUBSTANTIVE GROUNDED THEORY DEVELOPMENT ON SKILL-BASED STRATEGY AND ENTREPRENEURIAL LEADERSHIP

Protection of the Core Skills in the Face of Rapid Growth Requires Leadership and Strategic Assessment

The explosive growth of HCR in the last two years stems mostly from its success in natural product extraction, especially with regards to taxol manufacturing. This growth is changing the very fabric of the institutional character, as reflected in the comments in Section V.

While the need for more formal systems and programs to help HCR get a handle on the growth is important, what will be interesting to see in the future is how these systems develop relative to the skills and culture of the organization and what effects they will have on the long-term success of the firm to protect its core competence in problem-solving and technology transfer.

Scales is speaking as an operations man at Hauser, and is now at the core of the company in terms of money making. Given the history of the company and its established character as a home for generalists, he is not at the core of the institutional nature of the company. There seems to be something different in the way Scales speaks of the core of the firm. If this firm was to lose the skill-based edge in its problem-solving core, then when a product like taxol, or the process of extraction, becomes more competitive, HCR won't be able to craft its future and it will have lost is core competence to the process of over-specialization, which is the antithesis of the current business.

The institutional pressures on HCR are probably at their greatest now, and the demands for entrepreneurial leadership are also at their greatest. The company is beginning to fragment into cohorts and specialized functions and is beginning to lose its identity. Leadership is needed to protect the core and ensure that there is a process for generating new opportunities for the firm.

If the core of the company as a generalist problem-solving entity is to survive, HCR's leaders must make sure that its business development and new ventures functions are fed with the necessary resources to be able to evaluate and capitalize on new opportunities. A core skill-base of generalists must be nurtured and inspired to continue tackling the most challenging of problems. Any attempts to shift the resource-base substantially away from this core set of activities will lead to short-run profitability and long-run stagnation for this dynamic company.

IDIOGRAPHIC-LEVEL THEORY SUMMARY: HAUSER CHEMICAL RESEARCH

Growth of the Firm

(1) Pure skill-based growth is slow.
(2) Investments in technical and entrepreneurial capital are a means to quality growth.
(3) The merging of complementary skills both broadens and deepens the skill-base.

Problem-Solving Orientation: Skills, Systems, Technology, Structure and Value Creation

(4) Conceptualizing the firm as a problem-solving entity was a development strategy.
  (a) It is a value creation and appropriation strategy.
  (b) It is a basis for systems and structures development.
(5) Systems and structural developments must relate and feed the core skills.
(6) Skill-based strategy exploits economies of scope, which derive from economies of skill.
  (a) Technology transfer is a competence issue.
  (b) Systems and structure do matter.
  (c) The team formation process is at the heart of generating economies of scope, which derive from economies of skill.
(7) Problems are addressed by applications of technology, which are skill-driven.

Entrepreneurial Leadership and Institutional Competence

(8) Entrepreneurial leadership is first about skills, then recognizing opportunity and capitalizing on it.
(9) Protection of the core skills in the face of rapid growth requires leadership and strategic assessment.
(10) An incubating firm must be able to accept spin-offs and redevelop its skill-base

NOTES

1. Much of the information on the early years of Hauser Labs comes directly from *"Through the Wilderness with Hauser Laboratories"*, by Ray Hauser and Dean Stull, for the Division of Small Chemical Business, 15 April 1986.
2. *Ibid.*, p. 4
3. *Ibid.*, p. 12.

4. *Ibid.*, p. 13.
5. Hauser Chemical Research Document, Winter 1992.
6. Quote from Hauser Chemical Research Banks on Biomass Extraction, *The Boulder County Business Report*, Vol. 7, No. 10, October 1985, by John Kadlecek.
7. Hauser Chemical Research 1991 Annual Report, p. 7
8. Hauser, R., and Stull, D. (1991) *Into the Promised Land with Hauser Chemical Research, Inc.* Internal document, 4 December, p. 4
9. *Ibid.*, p. 7
10. Hauser Chemical Research 1990 Annual Report, p. 1.

8

ENTREPRENEURIAL LEADERSHIP AND SKILL-BASED STRATEGY IN A TECHNOLOGICALLY AND SKILL-INTENSIVE CHANGING SERVICE AND MANUFACTURING ENVIRONMENT IN A CORPORATE SETTING: THE CASE OF COMBUSTION ENGINEERING, 1982–1990

INTRODUCTION

Until it was acquired by Asea Brown Boveri (ABB) in 1990, the various business units of the Combustion Engineering Corporation (C-E) had been leading providers of engineered products, systems, and services to the worldwide power generation, process controls, and public sector and environmental waste and energy markets (see Table 8.2 for complete business segment information). Today (1992) Combustion Engineering is but one, albeit large and complex, part of what ABB's management has called "the world's most aggressive expansion strategy in the electrical machinery and power engineering sectors".

Combustion Engineering was considered a viable component of ABB's expansion in the late 1980s and early 1990s, in large part due to the turnaround and restructuring efforts of then CEO Charles F. Hugel and his management team, who inherited a Combustion Engineering company in 1982 that was in poor shape both financially and in terms of its capability at meeting customer needs.

This brief case history traces the reinvention of Combustion Engineering since 1982, which George Kimmel, President and Chief Operating Officer of Combustion Engineering under Hugel, says was based on a "skill-based strategy". It is a special case in the book in that it is a corporate strategy case. It is a case that focuses on how the company's leaders affected change, using a skill-based strategy, to enable the corporation, across a wide array of different business units, to provide new and valuable services that would meet the changing demands of its customers.

Because of the extremely complex nature of the organization, and the products and services that Combustion Engineering offers, this case focuses mainly on the process by which C-E's top management team were able to begin to reshape the company, especially in terms of its management of skills and human capital, and as such does not purport to be a comprehensive case history in any larger sense.

COMBUSTION ENGINEERING'S HISTORY, SECTION I: HUGEL INHERITS A DINOSAUR

In spring 1979 George S. Kimmel joined Combustion Engineering as Treasurer. He had left Lykes as Chief Financial Officer to join the world's largest manufacturer of steam generating power systems, used by electric utilities, and one of the market leaders in process industries. The company was organized along product-oriented business units, and as Kimmel observes, "this worked very well for a long time, and we made lots of money. Orders used to be allocated to the businesses, and prices were determined by what would clear the markets". When Kimmel came aboard in 1979:

> The company was really just finishing up a long period of growth and reasonable stability. It had a solid order book of nuclear plants and fossil fuel plants, and a good strong industrial minerals business. The company had begun to diversify into oil service equipment, with the underlying strategy that they would be a company that would serve the energy markets.
>
> And that was a strategy that seemed to make a lot of sense at that point of time. But then a lot of things started to happen in the early 1980s. One thing that happened in 1979 was Three Mile Island; the rest is history as to what happened in the nuclear business. There also was a change in the price of energy; people began to think more about conservation. Utilities became a different business, they no long had a free hand, they had to deal with Public Utility Commissions, so they stopped buying fossil fuel plants, there were also environmental reasons . . . the point is that they stopped ordering plants of any kind.

> Much of the C-E oil field services were based on high energy prices, deep gas for instance . . . that dried up, there just wasn't any business. Basically 80% of the C-E business just went into a nosedive about 1982, and that was about the time Charley Hugel joined Combustion Engineering from AT&T as President and Chief Operating Officer.

Before coming to Combustion Engineering, Hugel headed Bell Laboratories and Western Electric for AT&T, but when he joined Combustion in 1982 it is unlikely that he knew how tough the next three years would be; few people did. Kimmel continues, "He came to the company before any of these things became evident. This was a business that had grown in a stable industry, with major players, maybe even an oligopoly".

Combustion Engineering operated in an environment where, according to Kimmel, "you were dealing with clients you had a close relationship with, you could work things out in an iterative sort of way. You had long horizons to get things done". Combustion Engineering's position in an oligopoly did not require that it be as focused or as efficient as it would have had to be in the face of more difficult economic conditions and competition. By 1982, as Mark Altman of Paine Webber noted, Combustion Engineering had "no clearly defined strategic direction, a lack of management depth, poor management information systems and high costs, [which] were largely masked by the early 1980s energy boom".[1]

By the early 1980s, according to George Kimmel, "the markets changed rather quickly, and competition became tough . . . orders dropped off sharply. We had a problem . . . there were no long-term contracts, and the company's basic business was in trouble. It became apparent at that point that C-E had to change its focus, it clearly had to develop ways to use its expertise in the market differently than it had been applied before". In early 1984 Charles F. Hugel succeeded longtime Combustion Engineering leader Arthur J. Santry as Chief Executive Officer and began to lead the effort to change how Combustion Engineering would operate in the future.

SUMMARY OF PRIOR CONSTRUCTS RELEVANT TO SECTION I

The following theoretical constructs are relevant to Section I of Combustion Engineering's history:

Corporate strategy (Hofer and Schendel, 1978)
Human capital atrophy (Schultz, 1961, 1990)
Market-based portfolio management (Henderson, 1979; Hofer and Schendel, 1978)
Quantum change and crisis (Miller and Friesen, 1980; Miller, 1982)
Related diversification—resources (Farjoun, 1990)
Related diversification—product (Rumelt, 1974, 1982; Salter and Weinhold, 1979; Lecraw, 1984; Palepu, 1985)

See applications section for details

APPLYING EXTANT THEORY AND DEVELOPING FIELD-BASED SUBSTANTIVE THEORY FROM SECTION I

Extant Theory Application

Combustion Engineering operated as a business corporation with a central corporate staff and many different, relatively independent operating business units that spanned a variety of product and market segments. As such this case represents an example of corporate strategy concepts (Hofer and Schendel, 1978), especially those concepts that focus on related diversification, both from a product point of view (Rumelt, 1974, 1982; Salter and Weinhold, 1979; Lecraw, 1984; Palepu, 1985) and from a resource point of view (Farjoun, 1990).

During most of Combustion Engineering's corporate life it faced stable markets and benefited from having a large market share, with competition from just a few other large firms. It operated in a mature and oligopolistic environment (Bain, 1959; Porter, 1980). There were few pressures to perform in a time-based (Stalk and Hout, 1990) manner and few pressures to integrate the functions of the various business units or to fully exercise its skills (Schultz, 1961, 1990).

Acquisitions and C-E's business portfolio (Henderson, 1979) were opportunity and market driven, as witnessed by C-E's diversification into the oil service and equipment businesses in the 1970s.

Little formal attention was paid to internal operating characteristics and information flows across business units until the markets began to change dramatically in the late 1970s. The "market-quakes" forced the corporate leadership to begin to re-evaluate their strategy and the structure of the company, and illustrate how crisis precipitates the search for quantum shifts in strategy (Miller and Friesen, 1980; Miller, 1982).

SUBSTANTIVE GROUNDED THEORY DEVELOPMENT ON SKILL-BASED STRATEGY AND ENTREPRENEURIAL LEADERSHIP

In Mature and Oligopolistic Settings the Firm's Skill-Base is Likely to Atrophy, Especially in Terms of Integrative Competence

Given the relatively benign environment that Combustion Engineering operated within for most of its life (the energy and power markets were quite healthy until the late 1970s), and its immense size and market power, it was no surprise that C-E did not have the capabilities to respond to quick changes in the market-place in the early 1980s. In such an oligopolistic setting there is little need to focus managerial time and attention on developing time-based or integrative skills. In relatively non-competitive or non-dynamic markets the skill-base of the firm, especially in terms of its ability to blend and package a variety of skills and resources in real-time, is likely to atrophy.

Corporate attention in such an environment is more likely to focus on opportunity-based activities such as developing a portfolio of companies that may be related in a macro product sense (like energy), but not necessarily a skill or resource sense. Little managerial time is devoted to building core competences and internal flexibility.

COMBUSTION ENGINEERING'S HISTORY, SECTION II: TRANSFORMATION THROUGH ACQUISITION

Hugel and his managerial team at Combustion Engineering faced a war on two fronts in the mid-1980s: (1) externally, the demand for C-E's traditional products and services had all but dried up. In March 1985, Hugel commented that "those who are waiting hopefully for the markets to return to where they were five years ago are engaging in wishful thinking"; (2) internally, the company was fragmented and did not have good information or coordination systems, and was extremely product-oriented.

Hugel attacked both problems by opening up a discussion within the company on the way that Combustion Engineering should redefine itself. Smith remembers that "Hugel, who had a lot strength in being new, [could] ask the dumb question . . . why aren't you guys working with you [other] guys to meet the needs of this customer?". George Kimmel continues:

> So that [redefinition] necessitated a top-down, bottom-up approach, where the people at the top, starting with Charley Hugel and the senior management group, framed the question and began to frame the issues and demand of the organization a focus on changing markets and our own internal strengths and weaknesses in which to build. That was really kind of iterative.
>
> You build credibility for the reality from the top down. You also begin to change people out, and begin to make acquisitions to bring in skills and points of view or capabilities that complement what you have, or even substitute for what you have. So we did all of those things.

Indeed, under the direction of Hugel, Combustion Engineering attempted to remould itself into a customer-responsive technology and solutions company. In practical terms this meant that C-E had to shift from being primarily an equipment supplier to the power, oil and gas, and process industries, to becoming a service-oriented engineering, technical services and process controls company that also sold capital equipment.[2]

The challenge, according to Kimmel, was to find a way that Combustion Engineering "could take 'know-how' and technical skills from our different business units and pull them together to respond to the different needs of our customers. We wanted to get to a point where we thought of ourselves as one company working together to serve clients".

> More often than not we [began to think] of C-E in terms of skills and resources, in terms of know-how and technical skills, rather than in terms of products and markets. Know-how and technical skills that could serve customers . . . . Our analysis of that [in the power generation area] led us to conclude that the strength of the company was in its technical know-how, its basic technology, its access to the market.
>
> [In terms of] the petrochemical business . . . the opportunity for us there was in our knowledge of the process and of engineering management, and therefore we had to take those skills and bring them to the market in a way that dealt with the current needs. Those needs were for such things as improved process technology, process control and things of that sort, new engineering methods. We then moved into, among other things, the process control business with some acquisitions.

Beginning in 1982 the company began to transform itself through acquisition and divestiture. During the period 1982–1989 Combustion Engineering divested businesses with approximately $1.13 billion in sales and acquired businesses with roughly $945 million in sales. A summary of these transactions appears in Table 8.1. Table 8.2 provides information on the business segments that Combustion Engineering was competing in as of 1989.

The philosophy behind the transformation was consistent. Combustion Engineering moved out of non-core equipment making, divesting its operations in miscellaneous areas and moved into services, process controls and related software. A few selective acquisitions in core equipment businesses were made to protect or enhance existing franchises. In the Power Generation segment, C-E acquired assets to upgrade its fluidized bed combustion technology—a move necessary to protect its leadership in fossil boilers.

On the service and controls side of the power business, implementation of the strategy to rebuild the company began in 1982 when Combustion Engineering acquired Taylor Instrument Company, a manufacturer of analytical instruments and process controls, for $80 million. It also bought Allied Corporation's Bendix Process Analytic Instrument Division that year. Early in 1984 it acquired Jamesbury Corporation, a supplier of valves and valve actuators for processing plants for $99 million, and Impell Corporation, a provider of computer-based management and engineering services for power plants, for $105 million. The acquisitions of Impell added to C-E's expertise in consulting, engineering and maintenance services. Those companies, along with plant simulation and electronic components operations, would come to comprise the nucleus of Combustion's engineered systems and controls group, which Paine Webber analysts felt was a good way for Combustion to build its technology base.[3]

In the Process Industries segment, C-E moved out of its oil- and gas-related equipment businesses and a few other operations, and began a major move

Table 8.1. Combustion Engineering's transformation, 1982–1989. Source: Shearson Lehman Hutton, 8 October 1989, p. 30.

Area	Segment	1983 sales ($M)	1983 sales (%)	1982–1989 divestitures ($M); Companies	1982–1989 acquisitions ($M); Companies
Power generation	Equipment	910	29		150: Lurgi
	Services, controls	407	13		105: Impell
Process industries	Equipment	800	26	715: Vetco Offshore; Gray Tool; Tyler; Beaumont; Natco; Cast Products and Equip.; Randall; Invalco	125: Sprout-Waldron
	Controls, electronics	100	3		400: Taylor Instruments; AccuRay; Bendix Process Analytics; Afora
	Minerals	330	11	160: Refractories, Basic	
Other	Engineering and construction	300	10		
	Resource recovery	na	na		
	Operations services	65	2	65: Vetco Services	120: Bell Technical Operations; C.E. Jordan
	Building materials	180	6	180: Morgan Products	
Total		3,092	100		

Table 8.2. Combustion Engineering business segment information, 1986–1988.

Primary segments	(000)	1988	1987	1986
Power generation	Net sales	$1,664	$1,498	$1,274
Principal markets				
Electric utilities				
Independent power projects				
Process and other industries				
Principal services, products and equipment	Operating income	($90)	$179	$150
Engineering and management services				
Maintenance and project management				
Power plant control systems and operating software				
Fossil steam systems				
Nuclear steam systems, fuel and service				
Complete power generation plants				
Fluid bed combustors	Backlog	$1,951	$2,161	$1,537
Power plant emission control equipment				
Heat recovery equipment and services				
Work and maintenance management software				
Process industries	Net sales	$1,546	$1,283	$1,079
Principal markets				
Pulp and paper refining				
Petrochemical, chemical and pharmaceutical				
Food and beverage				
Primary metals				
Minerals and mining	Operating income	$55	$48	$9
Textiles				
Principal services, products and equipment				
Measurement and control systems and services				
Operating software and management services				
Instrumentation and valves				
Engineering, construction and project management services	Backlog	$976	$624	$452
Petrochemical and refinery technology				
Pulping equipment				
Specialty minerals and refractories				
Screening, grinding and milling equipment				
Engineering design and information management software				
Public sector and environmental	Net sales	$274	$260	$197
Principal markets				
Municipalities				
Government agencies				
Process and power industries				
Principal services, products and equipment	Operating income	($114)	($19)	$4
Municipal waste-to-energy systems				
Hazardous waste site cleanup				
Environmental consulting, assessment and monitoring				
Hazardous waste systems operation				
Maintenance and training services				
Mass transit engineering and construction services	Backlog	$777	$835	$777

into the process controls business, anchored by the Accuray and Taylor purchases. Shearson Lehman Hutton analysts commented that, "although C-E has paid a heavy price for this new profile, especially in its balance sheet, the [new] business mix probably makes it less vulnerable to cyclical capital spending downturns".

Kimmel comments that "what we did, was basically inventory what we had to work with, which were those skills, know-how and technology, and at the same time inventory what was happening in our markets, with our customers. And then we began to re-orient the business to capitalize on the strengths, which were the technology and the know-how".

SUMMARY OF PRIOR CONSTRUCTS RELEVANT TO SECTION II

The following theoretical constructs are relevant to Section II of Combustion Engineering's history:

Leadership—architect of purpose (Andrews, 1971, 1987; Wrapp, 1967; Peters, 1984)
Modularity of capacity (Ansoff, 1978, 1979)
Organizational change and adaptation (Chakravarthy, 1982; Miller, 1982; Ginsberg, 1988)
Organizational inertia (Miller, 1982)
Related diversification—resources (Farjoun, 1990)
Strategy as fit, building on strength (Hofer and Schendel, 1978)
Synergy (Ansoff, 1965; Hofer and Schendel, 1978; Clarke and Brennan, 1990)

See applications section for details.

APPLYING EXTANT THEORY AND DEVELOPING FIELD-BASED SUBSTANTIVE THEORY FROM SECTION II

Extant Theory Application

When Combustion Engineering's traditional markets began to dry up, there was a great deal of organizational inertia (Miller, 1982) that existed, and which had to be dealt with by C-E's leadership. The only way C-E was going to be able to change was if Hugel and the leaders could convince the middle managers that this was a new world and that it demanded new action. Hugel became the voice for change and an architect of purpose (Andrews, 1971, 1987; Wrapp, 1967; Peters, 1984), with a task to redefine the mission of Combustion Engineering.

One critical aspect of the process of inducing such radical change at the corporate strategy level (Hofer and Schendel, 1978) was within the direct control of Hugel and his top management team—acquisition and divestiture.

Combustion Engineering's leadership began to reshape the company from a skill-based or resource relatedness perspective (Farjoun, 1990), which was a departure from previous acquisition criteria. The goal was to build an organization that could exploit synergies of strength across business units (Ansoff, 1965, 1978; Hofer and Schendel, 1978; Clarke and Brennan, 1990), and provide rapid, quality service and equipment to its customers. The company thus moved out of the oil services businesses it had acquired in the days of opportunity-based acquisition, and moved into engineering services and process control technology companies.

SUBSTANTIVE GROUNDED THEORY DEVELOPMENT ON SKILL-BASED STRATEGY AND ENTREPRENEURIAL LEADERSHIP

(1) The entrepreneurial challenge at the corporate level is to learn how to package and repackage the skill-base of the company to create new value.

(2) A skill-based acquisition program must be customer oriented, not shareholder oriented.

The Entrepreneurial Challenge at the Corporate Level is to Learn How to Package and Repackage the Skill-Base of the Company to Create New Value

At the corporate level it is the role of the top management team, and of each business unit manager, to shape the premise for action and interaction within the firm. A significant part of the entrepreneurial leadership challenge, from a skill-based perspective, for corporate organizations is to learn to package and repackage the assets, particularly the human capital, of the entire organization in such a way as to bring new value to current markets or to create new markets. Kimmel comments:

> We thought of entrepreneurship more in terms of developing, bringing together skills from various parts of the company and forming and offering [a solution] to a client who had a specific need, and that was a principle way we could distinguish ourselves from our competition . . . .
>
> Our notion was that we must distinguish ourselves from our competition by having a better set of skills and the ability to deliver them quicker and responsibly and more competently than the competition. We believed that we could, and must, distinguish ourselves from our competitors through a stronger skill-set, and a kind of approach to business that I have described.

In the case of Combustion Engineering this challenge involved bringing into the company much more of a cooperative service element than had existed before, and required that both corporate and business-level executives learn new integration skills and styles of doing business. Interestingly, the skills and style of firms became important attributes of the acquisition decisions that were made by C-E executives during the 1982–1989 restructuring. The acquisition of Impell, in particular, stands out as a clear example of a skill-based human capital development-oriented acquisition. The managers in this company were eventually used to seed other C-E businesses, and became an agent of change for many business units.

The divestiture and acquisition program at C-E during the 1980s centred around the concept of building a new core set of corporate skills and competencies that could create value in the changing market-place. As Kimmel stated, the process was strategically oriented towards building upon key skills, strengths and technologies that C-E management felt it commanded, and which were most likely to allow it to meet the changing needs of the company's customers. In this case it was a classic example of building strategy on the strengths of the company and supplementing those strengths with new skills and perspectives on how to use those skills with new talent from outside the traditional business units. Kimmel is careful to note, however, that the process of examining skills relative to markets is recursive:

> . . . obviously when you think of something like this you have to think about two sides of it. You have to start, my theory is always, you start with the market-place and kind of work back. But, at the same time, you've got to look at what your skills are, and what your capability is, and if you have a mismatch, or a market is changing rapidly, you have to deal with that.

A Skill-Based Acquisition Program Must be Customer Oriented, not Shareholder Oriented

The new emphasis at Combustion Engineering was on developing a fit between corporate skills and capabilities and customer needs; with the knowledge that such a fit would eventually lead to superior shareholder value. Kimmel observes that

> it was our view that it's the knowledge or the know-how, or *the collective functioning of the organization, basically the skills element, the human resource element*, that is going to allow us to make a real success at the business, and make money. What we wanted to do was to build shareholder value, we believed that you did that through technology, know-how, what we call value-added, being able to understand the customer's needs, be responsive—build relationships that would bring value for the customer, and therefore the customer would be willing to pay for it, allowing us to get a price that would allow to get shareholders value. [our emphasis]

This skill-based customer-oriented corporate strategy, developed in the face of a changing and turbulent environment, appears to be more substantial and dynamic than the previous strategy, which was opportunity driven and which focused on developing a fit between corporate positions in markets and shareholder needs.

COMBUSTION ENGINEERING'S HISTORY, SECTION III: TRANSFORMATION FROM WITHIN

Concurrent with his reshuffling of the business portfolio, Hugel tried to reorient Combustion's internal dynamics. A matrix approach to businesses, encompassing client sectors and operating units, was adopted. The matrix was intended to make C-E more responsive to developments in its customer base and to enhance cross-selling of C-E products and services throughout the firm to clients of each operating unit. The key concept of this strategy was "bundling and unbundling" of products and services, based, according to Kimmel, on "know-how and technical skills that could serve customers". For instance, the relationship established between a paper client installing an industrial boiler and C-E's fossil systems group could be leveraged into orders for related process controls, engineering design work or even kaolin clay supplies.[4]

This shift to a customer-driven service orientation was problematic for the company from an internal perspective, as Dale Smith, former Combustion Engineering Vice-President of Human Resources comments: "a major issue, particularly when you move from selling products to solving customer problems, or selling systems, which is maybe another way to look at it, is that it is very hard just to get people to *think* that way". George Kimmel continues:

> The real impediment in the process was the people who were raised in C-E for their whole career in a stable environment, really didn't have a conviction early on that in fact the world had changed, that our capabilities had to be applied differently . . . organizations haven't caught up with the changes in the markets and competitive environment, management processes haven't caught up.

Information exchange within the company was minimal, and the sharing of knowledge across business units, and sometimes within business units, was difficult, as Kimmel notes: "there were many 'cottage industries', technical groups, within a single business unit—based on know-how—which didn't often share data or ideas with one another". The corporation began to work on ways to merge information networks, management and control systems and engineering automation software throughout the company so that C-E could strengthen its traditional capabilities and create new ones.[5] The process of change was slow and painful. According to Kimmel, "This is a process that takes some time. It obviously represents a change in thinking, which imposes a certain set of issues . . . it is never easy to get people to think about things like this".

As Kimmel, who had been named President and Chief Operating Officer of C-E in 1988, noted earlier, the involvement and commitment from top management was important to getting the process of internal change started. It was the responsibility of the people at the top, starting with

Charley Hugel and the senior management group, to frame the important questions and issues of organizational change. Unfortunately, much of the early impetus for change was top-heavy, as the analysts for Shearson Lehman Hutton noted in November, 1989:

> After several mis-starts, the matrix approach [to reorganization] may finally begin to pay off at Combustion Engineering. An organizational superstructure was initially built on top of individual operating units to coordinate business activities and leverage client relationships.
>
> This superstructure and associated staff levels have been ripped out and replaced with an emphasis on communications between segments, with only a few corporate level coordinators to facilitate this communication.[6]

The latter attempt to effect change at Combustion Engineering was based on an approach where operating decisions would remain decentralized, but where key functions such as finance and human resources would be relatively more centralized than before. Dale Smith elaborates:

> We viewed the Human Resources Function (HRF) as central to the strategy and its success. We gave a lot of support to the HR function. There was support from Hugel all the way through our senior staff. Our business units operated in a decentralized way as to operations, but in certain functions we had a key role: in finance and human resources. In the Human Resources Function we were very careful of the selection of each officer in the business units. Once we had the good people in place in the HRF in each unit, then they became part of a functioning HR network around the company, as well as part of senior operating management, all focused on trying to get this strategy into an implementation mode.

This focus on human resource development was an important factor in the success of Combustion's effort to redefine itself, and was an important factor in some of the acquisitions that were discussed earlier. As part of the strategy to build a new organization, Combustion Engineering had acquired Impell, which possessed expertise in computer-based management and engineering services for power plants, consulting and maintenance services. Kimmel remarks that, under the leadership of Ray Fortney, Impell had developed into "an engineering company that was superb at recruitment and development of its people in the way we needed. The managers were young and aggressive, and we used this company as a model and as a source to 'seed' other business units".

Reflecting on the process of change that took place at Combustion Engineering, which ultimately focused on human resource development and knowledge integration, Kimmel comments:

> Hugel provided the leadership . . . it was a long process, we had a cadre of people that believed in this notion [of decentralization and development of human resources]. I became a convert because it became absolutely clear to me, and still is, that you can't run a large organization in a rapidly

> changing environment in a monolithic way, institutionalized. Your only hope of success is to have good people, who have shared values, who understand what the objectives are, and have them organized in a way, with the right kind of leadership that keeps them focused and protects them. It protects their ability to do their jobs, to be creative and use initiative . . .

The goal of all this restructuring and redefinition of purpose was to allow Combustion Engineering to become more capable at responding to new customer needs in real-time. According to Kimmel, "we eventually were able to get to a point where we could put teams from different business units together to go after entire chunks of business, such as an entire process or pulp plant". The company introduced new compensation systems to encourage and reward this type of team action, and made a commitment to its stakeholders that the redevelopment of the company had solid foundations in human resource development, information technology application and new product technology development, as reflected in this statement from Hugel and Kimmel in C-E's 1989 Annual Report:

> Our commitment to clients is to be the best value producer—offering products and services that optimize quality, cost and functionality. We are achieving this in three principal ways. First, we are acquiring, developing and retaining the very best people. Second, we are making effective use of information technology to create advantages for C-E in the marketplace. Third, we are making sure that we understand our clients' changing requirements and are continuing to acquire and develop relevant technologies to meet those needs. We believe this operating approach will serve our shareholders, clients and employees well.

The progress that Combustion Engineering had made in reshaping itself for competition into the 21st century, during the years 1982–1990, was just beginning to be seen when, in 1990, the company was purchased by Asea Brown Boveri (ABB). In Kimmel's judgement, "we were having some real success with our strategy, we were distinguishing ourselves, we were picking up market share and we were beginning to do quite well financially before the ABB merger". Now, of course, under new management it remains to be seen whether Combustion Engineering will indeed be able to capitalize on the changes it has made.

SUMMARY OF PRIOR CONSTRUCTS RELEVANT TO SECTION III

Commitment (Ghemawhat, 1991)
Human capital (Becker, 1964; Schultz, 1961, 1990)
Human resources function (Butler, 1988; Lengnick-Hall, 1988)
Leadership (Kotter, 1990; Eastlack and McDonald, 1991; Hofer, 1992)

Leadership—architect of purpose (Andrews, 1971, 1987; Wrapp, 1967; Peters, 1984)
Loose–tight fit (Pascale and Athos, 1981)
Organizational unlearning (Hedberg & Starbuck, 1981)
Organizational inertia (Miller and Friesen, 1980)
Shared values (Peters and Waterman, 1982)
Skill-based competition (Naugle and Davies, 1987; Klein *et al.*, 1991)
Systems (Ansoff, 1965; Goldratt, 1990)

See applications section for details.

APPLYING EXTANT THEORY AND DEVELOPING FIELD-BASED SUBSTANTIVE THEORY FROM SECTION III

Extant Theory Application

The challenge to redefine Combustion Engineering through internal change was perhaps greater than the challenge to change the company through acquisition and divestiture. The company had developed, in a relatively stable and benign environment, an internal inertia (Miller and Friesen, 1980), or "way of doing business" over many years. There was little sense of shared value (Pascale and Athos, 1981), which put great pressure on Hugel and his management team to provide architectural purpose and competence (Andrews, 1971, 1987; Peters, 1984; Hofer & Bygrave, 1993; Teece *et al.*, 1992).

The recognition by Hugel that the company must utilize its human capital (Becker, 1964; Schultz, 1961; 1990) more effectively led to his integration of the Human Resource Function into the strategic fabric of the company (Butler, 1988; Lengnick-Hall, 1988).

Combustion Engineering used an operations-oriented Human Resource Function as a source for maintaining a loose–tight (Pascale and Athos, 1981) control of the company. The HR function served as an integrating tool in terms of shared values, and provided the company with a relatively centralized picture of the skill-base of the firm, thus putting the company on track to implement its strategy for skill-based competition (Naugle and Davies, 1987; Klein *et al.*, 1991).

SUBSTANTIVE GROUNDED THEORY DEVELOPMENT ON SKILL-BASED STRATEGY AND ENTREPRENEURIAL LEADERSHIP

(1) Integration of function and purpose is essential to effective skill-based strategies.
(2) An operations-oriented human resources function is an important tool in the crafting of skill-based strategies.

Integration of Function and Purpose is Essential to Effective Skill-Based Strategies

Many of the problems Combustion Engineering faced in responding to the changes in the market-place in the early 1980s were a result of its not being an integrated corporation, both from a skills perspective and from an institutional perspective. In terms of skills, even the various business units were fragmented into small pockets of know-how and technical expertise. Communication within, and especially between, business units was poor, so that the knowledge necessary to understand the firm's skill-base, and then utilize it effectively, was weak to non-existent. In terms of the institutional environment of Combustion Engineering, there was no unifying mission or shared sense of value within the company.

Given this internal environment, it is no wonder that acquisitions in the past were opportunity driven, there was little core building competence to integrate the company's skills at the corporate level, and even less institutional competence to provide C-E a shared sense of purpose and character. It was not "one company working together".

This began to change under the entrepreneurial leadership and commitment of Charles Hugel and his top management team, who realized that change was going to be slow and that it had to come from the top. Kimmel comments:

> Tradition in business limits the ability of firms to change. This is related to the age of the company, the building of formal systems. The word "culture" is too broad to be of meaning in the context of management change . . . I prefer "shared value", which is very important. Changes at the top can affect the shared values of the company, but it can take 5–10 years to significantly change the company's way of doing business.

Hugel and his corporate management team, along with a host of business level managers, provided the necessary entrepreneurial initiative to begin changing the context within which Combustion Engineering's businesses operated. The ultimate goal was to unite all the business units in terms of their ability and willingness to assemble a wide variety of skill packages, across businesses, that could provide value for customers; while at the same time giving the leaders of the business units as much operational autonomy as possible. Incentives to work in teams, to meet the needs of customers, were put into place and the company began to put a new focus and priority on Human Resource Management.

An Operations-Oriented Human Resources Function is an Important Tool in the Crafting of Skill-Based Strategies

An important ingredient in the efforts of C-E's top managers to change the company was the inclusion of the Human Resources Function into the strategic fabric of the corporation. As Kimmel noted earlier, "it's the knowledge or the know-how, or the collective functioning of the organization, basically the skills element, the human resource element, that is going to allow us to make a real success at the business and make money". Combustion Engineering's top management seeded change in business units through the development and strengthening of the Human Resource Functions within each unit. Dale Smith observes that "increasingly Human Resource positions at all levels are being filled by people who aren't HR professional, career people. We see a lot of operating people, and other technical people moving into Human Resources".

This focus on building a line or operations-oriented Human Resources Function is critical to building an effective understanding of the skill-base of a complex company. As Smith intimated in an interview, the traditional Human Resource Function at C-E and other large companies has not served the strategic needs of their companies well. The function of most HR departments has been staff-oriented and has been driven by personnel administration factors, few of which relate to tracking the skill-base of the company. According to Smith, there has been a general disconnect between the strategic needs of the firm, which relies heavily on its human capital, and the ability of non-operations-led HR departments to provide useful and timely information to meet those needs.

At the root of any effective skill-based strategy there must be a solid understanding of the specific skills that can be accessed by the firm, and how these skills can be assembled. The most likely storehouse of useful strategic information about the skill-base of the firm is the Human Resources Function, provided that it has operations and line-oriented people to lead and staff it. An operations-oriented HR staff, that is steeped in company and industry experience, can build an effective skills-related database that can inform top management on the profile and character of its human capital.

IDIOGRAPHIC-LEVEL THEORY SUMMARY: COMBUSTION ENGINEERING

Organizational Skill Obsolescence

(1) In mature and oligopolistic settings the firm's skill-base is likely to atrophy, especially in terms of integrative competence.

Value Creation in the Corporate Setting

(2) The entrepreneurial challenge at the corporate level is to learn how to package and repackage the skill-base of the company to create new value.

Skill-Based Acquisitions

(3) A skill-based acquisition program must be customer oriented, not shareholder oriented.
(4) Integration of function and purpose is essential to effective skill-based strategies.

HRM as an Operations Tool

(5) An operations-oriented human resources function is an important tool in the crafting of skill-based strategies.

SYNOPSIS OF PART TWO

In each of the four case studies presented in this book, three related "storylines" have been told that focus on the phenomenon of resource management and skill-based strategy: (1) the first storyline was a focused historical narrative that traced the evolution and development of each firm's skill-base and resource management capabilities, given different environmental conditions; (2) the second storyline was a brief summary and account that explored how extant academic constructs related to the narrative section; and (3) the third was a more concentrated and substantively based idiographic-level theory discussion on the details of the narrative with respect to emerging concepts about resource management and skill-based strategy.

Each of the four cases stands on its own merit as a practitioner extracted idiographic-level source of ideas about resource management and skill-based strategy. Up to this point, there has been little systematic attempt to move to a more general substantive-level of analysis. The research questions that were addressed in Part Two included:

Q1-A: What language do practitioners use in describing and understanding skills and resource-based phenomena?
Q1-B: What processes are involved in how managers build, nurture, change and deploy skill-based resources?

NOTES

1. From "Combustion Engineering Status Report", Paine Webber, 9 October 1985, Mark D. Altman.
2. From a Shearson Lehman Hutton Company Analysis Report on Combustion Engineering, 8 November 1989, p. 29.
3. Paine Webber Company Report, 9 October 1985, Mark Altman, p. 2.
4. Shearson Lehman Hutton Company Report, 8 November 1989, p. 30.
5. Combustion Engineering 1989 Annual Report, p. 1.
6. Shearson Lehman Hutton Company Report, 8 November 1989, p. 30.

PART THREE:
GROUNDED THEORY

9. A CROSS-CASE ANALYSIS OF SKILL-BASED STRATEGY AND ENTREPRENEURIAL LEADERSHIP: DEVELOPING A GROUNDED THEORY FROM IDIOGRAPHIC MODELS
10. ENFOLDING THE LITERATURE TO CREATE A SUBSTANTIVE-LEVEL, GROUNDED THEORY OF SKILL-BASED STRATEGY AND ENTREPRENEURIAL LEADERSHIP
11. SPECIFICATION OF A THEORY OF VALUE CREATION, SKILL-BASED STRATEGY AND ENTREPRENEURIAL LEADERSHIP: PROVIDING A TESTABLE THEORY

PREVIEW OF PART THREE

In Part Three, the idiographic-level theories and ideas presented in Part Two are discussed across the four sites and will serve as the base for the development of a more general, substantive-level, grounded theory of value creation from business activity, which examines how firms create value based on the exercise of entrepreneurial leadership and skill-based strategy. Part Three reflects an effort to make analysis, as Strauss (1987) and Glaser (1978) argue we must, not merely a tool for collecting and ordering, "a mass of data, but on *organizing many ideas* which have emerged from the analysis of the data (Strauss, 1987, p. 23)". The research questions to be explored and tasks accomplished during Part Three include:

Q2-A: What are the primary resource-based process determinants of firm performance and potential sustainable competitive advantage?

Q2-B: How are these determinants related to performance and one another, and why are they important?

T2-A: To develop an integrated, grounded, substantive-level theory of skill-based strategy.

T2-B: To develop an inventory of propositions that specifies the theory in more precise form.

T2-C: To develop a series of diagrams and a hierarchical spreadsheet summary of the theory.

9

A CROSS-CASE ANALYSIS OF SKILL-BASED STRATEGY AND ENTREPRENEURIAL LEADERSHIP: DEVELOPING A GROUNDED THEORY FROM IDIOGRAPHIC MODELS

INTRODUCTION

In this chapter, the process of building a more general, substantive-level theory begins with the presentation of the results of the cross-case analysis, whose methods are discussed in Appendix Two. This chapter, and the two that follow, reflect an effort to make analysis, as Strauss (1987) and Glaser (1978) argue we must, not merely a tool for collecting and ordering, "a mass of data, but on *organizing many ideas* which have emerged from the analysis of the data" (Strauss, 1987, p. 23).

The chapter begins with a brief review of the individual idiographic-level models that were developed from the data at each case site. The common elements and relationships of the models and cases that were identified during the reanalysis of the combined database, using an extension of the constant comparison analysis methods, are then discussed, and a model of the common elements is presented.

The unique elements of the idiographic-level models are then identified and evaluated as potential contingency variables for the basic "common elements" model. The practitioner-derived model presented in this chapter will serve as the base for enfolding the literature and developing a more formal theory.

A BRIEF REVIEW OF THE FOUR IDIOGRAPHIC MODELS

In this section the summary of the results of each of the idiographic-level case analyses is reviewed, as they provide the base for the cross-case analysis. The summary diagram and the summary statements of concepts that were presented at the end of each case are reproduced below.

The Precision Instruments Model and Idiographic-Level Theory

Summary

The Precision Instruments Company is one of the world's leading designers and manufacturers of hand-crafted surgical instruments. As of 1992, the company competes in five distinct product-market segments and is well positioned in each. The most important concepts at the Precision site were that its leaders maintained a focus on developing the firm's skill-base so that the company could "stay in the hunt" and "meet customer wants and needs".

The practitioners at Precision maintained that the process of meeting customer wants and needs should be customer driven, and that knowledge of the buying process and the product life cycle was critical to success.

The primary challenge at Precision was to integrate its knowledge and expertise about, and its experience in, specific customer markets, with specific material, design, and manufacturing skills. As Petrush summarized it, "we strive to achieve the highest levels of craftsmanship, combined with the most advanced metallurgical technologies, and overlaid with the most advanced designs (techniques) of surgery."

Summary of Idiographic-Level Theory

Figure 9.1 identifies the primary concepts that were discussed in the Precision Instruments case history and idiographic theory development chapter. This summary reflects the integration of both practitioner and academic-based concepts that took place during the analysis of the Precision case. This grounded integration was presented in the "Applying Extant Theory", and "Developing Field-Based Theory" sections of the case, and is reproduced in summary form.

The Custom Cabinets Model and Idiographic-Level Theory

Summary

Custom Cabinets Company is a small cabinet-manufacturing company in the South that has recently adopted "Synchronous Manufacturing" methods,

Firm Growth and Pace of Development
1. Building a craft-oriented skill-base takes time and commitment.
2. Even a short-term harvest can severely damage the skill-base.

Development of Systems to Protect Skills
3. Building skills requires consistent implementation of skill-based systems.

Entrepreneurial Leadership and Skill Development
4. Skill-based strategy enhances the effectiveness of entrepreneurial actions.
5. Economies of scope provide an effective mechanism for moving on emergent opportunity.
6. Core development often requires entrepreneurial competence.

Life Cycle Effects on Skills
7. Understanding and managing the skill and product life cycle of a company is a critical aspect of effective skill-based strategy.

Institutional Competence and Skills
8. There is a strong interaction between skill-base development and the development of institutional character.

Fig. 9.1. Idiographic-level theory summary from Precision.

based on the Theory of Constraints (Goldratt, 1990). These methods are skill and resource-based, and the firm has made a strong commitment to training its entire workforce in the methods. The firm has had an exemplary growth record since the late 1980s, and has developed the ability to deliver custom cabinet sets in 10 days in a business where the industry delivery average is between six and eight weeks. The company exercises unique craft skills and operates in a very cyclical business.

The most important concepts at Custom Cabinets were that the company meet customer expectations and strive for "making more money now and in the future" through a company-wide commitment to solving the customer's problems.

Custom Cabinet's management was concerned that the company be able to work together to solve customer problems and meet their customers' expectations in terms of quality, service, variety and reliability. The company had developed integrative systems to help employees work together in defining and solving a wide range of customer problems. Systems development, in terms of decision-making systems and in terms of work-flow, were important elements of the Custom Cabinet focus.

The development of human resources was critical to the company, as the human resources were the source of continuous improvement efforts.

Summary of Idiographic-Level Theory

Figure 9.2 identifies the primary concepts that were discussed in the Custom Cabinets Company case history and idiographic-theory development chapter. This summary reflects the integration of both practitioner and academic-based concepts that took place during the analysis of the Custom Cabinet case. This grounded integration was presented in the "Applying Extant

---

Deep Knowledge and Skill Types

1. Businesses emerge from what their founders know and enjoy – leveraging key skills.
   (a) Skills and opportunity.
   (b) Experience as knowledge.
   (c) Craft and knowledge skills.
   (d) Leveraging skills in making entrepreneurial choices.
   (e) Rewards: the positive effects of leveraging key skills.

Organizational Growth and Skills

2. Following up on entrepreneurial choices: innovations to work smarter.
   (a) Breaking barriers and serving the customer.
   (b) Developing new skills and competences.
3. The failure to understand the boundary of core skills can be fatal.
4. The need for a skill-based strategy that is rooted in competent systems.

Organizational Systems and their Relationship to Skills

5. Systems and skill.
   (a) Organizational competence develops with systems.
   (b) Skills drive systems and systems are the bridge between individual skills and organizational competencies.

Institutional competence: commitment

6. Commitment is the baseline for institutional competence.
   (a) Commitment to solving customer's problems: linking the core.
   (b) Commitment to employees: protecting the core.
7. The thinking organization: TOC and human resource evolution.
   (a) The development of institutional competence.

Theory of Constraints as a Resource-Based Theory of Management

8. The Theory of Constraints as a resource-based process theory.
9. TOC is also a theory of organizational competence.
   (a) The "thinking process" generates future competence maps.
10. The firm is a net generator of resources: this affects factor values.
    (a) Using slack as an investment.
11. Competitive advantage from commitment and competence: summary chart.

Fig. 9.2. Idiographic-level theory summary from Custom Cabinets.

Theory", and "Developing Field-Based Theory" sections of the case, and is reproduced in summary form.

The Hauser Chemical Research Model and Idiographic-Level Theory

Summary

Hauser Chemical Research, Inc. (HCR) is one of the world's leading manufacturers of extracted natural products and is the world's only current (1992) commercial-scale manufacturer of the anti-cancer agent, taxol. The company offers a wide range of chemical and engineering services and offers its talents internationally. The company is known for its fast turnaround times and quality service.

The company relies heavily on its base of highly trained scientists and technicians, its special kind of human capital, as a means of solving the particular problems of its customers. The company has no marketing capacity *per se* and operates in a "marketless" environment. That is, the company is primarily in the business of solving a vast array of problems, each of which has its own market context. It is a professional service company. The most important concepts at Hauser Chemical Research were that the company learn to maintain and enhance its ability to solve customers' problems through the interdisciplinary application of its employees' skills and "suite of knowledge".

Hauser is conceived of as being part of the "need fulfilment process" of its customers. The entrepreneurs and managers at Hauser have designed a flexible organization, built around the concept of "generalists", that can come up with unique and viable solutions to myriad problems in chemistry and engineering.

Hauser has also begun to learn, despite the "market-less" nature of its current business, how to tap into the value it creates for a particular customer and generalize its solutions.

Summary of Idiographic-Level Theory

Figure 9.3 identifies the primary concepts that were discussed in the Hauser Chemical Research case history and idiographic-theory development chapter. This summary reflects the integration of both practitioner and academic-based concepts that took place during the analysis of the Hauser case. This grounded integration was presented in the "Applying Extant Theory", and "Developing Field-Based Theory" sections of the case, and is reproduced in summary form.

Growth of the Firm

1. Pure skill-based growth is slow.
2. Investments in technical and entrepreneurial capital are a means to quality growth.
3. The merging of complementary skills both broadens and deepens the skill-base.

Problem-Solving Orientation: Skills, Systems, Technology, Structure and Value Creation

4. Conceptualizing the firm as a problem-solving entity was a development strategy.
   (a) It is a value creation and appropriation strategy.
   (b) It is a basis for systems and structures development.
5. Systems and structural developments must relate and feed the core skills.
6. Skill-based strategy exploits economies of scope, which derive from economies of skill.
   (a) Technology transfer is a competence issue.
   (b) Systems and structure do matter.
   (c) The team formation process is at the heart of generating economies of scope, which derive from economies of skill.
7. Problems are addressed by applications of technology, which are skill-driven.

Entrepreneurial Leadership and Institutional Competence

8. Entrepreneurial leadership is first about skills, then recognizing opportunity and capitalizing on it.
9. Protection of the core skills in the face of rapid growth requires leadership and strategic assessment.
10. An incubating firm must be able to accept spin-offs and redevelop its skill-base

Fig. 9.3. Idiographic-level theory summary from Hauser Chemical Research.

The Combustion Engineering Model and Idiographic-Level Theory

Summary

Until it was acquired by Asea Brown Boveri (ABB) in 1990, the various business units of the Combustion Engineering Corporation (C-E) had been one of the leading providers of engineered products, systems, and services to the worldwide power generation, process controls and public sector and environmental waste and energy markets. Today (1992) Combustion Engineering is but one, albeit large and complex, part of what ABB's management has called, "the world's most aggressive expansion strategy in the electrical machinery and power engineering sectors".

The case traces the reinvention of Combustion Engineering since 1982, which George Kimmel, President and Chief Operating Officer of Combustion Engineering under Hugel, says was based on a "skill-based strategy". It is a special case in the book in that it is a corporate strategy case. It is a case that focuses on how the company's leaders effected change using a skill-based strategy to enable the corporation across a wide array of different business units to provide new and valuable services that would meet the changing demands of its customers.

The most important concepts at Combustion Engineering involved the role that top management must play in reshaping and developing new sets of shared-values that are needed to support a change in strategy and operations. The challenge at C-E was to get the entire company, across many different strategic business units (SBU), to begin working together to serve clients needs and solve their problems.

The primary instrument to change at C-E was the Human Resources Function, which served as a catalyst and bridge between different business units with respect to pulling skills together across the company and providing systems to encourage a team approach to solving problems.

Summary of Idiographic-Level Theory

Figure 9.4 identifies the primary concepts that were discussed in the Combustion Engineering case history and idiographic-theory development chapter. This summary reflects the integration of both practitioner and academic-based concepts that took place during the analysis of the C-E case. This grounded integration was presented in the "Applying Extant

---

Organizational Skill Obsolescence

1. In mature and oligopolistic settings the firm's skill-base is likely to atrophy, especially in terms of integrative competence.

Value Creation in the Corporate Setting

2. The entrepreneurial challenge at the corporate level is to learn how to package and repackage the skill-base of the company to create new value.

Skill-Based Acquisitions

3. A skill-based acquisition program must be customer oriented, not shareholder oriented.
4. Integration of function and purpose is essential to effective skill-based strategies.

HRM as an Operations Tool

5. An operations-oriented human resources function is an important tool in the crafting of skill-based strategies.

Fig. 9.4. Idiographic-level theory summary from Combustion Engineering.

Theory" and "Developing Field-Based Theory" sections of the case, and is reproduced in summary form.

THE COMMON ELEMENTS IN THE FOUR IDIOGRAPHIC MODELS

In this section, the important conceptual elements that are common across all four cases are identified and an integrative model developed. The results presented in this section are the output of the cross-case analysis procedures discussed in the last chapter, which included a primary group-wise examination for similarities across a divergent set of firms, and a secondary examination of unique elements at each firm. Both cross-case analysis techniques were used to generate core categories, in the language of the practitioners, that could provide a systematically derived base for development of a more general theory of resource management and skill-based strategy.

The following discussion presents a general summary of what appear to be the most important concepts and ideas that emerged directly from the combined practitioner data collected and analyzed in the four case studies. This summary identifies the core categories that emerged during data analysis, using the language of the practitioners.

From Open to Selective Participant-Derived Codes

During the initial phase of the qualitative analysis of the case data in this book, the starting, or "open" codes for the concepts that were identified were based on the principle of *in vivo coding* (Strauss, 1987, p. 30), which reflects a preference for identifying concepts in the terms used by the people being studied, in this case the managers and employees in each firm. These participant-derived codes were the roots from which all subsequent analytical and selective coding emanated. Strauss (1987) and Glaser (1978) note that:

> selective coding pertains to coding *systematically* and concertedly for the core category [of the phenomenon under study] (Strauss, p. 33). "The other codes become subservient to the key code under focus. To code selectively, then, means that the analyst delimits coding only to those codes that relate to the core codes in sufficiently significant ways as to be used in a parsimonious theory" (Glaser, p. 61).[1]

In the next subsection, the core category that was established during the selective coding analysis process as the data were analyzed across all four cased is identified, as are the major common categories.

A Practitioner-Based Framework of Core and Major Categories

The idiographic-level theories presented in this book have focused on how the leaders of different firms have attempted to create value for their customers by assembling and leveraging the resources, especially the human resources, at their employ. The concept of value creation emerged consistently across all four cases as one of the most important concepts to the practitioners, and was identified under the various practitioner codes of: "*value-added*", "*solving the customer's problem*", "*meeting the customer's needs*", "*building relationships that would bring value for the customer*", "*we could bring things together and deliver them to the customer in a new way*", and "*I will do my best to make a customer happy*". The general managers' concern for meeting customer needs and expectations and solving their problems all reflected a value-added approach to managing their businesses.

The selected participant-generated code "*meeting customer needs, expectations and solving problems*" thus **emerged as the core category, across all the cases, around which other categories have been analyzed and subordinated.**

Figure 9.5 presents excerpted data, across all four cases, with respect to the core category. Note that this core category is also identified explicitly in all four summary diagrams that were presented above.

Figure 9.6 presents an example of excerpted data for the selected major categories "*SKILLS, EXPERTISE and KNOWLEDGE*" and "*SHARED VALUE and CULTURE*", which were also major common themes across all four cases. Both Figs. 9.5 and 9.6 have been included to show the reader how the data were identified in the combined raw data file, how theory notes were related to the data and to show that evidence does exist in the cases to provide a grounding for the model that is presented.

Note: Comments offset with the "##" symbol are theory comments that were made in the master theory journal.

CORE CATEGORY: PROBLEM-SOLVING, OPPORTUNITY, and MEETING CUSTOMER NEEDS

Precision Codes and Comments:
LP: ** **01.2 END-USER** "You have to define the business by what the customer wants and needs, not by some internal examination of technology or manufacturing".
This clearly shows that opportunity is perceived as external to the firm, as Hofer argues. The skill involved is in knowing what customer to serve... this can only be done with reference to the skill and technology base. In some sense that is entrepreneurial choice... looking to match customer with skill that can create and add value... it takes skill itself (perception, search)

** **05 CUSTOMER AS INNOVATION SOURCE** "You need to get into the surgery room to *stay in the hunt*".
This comment led to the "Oh, shit!" product and market development approach. Watch and wait until a surgeon says "Oh, shit!", in surgery, then figure how to make an instrument that can reduce fatigue, chances for error – to correct the "problem".

** **04.1 FEATURES** It is critical that businessmen know the difference between features and benefits and develop skills ** **04.2 BENEFITS** needed to support benefits. You need to understand the buying process of your customers. You should identify the leading edge customers [early adopters] that others look to for buying behavior [by product].

His scanning of his customer environment is really quite ethnographic – he goes into the field, listens, talks, watches. This is part of Kirzner's entrepreneur, and of the skill-set of entrepreneurs. See Herron (1990).

Benefits get to the heart of adding or creating value. Substance, not fluff... to get substance requires a more skilled approach.

RH: I will do my best to make a customer happy. So the bottom line is that we try to produce a good product and be extremely knowledgeable about it. ** **05 CUSTOMER AS INNOVATION SOURCE** Part of that has led us into where we are now into what we call "procedural specific instruments". It is not that we can make something that can make everything, we make something that can do something very specialized, which will lead back into something that will do everything.

I don't think we will ever be real big... we are not going after the major market share... most people want to buy the vanilla, not the real spices. The real spices is where you make the money.

Custom Cabinet Codes and Comments:
MV: ** **03.1 SOLVING THE CUSTOMER PROBLEM** Well... I think any company that is successful is basically solving a problem for somebody... and what I see our company doing... is... we provide something that is necessary in the end product... uhhh... and there are quite a few companies who also do that... what I think helps distinguish us is that we try to... uhh... solve the problem... ** **03.2 TAKING ON RESPONSIBILITY** we basically take on the problem that the superintendent or the builder has... and provide him something's that is appealing to the market. ** **07.2.1 CUSTOM SHOP** But... if he's just trying to fill the space he can do it much less expensively than using us.
** **03.1.1 APPEALING** So we've got to provide something that has appeal... that is less of a headache for him than he's had in the past. You know, I think that's what almost any company is basically doing.

CCC is a service company, that has a manufacturing component... the service is described in terms of problem-solving for the customer, which is an extremely broad way to view the business... and implies a higher order of thinking about the skills that are/may be required to run the company... this is an important point to compare..

MV: ** **10.2 CUSTOMER EXP/REPUTATION** And this has probably, for me in ** **39.3 LONG-RUN** the long run, the most important thing, to be known as being reliable, whether I made or lost money on that individual job wasn't important... was my reputation for reliability... and that's what we guard against now. + ** **07.4.1 PACE and GROWTH** We don't want to take on more business than we can be reliable with.

The pace at which the firm grows is limited by the reliability question, and delivery schedule measures reliability. Reliability is the stock resource borne from the competencies of the "system" he has built.

MV: ** **29.8 COMMITMENT** ** **29.8.1 WHAT WE DO** I think... the basic thing is the commitment from management that is what we do. If what we do... was build a cabinet, put it in a box and deliver it to a job-site, we would have to produce a different kind of product. ** **10.3.1 CUSTOMERS AS PARTNERS** But we've basically taken the attitude in this that we are a partner in this, we want to be accommodating to the customer... we want to provide what they tell us they need. ** **07.7 DIFF/SEGMENTATION** We try to differentiate our product by making it more of a service than just a commodity.

Mike talks about the management teams commitment to the service the company offers... that commitment is unique and important to the process of offering a custom product... and is important in keeping communication, etc open between the firm and the customers. This reflects a bit of the word Ghemawat uses.

** **44 COMPETITIVE ADVANTAGE** so we've really got a competitive advantage with the fact that we've got a custom product that is high quality, we have a short lead time, and we just cover them up with service. As soon as its delivered, the next day it's installed and service... [unintell]...

So, the mix of custom product, high quality and service, in terms of delivery schedule, installation, guarantee... together, provide CCC with a degree of competitive advantage. They have developed a strategic competence in that their strategy leverages their core skills and competencies to a customer arena that recognizes the value of the CC product/service mix, and is willing to pay for it.

Hauser Codes and Comments:
So we were just brainstorming, and it occurred to me anyway,
** **03.4 SKILLS AS RESOURCES** that we had an interesting set of skills ** **04 PROBLEM SOLVING** as problem solvers. ** **02.2 MOTIVATION** We were people who enjoyed taking other peoples' problems and finding solutions. In particular chemical solutions to chemical problems.
** **04 PROBLEM SOLVING** So we've looked around to find a way to start a business that would do that for people, and in the process of solving their problems create business opportunities downstream. So we really sort of did have a business, but we didn't know what is was yet. ** **05 BUSINESS**

DEFINITION But problem solving was the orientation we took, and we were going to look for the actual business we would do after we got into problem solving.

Solving the customer's problems is really a form of adding value, a way to think about it. What these guys have done is link the problem solving perspective as an engine to drive towards new opportunities to keep some of the long term residual value from their "generalizable" experience.

TZ: ** **02.5 DRIVING FORCE** Well, in my view you can't separate those. In one sense you can say that it is opportunity driven, but realize that those opportunities are only realized because of the skills of the people. The same thing can be told to people who don't have the appropriate backgrounds and they don't see the opportunity in it.

No, we don't study the skills of our people and then go out and search for an opportunity that fits the people. They naturally come together if you have... realize that our labs group, for example, performs about 1500 projects per year... and they have a huge variety of types of things that they do in engineering and chemistry.

** **02 OPPORTUNITY** Opportunity is in the eyes of the beholder, and if you have people attuned to being able to see the bigger picture, and realizing the potential opportunity when they run across it... they kind of pop up.

This reflects a bridge between the Schumpeterian and Kirznerian view of entrepreneurship and innovation... one has or makes choices about the skills and capabilities the organization should have, then is opportunistic in applying those skills and capabilities, and building new ones.

Combustion Engineering Codes and Comments:

** **07 MARKET CHANGES** CE is in process businesses and power generation....initially we had product-oriented business units. This worked very well for a long time, and we made lots of money. Orders used to be "allocated" to the businesses, and prices were determined by what would clear the markets. But the markets changed rather quickly, and competition became tough... orders dropped off sharply. We had a problem... ** **06.1 SKILL TYPES** how could we take "know-how" and technical skills from our different business units and pull them together to respond to the different needs of our customers? We wanted to get to a point where we thought of ourselves as "one company working together to serve clients". ** **08 COMPETENCE** We needed to work across lines: product, function, business lines; systems get in the way, compensation, incentive, social.

Fig. 9.5. Data from cases on the core category: solving customer problems and meeting their needs.

MAJOR CATEGORY: SKILLS, EXPERTISE and KNOWLEDGE

Precision Codes and Comments:

LP: ** **08.1 SKILLS AT CORE** Well... if you will recall, I did this, in what I call our quality... and our corporate definition. Which I wrap into two. Which is, we are bringing to markets, the products... the latest designs... now you say latest, I'm going to digress on that... but understand that I am in a competitive posture on latest... so that is relative to competition. Then I'm overlaying that design with manufacturing capability, utilizing highest levels of hand-craftsmanship. Now, the third item, which is actually a technical item, that overlays that, is using the most advanced metallurgy, and engineering.

So, there is a solid, but general description of the core competence of this firm... its ability to blend the various technologies, assemble the resources (again in an ad hoc manner)... emergent strategy... skills really drive the ability to exploit emergent processes... if they are general in nature. If they are specific, and rigid systems develop... then there is an inertia to strategy (think about this more in terms of Mintzberg)

LP:** **02 FULL MARKET SERVICE** We are going to "full market" in the plastic surgery and endoscopic markets. We think the endoscopic market will eventually cannibalize the general market, and eventually the others. "Prior to my arrival, TPC manufactured 98% of what we sold, but in order to compete today, and be full market oriented, we are beginning to outsource non-critical products.... ** **08 CRAFTSMANSHIP** we are keeping the products and processes that require the skilled craftsman we have."

This is Quinn's hollowing out... around core skills. The competence idea, of blending resources and assembly skills, goes beyond owning the resources... it is coordinating them and linking this to the ability to offer full service to the customer.

LP: In the mid 80s the competition in our business was 60% German, and 40% US/UK... we felt we needed to re-position ourselves vs Germany:

We pursued this re-positioning by:

(1) Beginning to improve our technology and products. (We had actually considered outsourcing to the Germans and having our name put on the instruments, as our name was highly regarded.) (2) In the long-range we wanted to broaden into specialty segments, and split production into domestic and international components.** **08.1.2 BREADTH and DEPTH** (3) We needed to re-train and work our "skill-base", in terms of both its "depth" and "breadth".

** **06.1 SYSTEMS EVOLUTION** We needed to change our bonus system to reward skill breadth and functional abilities... raises based on skills. We were/are looking for people with long-term interest in being in our workforce.

This illustrates how Prec leveraged the scope economies of his craft skills to get into niche markets... niche market, high-skill strategy... even in labor-intensive markets (CCC holds to this too... skilled labor as a fixed cost).**Custom**

Cabinets Codes and Comments:
MV: ** **45.1 SALES STAFF ** 27.3.3 BOUNDARY SKILLS** Well... one element in that is our sales staff... they are really the link between what the customer wants and our manufacturing capability... ** **27.3.3 BOUNDARY SKILLS** they've got to describe to manufacturing what the customer wants as quickly and accurately as possible. And that's a big undertaking for us because we have such a wide set of options available. ** **02.4 EXPERTISE** So our sales position requires a pretty good level of expertise between know what we can do manufacturing-wise and trying to make that match for what the customer wants. ** **27.3.3.1 INTERNAL SPANNING** Drafting sits right in between sales and manufacturing... they've got to take that information and convert it into language that manufacturing can produce a product with. ** **27.3.7 SKILL TYPES COMMUNICATION ** 27.3.4 SKILL TYPES TRANSLATION** ** **27.3.5 SKILL TYPES CAPTURING DETAILS** So I see that we have... we should have people in sales that are very good a communicating and capturing details... in drafting we should have... and obviously we do... not to the level I would like to have, but we have very talented people that now take that information from one format and turn it into another format that the computer can understand, that then produces the list.
** **07.2.1 CUSTOM SHOP** Some of the people we compete with... are providing . . they may not do the design, they may not do the sales, they may just do the manufacturing, and somebody else is selling the product for them... ** **27.1 SKILL LEVEL** so we have... because we are doing the whole thing, we have a lot more people involved, and a lot more skill and talent required to accomplish that. . .

This entire passage reflects the essence of the core skills CCC has built... and which have provided the base for their core competencies... the translation skills are important and fit into the "interpretation" logic of Weick.... Again the "systems" are driving the need for particular skills at the individual level... and the coordination of these elements are dictated by the "system"... the "system" in Viera's words, is really the key to the skill-assembly process, or the building of competencies.

Viera sees the system, and his firm's evolution competence in dealing with it, in terms of what leverage it gives him in terms of his customers (to solve their problem), competitors (Business definition), and employees.

Hauser Codes and Comments:
DS: ** **15 COMPETITIVE ADVANTAGE** Well the... we are worried about protecting our competitive advantage from our proprietary technology viewpoint. ** **09.3 TECHNOLOGY CHANGE** But not nearly as concerned as we are about keeping our people abreast of technology changes, ** **15.1 GENERALIST** and working as generalists. ** **15.2 DIFFERENTIATION** Which I think is a key differentiation we have from a lot of other companies. Here, if you are not able to apply a variety of disciplines, or technical disciplines to solve a problem, you are probably not going to be very happy. We want people that can tune their cars, that can fix a turntable that doesn't work, don't mind plumbing... might even go out with a nail and a hammer to build their own house... and they may have a technical degree to go with it.

The success of the company, in terms of being able to solve problems and benefit from them, is in large part due to the broad-based scope and skills of the people in the firm. This is an attribute built into the company by the founders, who are somewhat general in their skill-base. The focus is not on technology, per se, but on the capability of the people to understand the new technologies and to adapt and change the technologies as needed, given customer problems.

DS:** **03.3.2 COMMUNICATIONS SKILLS** One thing that we think makes us a better company is that we have really great internal communications. If we have a meeting here, and a customer comes in on Tuesday, and they are gone at two o'clock, we will generally meet the same day to de-brief on that meeting. We have both e-mail and v-mail internally, and there are 130 stations... people are constantly communicating.

Open and flexible communications (and informal) seems to support the team building process... the focal point concept is a good one to look at in more detail. Note that there are physical systems that link the humans... computers and phone systems that facilitate the integration process.

Combustion Engineering Codes and Comments:
** **06.1 SKILL TYPES** Our analysis of that led us to conclude that the strength of the company was in its technical know-how, its basic technology, its access to the market, and, obviously an ability to be responsive to their shorter term needs now. Their needs were changing, instead of needing a new plant, they needed a way to make the old plant better.
That is quite a different mentality than goes with the construction piece of it. The same could be said of the petrochemical business... the opportunity there for us was in our knowledge of the process and of engineering management, and therefore we had to take those skills and bring them to the market in a way that dealt with the current needs. And those needs were for such things as improved process technology, process control and things of that sort, new engineering methods. We then moved into, among other things, the process control business with some acquisitions.

** **08 COMPETENCE** We eventually were able to get to a point where we could put teams from different business units together to go after entire chunks of business, such as an entire process or pulp plant. We had to develop new compensation systems to encourage and reward this type of team action.

MAJOR CATEGORY: SHARED VALUES and CULTURE

Precision Codes and Comments:
LP:** **10 LEADERSHIP** PRES: They (the workers) know I love this business, and that I respect them... I bonus them... I can't deny them the ability to achieve... they have an ownership position.
Again, there is a strong institutional element... leadership and commitment to the people, the culture and the "way of doing things". Institutional competence!

LP: ** **03 SHARED VALUES** So, I think if I were to define Precision today, I would not define it by products, I would define it by the people that are allowing us to come to the market with better designed products, of a higher quality, faster than our competitors. And anybody who looks at this business is going to understand that.
A good test to see if you have a skill-based strategy or perspective... do you think of the firm as a set of people with abilities to be tapped and utilized... to create value... or as an output function, with products and positioning in the market?

LP: ** **03 SHARED VALUES** These people, to the person, have allowed us to take this business where we've taken it. From the office staff... everybody in this organization is willing to go the extra mile, and put out the extra effort to support us.

Custom Cabinet Codes and Comments:
** **29 CULTURE** – "the environment of CCC makes one want to come back (from training/continuing education) and apply it for the good of this company, and not leave for other work".
This is an interesting statement that supports Viera's assertions of the efficacy of training, and his commitment to it.

** **29 CULTURE**
JM: In some aspects it is the same... in some aspects it's different... for example I think we're more dynamic now... Mike has always been a person of change... he's always been a visionary that could look down the road... I always thought we'd get to a plateau and stop... ** **29.5 HONESTY** but it's also impeccable integrity... I think Mike has tried to surround himself with people that want to have an honest organization... that people are not scared of work... that they want to put in a day's work... not just brain power... he wants brain power... he sets a good example... do as I say, and do as I do... he expects a lot out of you. We like to knock obstacles down... that is one thing that Mike has really stressed.

Hauser Codes and Comments:
DS: Our production facility here for taxol... we did not have anything in this building in the first of September... which was 10 months ago. Today this plant produces product... ** **04.3 TIME-BASED** it would have taken a normal pharmaceutical company 18-24 months to reproduce that process. ** **05.8.1 EMPOWERMENT** We did that because all the people in the plant that were helping to design, build and install it, had full authority to make all the decisions necessary to get to the end. And they made a lot of mistakes. But we told them all along that "if you make a mistake, I'll take the blame for it"... just keep making the decisions to get the plant running... and they did... and the contractors here, bar none, would tell you that they would rather work here than anyplace else.
In order for this broad skill-based strategy to work, the structures of the organization that are needed are inherently open and "inverted"... they are also quite informal, which allows and encourages cross-pollination of ideas, a more shared knowledge of the overall skill-base of the company, which is something that each scientist must know about, or be able to learn about, once he/she is confronted with a new problem from a customer.

Combustion Engineering Codes and Comments:
** **03 SHARED VALUES** "Tradition in business limits ability of firms to change. This is related to the age of the company... building of formal systems... the word "culture" is too broad to be of meaning in the context of management change, process, I prefer "shared value", which is very important. Changes at the top can affect the shared values of the company, but it can take 5-10 years to significantly change the company's "way of doing business".

** **05 CHANGING THE MIDDLE** Problem in change processes – mid-level and first general manager report levels. These people are very entrenched in how they do business. At CE (Combustion Engineering), inducing change: changing "key people", "seeding" in key positions. ** **04 TEAM BUILDING** Developing mission, "consensus building", building "shared values" is the key. This shared value has to translate in human resource policies, hiring, and selection process, promotion... and talking to customers.

Fig. 9.6. Examples of theoretical codes from combined master file on "Core Skills" and "Shared Values and Culture".

A Synthesized Summary Model and Analogy

Figure 9.7 presents a highly synthesized summary model of the primary common elements that were identified during the group-wise examination for similarities across the cases. This is a practitioner-derived model which, in a most general sense, illustrates how other major practitioner codes (which were derived from the cross-case analysis), such as "experience", "know-how", "technology", "expertise", "resources", "systems and integration", "speed", "flexibility", "quality" and "competitive advantage", relate to the core category within the context of the four cases.[2]

What is suggested by the framework is that the leaders of these organizations have each recognized that they must take a resource-based approach to the business in order to meet customer needs and expectations, to solve their problems and to add value for the customer. The challenge to the leadership is to know which customers to serve (this is a function of entrepreneurial leadership), which is a function of the firm's ability to serve them, based on the human and material resources at their employ.

Analogy as Model

The pictorial analogy that models this process is that of a liquid mixer and strainer, where various liquids (material and human resources) are mixed to create a composite fluid (speed, flexibility etc.) that, once it is properly

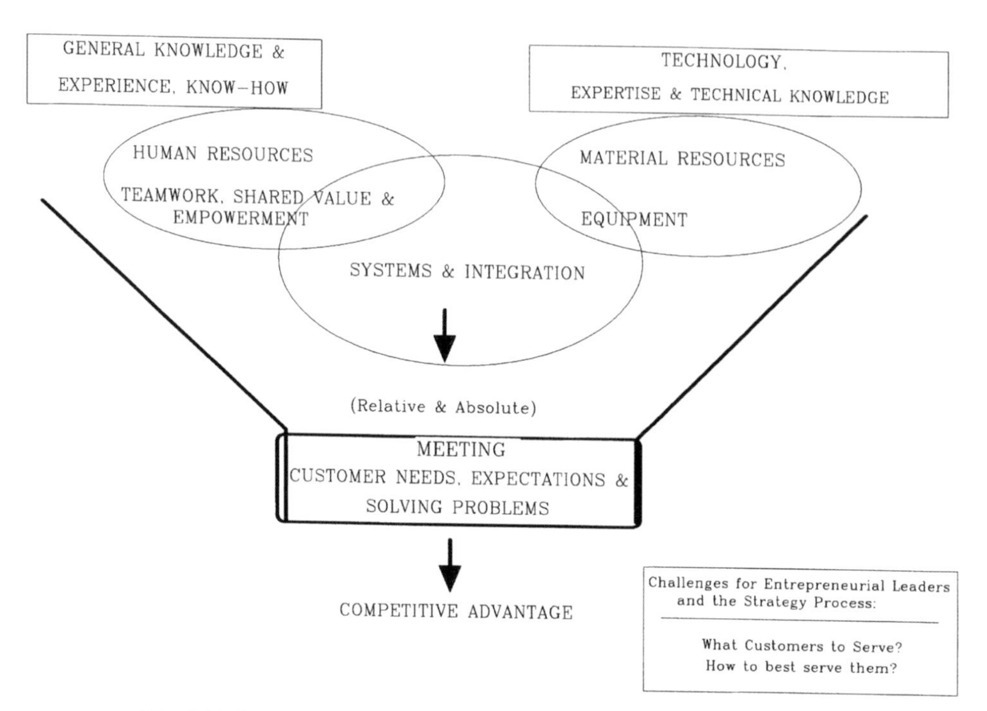

Fig. 9.7. A practitioner-based framework of core categories.

mixed (integration), will flow through a specific type of filter (customer needs), leaving a purified composite (competitive advantage) as the end-product. Mike Viera talks of "systems" as those integrating and mixing components:

> I think that . . . when I look at our company [in terms of resources], and of any company, you've got a building, a lot of equipment, and you've got a lot of people . . . and raw materials . . . by themselves none of those things really accomplish anything. What makes all that work is a system . . . and we, together the management team and the staff, we've got twenty years in developing our system. To extend the analogy, the leaders of a firm can tap two fundamental sources of knowledge which they can attempt to transform into value for the customer: (1) general knowledge, experience, know-how and creativity, borne in its human resources, and (2) technical knowledge and expertise, which is embedded in both human and material resources, and always in relation to some technology.

Kimmel, of Combustion Engineering, shows that one of the big challenges at his company was to tap into, and mix, appropriate resources, and particularly skills, to meet the needs of his firm's customers:

> We had a problem . . . how could we take "know-how" and technical skills from our different business units and pull them together to respond to the different needs of our customers? We wanted to get to a point where we thought of ourselves as "one company working together to serve clients".

Ziebarth, of Hauser, also talks about tapping into the knowledge-base of the company to solve customer problems and create value for the company:

> We're technologists. We are not fundamental discovers . . . . We try to know as much technology as we can so when problems or opportunities come up, we can draw on a bunch of resources to come up with a high-value solution that we can then do something with.

To the managers of the firms studied, it is not enough to know what resources are at the employ of a firm. The leaders of a firm must know what types of knowledge and other resources they can mix together to create a composite that will be of value in the market-place, and which will thus generate returns.

Since, according to the practitioners in this book, the customer is the ultimate filter in determining value, this implies that leaders must know the specific needs and problems of customers, and must also recognize, in detail, how they can mix their resources to create an appropriate valued composite.

In the next two chapters, a substantive-level grounded theory is presented that explores, in detail, the process of how firms create and add value for their customers through the discipline of entrepreneurial leadership and the exercise of skill-based strategies. It is argued in this theory that firms are most likely to achieve sustainable competitive advantage (and it is explained why) when they have entrepreneurial leaders and pursue skill-based strategies. The theory that is presented integrates the perspectives and ideas discovered in the field with relevant academic theory.

UNIQUE AND MAJOR ELEMENTS OF THE FOUR IDIOGRAPHIC MODELS AND THEIR PURPOSE

While the process model illustrated in Fig. 9.7 identifies general concepts and categories that are common to all the cases studied, some of the unique elements of individual cases are also important to analyze and discuss.

It is important to identify the major and unique elements of the cases because this "unique element" analysis puts the cross-case analysis into more of a contingency theory context, which Hofer (1978) and others (Harrigan, 1983; Hambrick and Lei, 1985) suggest is important in strategic management research.

The Unique and Major Elements of the Precision Company Case

One of the unique elements of the Precision Case was the importance that both the *product and skill life cycle* concepts had to the general managers of the company. According to Petrush,

> the life cycle concept is critical to running the business, in terms of the *product and the skills* of the people. You need to know the life cycle of each instrument and each person .... Everyone at this company is on the growth or development stage of the [skill] life cycle.

One of the important tasks of the general managers is to ensure that the company has a range of products that push the demand for internal skills beyond what it might be at present. In the Precision case, the company competes in five distinct product markets, each of which requires a different bundle of craft and knowledge skills.

The Unique and Major Elements of the Custom Cabinets Company Case

One of the unique elements of the Custom Cabinets case is that the management has committed the company to utilizing the "theory of constraints" (TOC) as its operating and management philosophy.

The "thinking process" of TOC, which includes "Effect–Cause–Effect Analysis", is a technique the firm uses to integrate its functional activities and to manage the organization as a whole as a system. As one manager at Custom Cabinets comments:

> What we call our Internal Jonah meeting ... [where the E–C–E analysis of each management team member is presented], each of us present our trees and help each other solve our problems ... and I think it does a couple of very important things. One, it gets everyone together and helping each other. It gives each person an understanding of what problems the others are having ... and in some instances one department can formulate a solution that the guy is just stumped on ... by changing something he

> does. It puts everybody on the same road . . . we're all, we all know each other's problems and are thinking in the same direction about what the solutions are. The biggest thing is that is helps us plan . . . . Plan and solve problems . . . .

The theory of constraints is a resource-based theory of management, with a very particular set of assumptions and thinking processes. As such, it separates the Custom Cabinets Company from the others in the case in terms of decision-making processes and systems development.

The Unique and Major Elements of the Hauser Chemical Research Case

One of the unique elements of the Hauser Chemical Research case is that the company has established both a new venture and a business development function in the firm to assist the company in tapping into general value-added solutions that might become apparent in the attempt to solve a particular client's problems. The company is really both an operations company, with the labs and the extraction–manufacturing functions, and a technology and intellectual holding company (Quinn, 1990).

Another unique aspect of the Hauser case is that the company does not have a separate marketing and sales function. Each scientist acts as a market representative, and the company as a whole mobilizes to meet the needs of specific clients as they become known.

The Unique and Major Elements of the Combustion Engineering Case

One of the unique elements of the Combustion Engineering case is that it is a corporate strategy case. The issues at C-E revolve around the integration of business units, rather than functional units.

Despite this unique element, however, the basic processes of systems integration, meeting customer needs, building skills and teams is very similar to the other cases. This suggests that, from a skill-based strategy perspective, there may not be as many differences in how to manage resources at a corporate level, as opposed to the business level, as one might think.

ACCOUNTING FOR THE UNIQUE ELEMENTS IN THE BASIC MODEL: SPECIFYING CONTINGENCIES

Given that there are some unique elements that are important to each case, such as those discussed above, it is important to see if and how they might relate to the integrated model presented in Fig. 9.7 earlier. With respect to the life cycle concept, although no other managers discussed it, there are theoretical reasons (Hofer & Schendel, 1978) and empirical evidence (Hambrick and Lei, 1985) to believe that life cycle effects, in terms of both products and skills, is a general contingency factor that affects the model. That is, the way in which skills and material resources are assembled and integrated may be affected by life cycle effects. This will be discussed further in Part Four of the book, the implications and future research section.

With respect to the unique aspect of the theory of constraints, this theory is based on general marginal analysis principles of microeconomics, and shares many other principles with other general decision-making models, such as those discussed by Deming (1986) and others. Therefore, it is unlikely that implementation of the theory of constraints would *per se* be a general contingency factor. However, it does raise the question as to how much of a contingency factor the decision-making systems and style of general management teams may be.

With respect to the level of strategy (corporate, business, functional), the analysis of the cases suggests that the basic processes identified may in fact be ubiquitous across the levels. The magnitude resource management activities may be different, but not the basic processes.

CHAPTER SUMMARY

In this chapter the results of the cross-case analysis have been discussed, and evidence has been presented to support a general practitioner-derived cross-case model of resource management processes.

Each of the idiographic-level models was reviewed, and then the core category, "solving problems and meeting customer needs", which integrates across all the cases, was identified. Other major categories that united the cases, and which were related to the core category, were identified, and examples of the chain of evidence were illustrated (Yin, 1989).

A synthesized integrated process model was introduced and explained in terms of an analogy—using evidence to support the use of the analogy. This practitioner-derived model will serve as the base from which to begin building a more general theory that will encompass extant literatures and provide a base for future empirical tests about the efficacy of skill-based strategic processes.

Finally, the unique elements of the cases were examined in light of their potential for being included as general contingency variables that may affect the basic process model.

NOTES

1. This entire quotation was taken from Strauss (1987, p. 33), where Strauss quotes directly from the Glaser (1978) work.
2. Note: For each of the categories identified in the summary model, such as technology, expertise and systems, the process of tagging and collecting evidence

across the four cases that has been illustrated in the last chapter with respect to "flexibility", and in this chapter with respect to "solving customer problems", "core skills" and "shared values and culture", was also completed. See the indexed summary of codes, in alphabetical order, in Appendix Two for a complete list of codes and their locations in the raw combined data file.

10

ENFOLDING THE LITERATURE TO CREATE A SUBSTANTIVE-LEVEL, GROUNDED THEORY OF SKILL-BASED STRATEGY AND ENTREPRENEURIAL LEADERSHIP

MOVING TO MORE FORMAL THEORY: THE LITERATURE ENFOLDMENT PROCESS COMPLETED

In this chapter, the final steps in the literature enfoldment process used in the book are discussed. The practitioner-based model discussed in the last chapter serves as the skeleton for the creation of a more formal, but still substantive-level theory. The primary goal of the enfolding process was to generalize the practitioner-based concepts and models within a more formal theoretical framework through a systematic process of comparison.

The enfoldment analysis process was begun in Phase Three, when the primary researcher was close to the idiographic-level data. As was mentioned earlier, specification of academic concepts and codes occurred during the analysis of data from each site, and accompanied the historical narratives. Much of that initial enfoldment data was then drawn upon and extended in the final enfoldment process, as developed in this chapter.

Specification of more formal theory, via enfoldment, involved three steps:

(1) The linking of the major practitioner-driven themes and categories with extant academic constructs and/or variables.
(2) The establishment of the definition and meaning of the linked constructs through critical analysis. That is, definitions of extant constructs were

compared with practitioner-use and the data as a validity check, and used if appropriate. If such a check indicated that a definition should be altered from extant use, a careful exegesis of the term was conducted and a grounded definition was provided.

(3) The establishment of relationships among the constructs. After specification of the meaning of constructs, the relationships among the constructs was examined and developed through a similar critical analysis of data.

In the next section, the practitioner-derived model developed in Chapter 9 is reviewed. Following the brief review, the elements of the practitioner-based model are then analyzed with respect to extant literatures and a revised integrated model is presented at the end of the chapter. This revised model is the base from which a more formal theory of skill-based strategy is developed in the next chapter.

REVIEW OF THE ELEMENTS OF THE PRACTITIONER-BASED MODEL

Figure 10.1 illustrates the practitioner-based model that was developed in the cross-case analysis chapter. The core category was *"meeting customer needs, expectations and solving problems"*. The other major categories that were common to all the cases are identified in Fig. 10.1; they include:

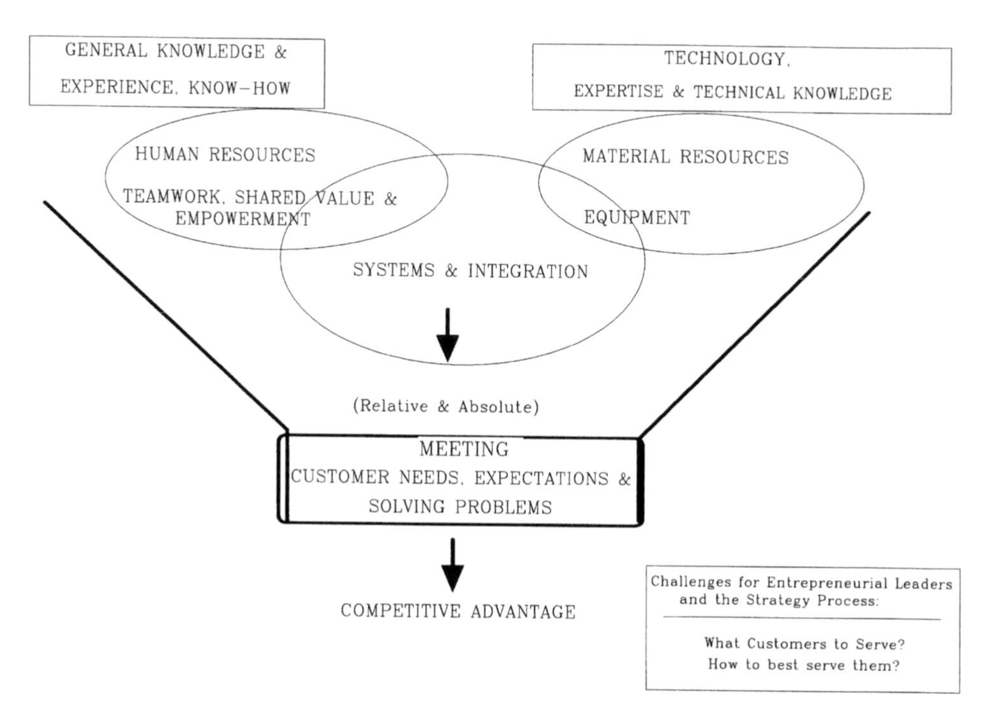

Fig. 10.1. Review of the practitioner-based model.

"general knowledge and experience, and know-how", "technology, expertise and technical knowledge", "human resources", "teamwork, shared values and empowerment", "material resources, facilities and equipment", "systems and integration" and "price, speed, flexibility and quality".

This model, as presented, is quite simplistic and general, and does not provide one with much of an idea about how the core and major categories are related to one another. The following, however, are important themes, as shown in the data presented in both the idiographic theories, and in the last chapter:

(1) Meeting the customers' needs is a function of knowing what the customers want and need.
(2) Resources in and of themselves are not as important as what management does in assembling them and developing integrative systems that can allow the company to produce quality products and be flexible, fast and cost efficient.
(3) The challenge for general managers is to keep the company "in the hunt" by knowing both what the customers want and what the company is capable of doing with its resources and skills.

In the next three sections, each of these themes is discussed in terms of the relevant concepts identified in the practitioner model, and their academic counterparts, where they exist, are enfolded in the discussion.

Meeting Customer Needs is a Function of Knowing what the Customer Wants

One of the key points that each of the practitioners emphasized was that they, and their companies, had to be attentive to what the market-place was telling them. Each general manager was concerned that they had experienced people "in the field" who could interpret and even anticipate customer needs, and who could identify potential solutions for the customer.

Each manager had a different approach to identifying and listening to their customers, but the common element was that they felt that it was important to build capabilities in their firm that would allow them to better understand their customer. In the next section, the academic literature relevant to the process of "knowing what the customer wants" is enfolded into the practitioner-model.

Recognizing and Interpreting Customer Needs and Values: An "Outside-In" Process of Entrepreneurial Discovery and Leadership

One of the important phenomena in the practitioner-based model, which is related to the skill-base of the firm, is that the leaders and managers of a firm must have the requisite abilities to absorb and interpret relevant information about consumer needs and values if they are to develop effective skill-based strategies to satisfy them (Daft and Weick, 1984; Cohen and Levinthal, 1990). The recognition of need, which may present itself as a new opportunity, is at the heart of entrepreneurial theories and is one of the primary processes in the creation of value (Kirzner, 1979).

As discussed earlier, the leadership of each of the firms studied in this book all recognized that meeting customer needs and expectations and solving their problems was the primary challenge their firm faced. A central aspect of being able to solve a problem is being able to understand and define the problem or opportunity—or absorb its essence. Absorptive capacity, as Cohen and Levinthal (1990) define the term, is a critical stock level human resource that a firm accumulates, primarily through experience, from the skills and competence of it boundary spanning individuals, which should include top management (Thompson, 1967; Itami & Roehl, 1987). Cohen and Levinthal suggest that

> the ability to exploit external knowledge is . . . a critical component of innovative capabilities . . . . Prior related knowledge confers an ability to recognize the value of new information, assimilate it, and apply it to commercial ends. These abilities collectively constitute what we call a firm's *absorptive capacity*. (p. 128)

One of the themes that emerged from each of the business-level cases in this book was that if a firm wants to remain competitive and viable, over the long run the top leadership of the firm must be "in the hunt", as Len Petrush put it, and be central contributors to the firm's absorptive capacity. A firm that wishes to renew itself must have strong entrepreneurial leaders that can absorb relevant external knowledge, and then be in a position to know what to do with the capabilities of their firm.

Daft and Weick (1984) suggest that the leadership of firms must develop processes and mechanisms that allow it to detect relevant information about its environment *and interpret it*. Interpretation is defined as "the process of translating events and developing shared understandings and conceptual schemes among members of upper management. Interpretation gives meaning to data, but it occurs before organizational learning and action" (1984, p. 286). This perspective is important because, as Daft and Weick note:

> To survive, organizations must have mechanisms to interpret ambiguous events and to *provide meaning and direction for participants*. Organizations are meaning systems, and this distinguishes them from lower level systems. (1984, p. 293) [our emphasis]

These perspectives on organizational, and particularly general management, absorption and interpretation processes are important because they acknowledge the central role that general managers have in understanding and shaping the organizational context for potential action (Bower, 1970). Indeed, one of the major findings in the field was that each of the leaders felt it was their responsibility to make sure that the firm had future opportunities to pursue, and that the firm had the requisite shared values, skills and capabilities to pursue them effectively. As Len Petrush noted, "our challenge as managers is to ensure that our company is on the leading edge of the market . . . and that drags everybody into the vortex . . . [the employees] have to increase their levels of skill applying to new products. This obviates any obsolescence".

Each leader of the firms studied was intimately connected to the entrepreneurial process, which Bygrave and Hofer (1991, p. 3) describe as "all the functions, activities, and actions associated with the perceiving of opportunities and the creation of organizations to pursue them"—each was an entrepreneurial leader.

Entrepreneurial Leadership: Moving "Outside" Opportunity "In"

John Kotter (1990) has argued that there is an important difference between leadership and management in organizations, where, "management is about coping with complexity . . . good management brings a degree of order and consistency to key dimensions like the quality and profitability of products". On the other hand, leadership, "is about coping with change . . . the function of leadership is to produce change" (p. 18). As a summary of this thesis, which emerges in many settings in Syrett and Hogg's *Frontiers of Leadership* (1992), the editors note that

> Therein lies the fundamental difference between the tasks of leaders and managers. Managers do things right, leaders do the right things. Managers accept the *status quo*, leaders challenge it. Leaders create and articulate vision, managers ensure it is put into practice. (p. 5)

An empirical survey of 211 Chief Executive Officers, conducted by Eastlack and McDonald in 1970, suggests that the general approach to leadership and management that a CEO takes *can affect the performance of his or her organization*. The degree to which the firm's top person is a leader and/or a manager is an important factor that affects the entire range of organizational activities and outcomes.

Using statistical cluster analysis techniques they identified eight general CEO approaches to leadership and management style, based on the importance the CEO's assigned to 18 different managerial activities. Table 10.1 identifies the eight approaches, lists the primary high-value tasks that were associated with each general style, and presents performance information associated with each approach.

Overall there are three engaging aspects of the Eastlack and McDonald (1970) study from a value creation and skill-based strategy perspective:

(1) The top two approaches to leadership and management style that were associated with better performance, with regard to both sales growth

Table 10.1. Attributes of eight leadership styles (Eastlack and McDonald, 1979).

General approach to leadership and management	Primary activities that the CEOs were most personally involved in	Percentage median 4-year growth rate (rank)	Percentage median 4-year ROI (rank)	Overall rank
Growth entrepreneur	Sales of existing products; specifying areas for development; budget activities; selection of personnel; Go–no-go	80 (1)	13.0 (3)	1
R&D planner	Specifying development budgets	65 (4)	13.1 (2)	2
Growth director	Review of progress vs. plan	67 (3)	12.5 (4)	3
Product manager	Specifying areas for development	50 (7)	13.3 (1)	4
Aloof strategist	Review of progress vs. plan	70 (2)	11.7 (7)	5
Remote controller	Identification of, and negotiation with, candidates for acquisition	63 (6)	12.0 (5)	6
Acquirer	Identification of, and negotiation with, candidates for acquisition	65 (5)	11.9 (6)	7
Activist	Organization and motivation of personnel	50 (8)	9.5 (8)	8

and profitability, were both hands-on approaches and were entrepreneurial in nature.

(2) The focus of the *growth entrepreneur* and the *R&D Planner* approaches is on new products and services, the creation of new value and the active management of change—they are leadership-oriented approaches.

(3) Finally, both of the top performing approaches focused on internal development processes and activities.

This analysis suggests that "entrepreneurial leadership" approaches may be the most important with respect to the development, growth and long-term health of companies. Hofer (1990) finds evidence of this assessment in a comparative study of the leadership regimes of G.E.'s Reginald Jones and Jack Welsh. This conception of entrepreneurial leadership is about the proactive management of change and opportunity, especially with respect to the creation of value through the development of an organizational ethos that encourages growth rooted in ever-improving capability. It is an "outside-in" process of moving outside opportunity into the firm's skill-based domain.

The primary task of the entrepreneurial leader is to ensure that the firm can create new value through the deployment and exercise of resources and skills at their potential employ. This requires that leaders must not only be competent at recognizing opportunities, but that they must also have a deep knowledge of the resources, skills and capabilities of their organization, and of the process of transforming those resources into services valued by the customer.

Knowledge of the Firm's Resource Base and the Development of Integrative Systems is Important

All of the practitioners studied in this book recognized that knowing the customers's needs was not sufficient to succeed in their business—they knew that they were part of the customers' "need fulfillment process", as Dean Stull said, and that the company must be able to also fulfil its obligation.

A major theme that ran across the cases was that the development of systems that enabled the firm to integrate and assemble its resources in ways that would allow the company to be more competitive, in terms of price, flexibility, speed, and quality, was critical. As Mike Viera said:

> I think that . . . when I look at our company [in terms of resources], and of any company, you've got a building, a lot of equipment, and you've got a lot of people . . . and raw materials . . . by themselves none of those things really accomplish anything. What makes all that work is a system . . . and we, together, the management team and the staff, we've got twenty years in developing our system.

The process of converting resources, of mixing them together and integrating them into systems, was one that the general managers studied in the book felt was important, and that they were personally involved in. In the next section, the relevant academic literature related to "systems development and resource integration" is enfolded into the basic practitioner model.

Transforming Basic Resources into Productive Services and Dynamic Capability: An "Inside-Out" Process of Resource and Skill-Based Management

Another important element that the general managers studied felt was important was that their company be able to grasp an opportunity once it was perceived—that the firm had the capability to deliver. The process of meeting perceived consumer needs and values involves: (1) the structuring of the firm's resources and the development of resource flexibility (Ansoff, 1979); (2) the development of core skills; (3) the development and management of technologies; and (4) the creation of dynamic capabilities, which involves skill and technology application, organizational learning and capability accumulation (Teece *et al.*, 1992).

Structuring Resources: Resource Flexibility

One of the specific themes that emerged in the cases was that the general managers were concerned with ensuring that their firms were flexible and able to respond to many different challenges from their customers. In this section, the concept of "*resource flexibility*" is explored. Resource flexibility represents the way a firm chooses to configure its resources in an effort to achieve flexibility which can be used to broaden the set of entrepreneurial choices a firm might have (Ansoff, 1978; Teece, 1980; Porter, 1985; Goldratt, 1990). As was noted in Chapter 2, Ansoff identified in 1978 that,

> unlike external flexibility, internal flexibility has received relatively little attention from strategic planners. *But recent history shows it to be a crucial ingredient in strategic preparedness . . . logistic [resource] flexibility has received even less attention than managerial flexibility*, and it is safe to predict that, in the coming years, logistic [resource] flexibility will be used increasingly as a measure of strategic preparedness. (pp. 64–66) [our emphasis]

Resource flexibility refers to the ability of a firm to structure its resources such that the firm possesses: (1) an adequate diversification of skills, (2) an adequate diversification of critical resources, (3) resource liquidity, (4) convertibility of capability, (5) elasticity of volume, (6) modularity of capacity, (7) multi-purpose capacity, and (8) speed of convertibility. These capabilities are a function of both the type and structure of the resources in question.

In most cases the skill-based resources, which are tied to human skills and competencies, offer the most flexibility in terms of liquidity, convertibility, elasticity, modularity and speed of convertibility. Human resources (know-how) and specialized, indivisible physical assets offer the best chance for a firm to achieve *economies of scope*. "As a general matter, economies of scope arise from inputs that are shared, or utilized jointly without complete congestion" (Teece, 1980, p. 226). Each of the general managers studied in the book, when asked about economies of scope, both acknowledged that they were familiar with the concept, and that it was an integral part of their business strategies.

As Rothwell and Whiston (1990), among others (Stalk and Hout, 1990), argue, economies of scope are becoming a critical element in competition:

> Today, while experience and scale effects remain important, there is a growing emphasis on strategies and structures designed to enhance corporate flexibility and adaptability in a rapidly changing world. As part of this shift, the notion of *functional integration* is being emphasized (p. 193).

They continue:

> Therefore it is of considerable importance to a manufacturer to maximize his process flexibly in order to encourage product-design flexibility, rapid product updating, short batch runs, modular design, etc . . . . Thus, "Economy of Scope" displaces Economy of Scale as the dominant issue. (p. 199)

Rothwell and Whiston (1990) cite the work of Lawrence and Lorsch (1970) in noting that high-performing organizations achieved more *effective integration around critical interdependencies* than their less competitive rivals. That is, they integrate around core skills and technologies.

Developing Core Skills

One of the important aspects of the argument above is that the adroit integration around key resources is vital for effective organizational action, and that the development of competencies to manage such core activities is a key aspect of the resource management process. Prahalad and Hamel (1990) emphasize that:

> core competencies are the collective learning in the organization, especially how to coordinate diverse production skills and integrate multiple streams of technologies (p. 82). If core competence is about harmonizing streams of technology, it is also about the organization of work and the delivery of value.

There is still the question, however, of establishing which skills and resources are most critical to build around. The particulars for a firm will depend on many contingencies (Hofer, 1975), but in general a firm has an incentive to use the most valuable and specialized of its resources as fully as possible (Penrose, 1959, p. 71) and to pace its growth in terms of its critical, least-reducible resources (Thompson, 1967). This line of reasoning is interesting if we connect it to the driving force and theory of constraints concepts and recognize that, from a strategic perspective, the firm should choose the location of the least reducible element, or the primary internal constraint (Goldratt, 1990), which paces the growth and scope of the firm's activities in the area it is most likely to be able to develop core competencies in.

The existence of a core competence implies that the firm will be able to get the most out of its critical strategic resource areas (which are comprised of skills and technologies) and that protective capacity exists in all other areas to allow smooth flow of throughput away from the core area (Goldratt, 1990). The core competences of the firm also mirror the central skills and skill-sets that the firm has acquired and built, and reflects the strategic commitments that general managers have made for their firms through time (Ghemawat, 1991).

These commitments are dynamic constraints on strategy and are cumulative, but they can provide a firm with the potential for building isolating mechanisms based on organizational skills and technologies (Rumelt, 1987; Ghemawat, 1990, p. 14). According to Prahalad and Hamel (1990), "since core competencies are built through a process of continuous improvement and enhancement that may span a decade or longer, a company that has failed to invest in core competence building will find it very difficult to enter an emerging market" (p. 85).

The general managers studied in the book were concerned about identifying their "core of competency", as Petrush called it, and were able to describe the core elements of their business in terms of competency.[1]

The Development of Technology: An Important Process Aspect of Resource-Management

Technology is one of the fundamental ingredients to understanding organizations from a resource perspective, and it was another theme that was important to the practitioners studied in the book. Based on the analysis of data from the different firms studied in the book, technology is an inseparable aspect of the skill-base development process of a firm. One theme that emerged in all the cases studied, and in the work of Teece *et al.* (1992) was that the operational performance of a firm is tied to that firm's ability to blend its human capital (skill-base) with technological opportunities and requirements, and that this tie-in of skills and technology, with market opportunity, is at the centre of the innovation process. In order to understand this very close link between skills and technology better, let's look at what we mean by technology in the context of this work – for as often as the term is used in the strategic management and innovation literature, it is rarely defined[2].

Defining What is Meant by "Technology": State and Process Descriptions

Burgleman *et al.* (1988), however, offer the following description of technology, and suggest that technology is, generally speaking,

> the practical knowledge, know-how, skills, and artifacts that can be used to develop a new product/service and or new production/ delivery system. Technology can be embodied in people, materials, cognitive and physical processes, plant, equipment and tools. (p. 32)

While this is a useful definition, this conception limits technology to "new" applications and developments, which is unnecessarily restrictive, and it does not refer to any process element. That is, the definition is a classic an example of what Herbert Simon (1981) calls a *state description*, which,

> provide[s] the criteria for identifying objects . . . they characterize the world as sensed. For example:
>
> *A circle is the locus of all points equidistant from a given point* is an example of a state description of a circle. (p. 221)

From state descriptions one can identify an object or concept, but not know anything about how the object is created or how it functions. A better description of technology, in that it is more general and does include process elements, is that suggested by William Siffin (1983):

> Technology—any combination of rules, roles and resources which, appropriately applied, consistently produce predictable outcomes. (p. 1)

Like Burgleman *et al.*'s definition, Siffin conceptualizes technology as a means for accomplishing an end, however, technology is not limited to the domain of "new product/service" development. Siffin is much more general and specifies that when *rules* (regulating mechanisms for proscribed actions), *roles* (which define the authoritative actions of individuals in organized settings) and *resources* (like skill, practical knowledge and know-how) are combined *appropriately* (to produce consistent and predictable outcomes)—you have a technology. Siffin's definition implies that processes which are not *understood* are not technologies. A "black box" understanding of an organizational process or tool does not allow one to asses the "appropriateness" of its use, which can have profound strategic consequences for managers.

In order for strategists to have a sufficient basis for managing technology, technology must be understood in terms of *both state and process* description. According to Herbert Simon, process descriptions:

> characterize the world as acted upon; they provide the means for producing or generating objects having desired characteristics . . . . Recipes, differential equations, and equations for chemical reactions are process descriptions (p. 222). An example would be:
>
> *To construct a circle, rotate a compass with one arm fixed until the other arm has returned to its starting point.*

If one considers that rules, roles and hierarchies, and the means for acquiring resources in a specific situation, are all "*techniques*" that can be used to establish the context for the appropriate exercise of resources, then a basis for understanding technology in both state and process description is established. William Barrett, in *The Illusion of Technique*, elaborates on the meaning of "technique":

> A technique is a standard method that can be taught. It is a recipe that can be fully conveyed from one person to another. A recipe always lays down a certain number of steps which, if followed to the letter, ought to lead invariably to the desired end. The logicians call this a decision procedure. *Technology is embodied technique*. (p. 22)

For Barrett, technology is "embodied technique", it is a state descriptor which encompasses the "processes of technique". Technology is both form and substance, with form being a complete process function of human skill and volition. Most definitions of technology fail to recognize the vital importance of the process element. Information Technology is defined as: "the technology of sensing, coding, transmitting and transforming information". In this form this is a simple state description. However, the components of information technology are all "techniques" which may be characterized in detail by process descriptors. So the technique of translating information might be described as the "process where one takes an input series of data in one language and assigns binary values to each input, according to an appropriate codebook, in order to produce a translated machine-readable text".

The combination and appropriate blending of techniques, including translating, comprise "the technology of information". This is an important point to have laboured over because it is necessary to understand both the state and process descriptions of information technology if one is going to *successfully manage* it, for the essence of management is about process and context. In the most general sense, then, technology can be defined as:

> *A system of techniques which, when appropriately applied, consistently produces predictable outcomes.*

In an organizational context, technologies are systemic aggregations of techniques which exist in organized settings as a means for achieving desired end states. They are part of the tools of strategy and part of the process of strategy. Technologies, if they are seen in light of the process elements, affect the premises for decisions; have the potential for changing the balance of power and perspective in organizations; and are great potential sources of competitive advantage because they are really where human skills meet application, and are at the heart of where value is actually created in organizations, as Lenz (1980) notes,

> The proximate source of value creation in an enterprise resides, therefore, at the confluence of *knowledge about the creation of value and technical facilities and processes*. These elements in functional integration may be termed a knowledge–technique base for value creation. This base constitutes the operational core of an enterprise and is similar to Thompson's (1967) notion of a "core technology". (pp. 226–227) [our emphasis]

Perhaps none of the cases illustrates the ideas of technologies as both knowledge and process, and that they are indeed embodied in people, better than the Hauser case. At Hauser, which Dean Stull calls a "technology-based company", the focus is not on technology *per se* but on the capability of the people to understand the new technologies and to adapt and change the technologies as needed, given customer problems. Stull continues,

> we are worried about protecting our competitive advantage from our proprietary technology viewpoint. *But not nearly as concerned as we are about keeping our people abreast of technology changes, and working as generalists.*

The challenge in leading and managing Hauser is in keeping the skill and knowledge base of the human capital fresh and flexible. The concept of managing the firm as a problem-solving entity puts a direct emphasis on the need to have a process which can generate solutions for customers and for entrepreneurial downstream applications for Hauser. An interesting aspect of this problem-solving focus is the team formation process at Hauser, which was discussed at some length in the Hauser case. The process is largely *ad hoc* and informal, and is determined by the nature of the problem at hand and the underlying nature of the techniques involved in the technologies that might be applied to solve the problem. The teams are really complementary skill packages that are assembled to broaden the knowledge–technique base enough to potentially find a creative solution. This means that Hauser is an example of what Quinn and Paquette (1990) call an infinitely flat organization—any number of myriad teams may form at Hauser, making an operational organizational chart difficult to track at best.

Hauser is a storehouse of technology in both its process and content meaning—it is a coordinated system of interaction whose people know how and when to apply the techniques of discrete technologies to meet the needs of specific clients. Dr Tim Ziebarth comments,

> Most of our clients . . . are unable to solve their own problems because their views are too narrow. They don't really see the problem, they don't have a suite of knowledge that allows them to apply very many technologies to solving a problem. *Our people are very broad-based technologists, who are good problem-solvers*. They know how to ferret out the true problem, and draw in a lot more than one narrow discipline to solve these problems. [our emphasis]

In this passage Ziebarth uses an interesting word to describe his people: he refers to them as technologists. "We try to know as much technology as we can so when problems or opportunities come up, we can draw on a bunch of resources to come up with a high-value solution that we can then do something with."

This is a skill-based viewpoint, to the extent that the company sees itself as a cooperative of technologists; the technology drive is really a skill drive. Technology is embodied skill at Hauser, which is consistent with Rosenberg's (1985) view of technology as "a quantum of knowledge retained by individual teams of specialized personnel". It is also consistent with Dutton and Thomas's (1985) contention that, "technologies embed knowledge and are reflections of a growing stock of human knowledge. Firms' technologies therefore reflect knowledge acquired from their internal and external sources" (p. 188). Clearly, in such an environment the human capital base is most important. As Amendola and Bruno (1990) suggest:

> The human resource is the key of innovative outlooks: it is the reservoir of potential capabilities which can create new and different productive options. In this light, it does not appear as an input to be passively associated with given productive structures, and whose relevant aspect is

> then availability in the right amount and proportion, but as an asset made up of skills which are the end result of a sequential process of specialization. *As an asset rather than an input, labor is no longer something to minimize or get rid of, but to maintain and possibly enrich.* (p. 428) [our emphasis]

Indeed, the entrepreneurial leaders of all the companies studied in this book, including Mike Viera, who managed in a relatively technologically simple environment,[3] managed their companies with the Amendola and Bruno comments in mind—the skills and knowledge of human beings is what drives technology and innovation, which is at the heart of the process of value creation (Lenz, 1980).

The ability to acquire, generate and regenerate needed technologies and technical knowledge in a firm is closely tied to the nature and development of a firm's skill-base of human capital (Schultz, 1990). As the need for technological advancement increases, either through demand-pull or business-push mechanisms, there is a concomitant need for a more sophisticated skill-base in terms of both skill depth and breadth (Burgleman and Sayles, 1986). The leaders of a firm must continually choose an investment path in either the acquisition of technology and skilled capital or their internal development, or some combination, if they are to continue developing more broad or deep capabilities in their firms.

Creating Capabilities: Dynamic Capability and Organizational Learning

The development of systems that would allow the firms of the general managers studied in the book to integrate their resources at the most general level was also an important theme that emerged in the field. As Kimmel said:

> We wanted to get to a point where we thought of ourselves as "one company working together to serve clients". We needed to learn to work across lines: product, function, business lines.

The general capabilities a firm develops are essential to its ability to react to opportunities in the market-place. Firms which are to survive in a turbulent business world must be equipped, especially in terms of their leadership capacity, to change rapidly and they must be able to apply their productive services across traditional product industry boundaries—that is possess dynamic capability.

The concept of dynamic capability, in this book, builds on the extant concepts of transilience, which deals with the firm's ability to innovate and change with respect to its internal competencies and external market linkages (Abernathy and Clark, 1985), and strategic capability. Abernathy and Clark suggest that

> The foundation of a firm's position rests on a set of material resources, human skills and relationships, and relevant knowledge. These are the

> competencies or competitive ingredients from which the firm builds the product features that appeal to the market-place. Thus, *the significance of innovation for competition depends on what we shall call transilience*—that is, its capacity to influence the firm's existing resources, skills and knowledge. (p. 57) [our emphasis]

Lenz (1980) describes strategic capability as

> the capability of an enterprise to successfully undertake action that is intended to affect its long-term growth and development. It refers to an *organization's total capability*, which includes support that may be summoned from environmental aggregates and projected in pursuit of strategy. (p. 226) [our emphasis]

Lenz distilled his conceptualization of strategic capability from previous theoretical developments and identifies three dimensions of an organization's strategic capability: (1) its knowledge–technique base for value creation, which we have seen is a direct function of the firm's human and technological capital stock; (2) its capacity to generate and acquire resources; and (3) its general management technology (1980, p. 226). The concept of a knowledge–technique base for value creation is really what we have termed a firm's core competence, and is comprised of those skill-sets that can be called core skills, which are indeed at the confluence of knowledge about creating value and technical facility. Andrews (1987) also addresses the sources of capabilities for a firm:

> *Sources of capabilities*. The powers of a company constituting resources for growth and diversification accrue primarily from experience in making and marketing a product line or providing a service. They inhere as well in (1) the developing strengths and weaknesses of the individuals comprising the organization, (2) the degree to which individual capability is effectively applied to the common task, and (3) the quality of coordination of individual and group effort (p. 47) . . . in this connection, it seems important to remember that individual and unsupported flashes of strength are not as dependable as the gradually accumulated product and market-related fruits of experience (p. 48).

In this passage Andrews is indirectly pointing to the fact that individual skills are the basis for capability (Ulrich and Lake, 1991). This, combined with the ability to apply these skills towards a common task (competence), and to coordinate skills and skill-sets in the pursuit of opportunity (core competence) leads to firm capability. Additionally, he argues, like Nelson and Winter (1982), Ansoff (1984), Peters (1984), Itami & Roehl(1987), Winter (1987), Dierickx and Cool (1989), Badarocco (1990) and Collis (1991), that capability *evolves through the exercise and implementation of day-to-day operational activities*.

Dynamic capability is neither easily acquired nor easily maintained. Ansoff (1983) suggests that the capability of an organization is determined by three components: (1) the quality of human, physical and technological, scientific, informational and monetary resources; (2) the manner in which these resources are organized into work systems; and (3) the social system of values and expectations. Ansoff also notes that capability is a function of timing and spatial relations, and the speed of build-up in capabilities is dependent on sequences and that the total process of strategy and capability adaptation may take as long as 10 years. As such, capabilities are not easily imitated, and firms which possess generalized capabilities should be able to appropriate both Ricardian and entrepreneurial rents more readily than firms which do not possess such capabilities.

Dynamic capability conceptually links strategy formulation processes to strategy implementation processes. Yet, as Ansoff (1984) argues, despite its importance there has been too little work on the implementation end of strategic management. Ansoff suggests that *strategic evolution* (parallel strategy formulation and implementation processes) and *capability evolution*, which describes the process by which the organizational configuration and its dynamics evolve over time, will become the dominant paradigmatic keys to future strategy research (1984, p. 508).

Capability evolution, as Ansoff (1984) defined the term, describes the process by which the organizational configuration, or structure, changes over time to support and constrain strategic directions. Capability evolution is therefore a function of two major processes: (1) organizational configuration, or the structure of its resources, which was discussed earlier, and (2) organizational learning.

Organizational Learning

The development of systems to encourage organizational learning was an important aspect of the efforts of each of the general managers studied in this book to ensure that their companies would be able to deliver value to customers now and in the future. Mike Viera invested a great deal of time and money in the development of an education room, and hired an "educator" to help his employees learn about the theory of constraints and the thinking process. Each of the other managers made tangible investments in training, and in developing the abilities in their people to learn about their role in the business.

Organizational learning and unlearning (Hedberg & Starbuck, 1981) is a critical aspect of a "firm's *ability to develop insights and knowledge* between past actions, the effectiveness of those actions, and future actions" (Fiol and Lyles, 1985, p. 811) in order to: (1) maintain or improve its current position in an environment (Jelinek, 1979; Daft and Weick, 1984; Nonaka and Johansson, 1985) or (2) to respond to threats from its environment (Cyert and March, 1963; Cangelosi and Dill, 1965). Organizational learning is a separate and more dynamic concept than *organizational adaptation*, which only reflects the "firm's ability to make *incremental adjustments as a result of environmental changes*, goal structure changes, or other changes" (Fiol and Lyles, 1985, p. 811; Miller and Friesen, 1980; Chakravarthy, 1982; Miller, 1982; Hrebiniak and Joyce, 1985).

Organizational learning is a vital component of resource management because it represents important feedback (control), feed-forward, and memory mechanisms that link implementation-oriented processes of skill-development and utilization with the more formulation-oriented processes of environmental scanning and entrepreneurial choice (Walsh and Ungson, 1991). Hedberg & Starbuck (1981) refers to organizational learning as the link between individual skills and organizational capability in the following way:

> Organizations do not drift passively with their members' learning: organizations influence their members' learning, and they retain the sediments of past learning after the original learners have left. *Organizations can be thought of as stages where repertoires of plays are performed by individual actors.* (p. 6) [our emphasis]

An organization, in this sense, is a theatre where the "repertoire of plays" is actually a "repertoire of skills and routines" (Nelson and Winter, 1982; Peters, 1984) that are brought to bear in order to "entertain" and add value to targeted "audiences" of customers and external stakeholders. General managers have the vital role, in this "production company", of knowing what "plays" are the proper ones to stage, both in terms of timing and content, and they must be able to provide the context for the individual "actors" to "learn" to perform well.

Fiol and Lyles (1985) describe four contextual factors that affect the probability that learning will occur in organizations—or the probability that the "actors will know their parts":

(1) There must be a *corporate culture conducive to learning*. In terms of the role organizational culture plays in the learning and memory processes of an organization, Walsh and Ungson state that:

> Organizational culture has been the subject of increasing interest. It has been defined as a learned way of perceiving, thinking, and feeling about problems that is transmitted to members in the organization. (Schein, 1984). The words *learned and transmitted* are central to this definition and our purpose. Culture embodies past experience that can be useful for dealing with the future. It is, therefore, one of organizational memory's retention facilities (pp. 64–65).

The notion of organizational memory is important in that, especially as it relates to culture, it is an intangible resource that cannot be replicated easily, but can be managed within the context of an individual firm. As Wernerfelt (1984) says, "you cannot expect superior performance in a fair race against equals. Instead, you need to look for races where you have an advantage. You cannot expect to make above-average returns on investments in physical assets or blueprints

... with resources of the cultural variety, things are a bit different. Once you have them, you can 'grow' them at a cost way below the cost of imitation" (p. 11).

(2) There must be *strategies that allow flexibility* and learning. "Strategy influences learning by providing a boundary to decision making and a context for the perception and interpretation of the environment ... the strategic posture also creates a momentum to organizational learning ... similarly, the strategic options that are perceived are a function of the learning capacity within the organization" (Fiol and Lyles, 1985, p. 804).
(3) There must be an organizational structure that allows both *innovativeness and new insights*. Teece *et al.*, (1992, 1997) discuss the idea of a dynamic capabilities approach to strategy. In that approach they suggest that the creation and maintenance of *dynamic routines* is vital to organizational effectiveness over the long term. Dynamic routines are directed at learning and continuous new process and product developments. Like the approach to continuous improvement in the Theory of Constraints (Goldratt, 1990; Dettmer, 1997), the dynamic capabilities approach fosters innovative, insightful learning by individuals and the organization.
(4) Both the internal and external environments of the firm must not be too complex so that overload occurs, nor too simple that boredom sets in.

Meyers (1990) identified four primary organizational *"learning modes"* in her in-depth examination of the organizational learning changes in one exemplar, the Xerox Corporation, over the development cycle of its core technology. She collected longitudinal data that spans over 40 years, and she interviewed 130 managers. Meyers recognized a "progression of learning modes over the cycle" (p. 97):

(1) *Maintenance learning*—Involves the establishment of standard operating procedures and written guidelines for action.
(2) *Adaptive learning*—Incremental, but systemic rule changes occur.
(3) *Transitional learning*—Involves a shifting of strategic thrust in response to extreme environmental changes—experiments, rules of game changing, unlearning occurs.
(4) *Creative learning*—Involves radical redefining activities, constructive conflict (Pascale 1991), and the inventing or adopting of new technology.

The types of learning mode are listed in ascending order of complexity and degree of innovation and entrepreneurial choice that is involved. Each demands a different set of organizational skill repertoires, competencies, and to some extent, cultural orientation.

Organizational Learning and the Development of Systems

The process of organizational learning is a pivotal aspect of the development of dynamic capability (Teece *et al.*, 1992, 1997) and is the bridge to the development of organizational systems, which are often the result of the "routinization of skills and knowledge"—i.e. they are memory enhancing mechanisms (Walsh and Ungson, 1991). As a firm's systems evolve, they provide the base for the firm to exercise its "dynamic capabilities", which become a source of further development and *also of potential competitive advantage*. Here then is the real link between skill-base development and strategy—skills and technology are the building blocks of the value creation process, and if a firm can develop its skill-base more quickly and effectively than competitors, they will potentially benefit from opportunities that only they are ready to mine. At a corporate strategy level, as Prahalad and Hamel (1990) note, "the real sources of advantage are to be found in management's ability to consolidate corporate-wide technologies and production skills into competencies that then empower individual businesses to adapt quickly to changing opportunities" (p. 81), which was a central concern of Charles Hugel and the Combustion Engineering management during their turnaround efforts.

Recapitulation of the Discussion on Dynamic Capability

Figure 10.2 recapitulates the previous discussion on the development of dynamic capability. The wheel of dynamic capability is driven by three

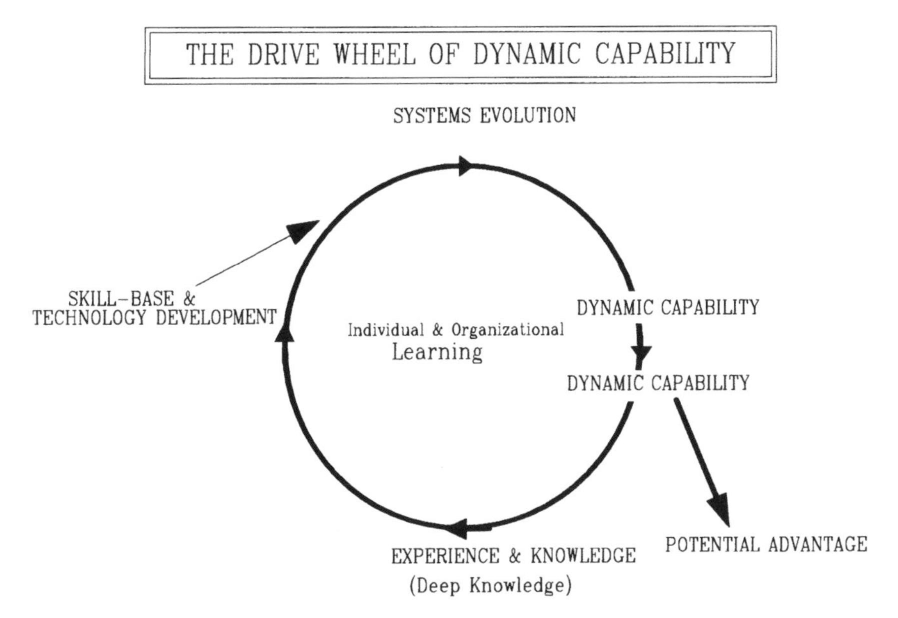

Fig. 10.2. The drive wheel of dynamic capability.

major forces and processes: (1) the experience and knowledge-base of the employees of the firm; (2) the process of skill and technology development; and (3) the related development of organizational systems. At the hub, or bearing, of this wheel is individual and organizational learning—the more learning that takes place, the faster the wheel can rotate, given the other forces.

Without a strong institutional commitment (Selznick, 1957) to an environment that encourages learning, the hub will actually inhibit the progress of dynamic capability development.

KEEPING THE COMPANY "IN THE HUNT": MATCHING SKILLS AND OPPORTUNITY

One of the important challenges that the general managers of the firms studied in this book identified was in making sure that their companies both knew what the customer wanted, and that the firms had the capability to do something about it—that the companies could match skills with opportunity. Hugel and Kimmel, in helping to turn round Combustion Engineering, actually conceived of a "skill-based strategy" to ensure that the various problems of the firm's customers could be met with the resources the company controlled—even though it meant major restructuring and a change in corporate culture. In this section we explore academic literature relevant to the concept of "integrating resources with opportunities—skill-based strategy".

Skill-Based Strategic Thinking—Integration of the Outside-In–Inside-Out Processes: Integrating Resources with Opportunities

One of the important elements to each of the practitioners was that the strategy of their firm must include both a detailed and systematic understanding of their customers' needs and their firm's capabilities. When asked if either the "customer" or "resource" aspects of their business dominated their strategic thinking, a typical response was like Mike Viera's, who said that, "they are like two wings of an airplane", and that there must be a balance in developing strategy—there must be a mastery in the process of being able to effectively match and integrate capability with opportunity. There must be a proficiency in both skill-based strategic thinking, formulation and implementation, and market-based strategic thinking.

As Rumelt (1979) notes, "a principal function of [strategic thinking] is to structure a situation—to separate the important from the unimportant and to define the critical subproblems to be dealt with . . . " (p. 199). What becomes most important in the skill-based strategic thinking process is knowing, as precisely as possible, what capabilities a firm has—and hence what set of opportunities it might feasibly pursue. The essence of skill-based strategic thinking, and what makes it different from opportunity-based strategic thinking (Porter, 1980), is that it focuses on developing a general proficiency in the strategist's ability to understand her firm in terms of generalizable capabilities rather than product markets. That is, strategy is less about product positioning in specific markets, than it is about developing superior skills, material resources and capabilities that can be deployed in a number of market settings.

The Skill-Based Strategy Process as Making Entrepreneurial Choices and Mobilizing Capability

Skill-based strategies are those designed to translate the specific skills of an organization into sustainable competitive advantages. *Skill-based strategies focus on the continuous process of creating and adding value in the market-place through the making of entrepreneurial choices which exercise relevant and critical organizational skills, competences and capabilities*. As such, skill-based strategies are inherently dynamic and require an approach to leadership and decision-making that is inherently entrepreneurial. The primary challenge to leadership, from a skill-based strategy perspective, is to find new applications for current skills and to build new skills for new applications—both of which involve creating new value and potentially new organizational forms.

Entrepreneurship, from a resource-based perspective, is essentially a dynamic process of creative destruction (Schumpeter, 1934) that is embedded in the process by which entrepreneurs, as leaders and catalysts, seek to link a firm's unique resource and skill capabilities with opportunities that they create or perceive in the market-place. Rumelt (1987) reinforces this idea:

> innovation and entrepreneurship are really about novelty and differentiation: models of commodity-producing collectives [Neo-classical and most IO models] may not be the best approach to their study. An alternative viewpoint, one that emphasizes the uniqueness of firms and identifies profits with resource bundles rather than with collectives, is offered by the strategy field. (p. 140)

In this statement Rumelt is recognizing that a theory of strategy, at least one that professes to analyze the long term, is highly related to a theory of entrepreneurship and innovation. Innovations, which are the end result of the process of mobilizing resources to bring new ideas and products to the market, are the byproducts of *entrepreneurial choices*, which reflect the decisions of general managers to seek value-added opportunities that leverage their firm's present or planned skill, capability and material resource endowments.

The concept of skill-based strategy rests squarely on the concept of entrepreneurial choice—which is all about the strategist being able to define and mobilize dynamic capabilities in a firm *and match them* with opportunities that she recognizes in the environment of customers and stakeholders. It is about building and maintaining innovative organizations. Innovative organizations are organizations designed to exploit change (Drucker, 1985). Whenever a business can alter the conditions for effectiveness, or respond to such changes by like action, the chances for success in the competitive market-place are enhanced and long-term survival is increased: this "altering of the conditions of effectiveness" results in *strategic change*—which is the systematic exploitation of the opportunities and problems that arise for a firm in an attempt to ensure effectiveness. A strategist who fully understands a firm's resource-base and capabilities, who crafts skill-based strategies, should be able to induce strategic change, or change the rules of competition if you will, more frequently than one who is more concerned with market position or structure.

REVIEW

In this chapter three general issues that emerged in the field research were examined in light of relevant academic literatures. They were:

(1) Meeting the customers' needs is a function of knowing what the customers want and need.
(2) Resources in and of themselves are not as important as what management does in assembling them and developing integrative systems that can allow the company to produce quality products and be flexible, fast and cost efficient.
(3) The challenge for general managers is to keep the company "in the hunt" by knowing both what the customers want and what the company is capable of doing with its resources and skills.

In the next section, these three issues are re-cast in the terms of three organizational processes which are part of the general process of value creation.

THREE BASIC PROCESSES THAT BRIDGE THE DISTANCE FROM BASIC RESOURCES TO MEETING CUSTOMER NEEDS: THE MAJOR ELEMENTS THAT COMPRISE A THEORY OF ENTREPRENEURIAL LEADERSHIP, SKILL-BASED STRATEGIC THINKING AND VALUE CREATION

Figure 10.3 outlines the four main elements that comprise the basic structure of a model, grounded in both practitioner and academic thought, that will serve as the base for a substantive-level theory on skill-based strategy and entrepreneurial leadership. The four core concepts of this model include: (A) the *actual and perceived needs and values of customers*,

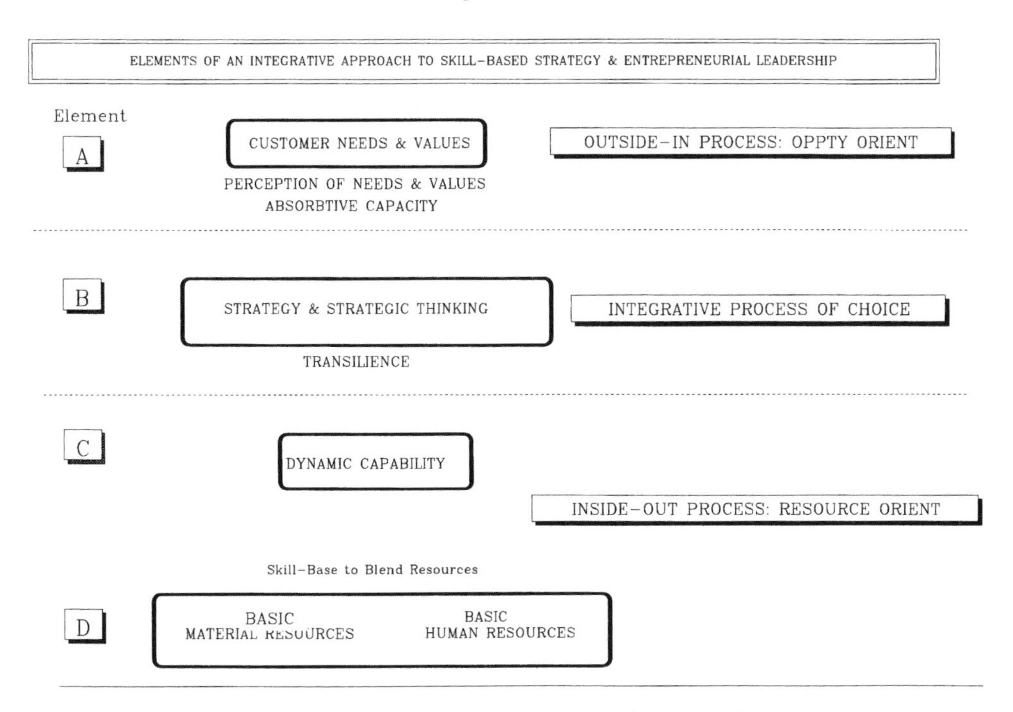

Fig. 10.3. An integrative diagram of the value creation process.

which are a function of individual tastes and broad socio-economic and political forces—and the entrepreneurial process that is involved in identifying and interpreting those needs; (B) the process of *skill-based strategic thinking and strategy*, which is a function of entrepreneurial choice activities and which involves managing the firm's ability to deal with changes in both consumer (market) linkages and internal competencies; (C) the process of development of a firm's *dynamic capability*, which is a function of the firm's ability to continuously reshape itself with respect to its knowledge–technique base (Lenz, 1980), which is itself a function of the firm's human and technological capital stocks; and (D) the firm's *resource base*, which is a function of management's ability to identify, acquire and build appropriate assets, given their entrepreneurial or strategic choices, that can be transformed into valuable goods and services.

The four elements are divided into three separate regions in the figure – *each of which addresses a different fundamental process of leadership and management*. The top region (containing element "A") of the figure addresses issues related to the perception and interpretation of customer needs and values, and reflects an "*outside-in, opportunity-oriented*" process of entrepreneurial discovery and alertness (Kirzner, 1979). The lower region of the figure (containing elements "C" and "D") addresses issues related to the skilled transformation of basic resources into dynamic capability for the organization. This is an "*inside-out, resource-oriented*" process of management.

The middle region (containing element "B") addresses strategic issues related to establishing a fit between capability and opportunity—it is where resource management meets entrepreneurial discovery. It is a holistic integrative process of choice.

The three basic processes identified above are part of the more general process of value creation and delivery. That is, the processes of entrepreneurial leadership, resource management and strategic choice are all integral parts of the process of creating value in the market-place through organized activity. In the next two chapters we will explore in more detail how value is created through the effective and strategic creation and deployment of resources in the pursuit of opportunity. It will be argued that if a firm is to sustain advantage in the market-place, as an organization it must be adept at all three primary processes discussed above: (1) it must be able to perceive opportunities; (2) it must be able to understand the specific resource implications the pursuit of a particular opportunity will have, especially in terms of skills and capabilities; and (3) it must be able to effectively mobilize and remobilize its resources as conditions change in both product and factor supply markets.

CHAPTER SUMMARY

In this chapter we have moved to the substantive level of theorizing about skill-based strategy and entrepreneurial leadership, and have begun to explore how these concepts are part of the general process of value creation.

Three major leadership and management processes were identified that are part of the more general process of value creation: (1) an "outside-in, opportunity-oriented" process of entrepreneurial discovery and alertness; (2) an "inside-out, resource-oriented" process of management, which addresses issues related to the skilled transformation of basic resources into dynamic capability for the organization; and (3) an integrative strategic thinking and decision-making process that centres on establishing a fit between capability and opportunity—it is where resource management meets entrepreneurial discovery.

NOTES

1. Each of these discussions is related in the text of the "core skills" section of the data presented in Chapter 9, and in the idiographic cases.
2. In a perusal of Tushman and Moore's second edition of Readings in the Management of Innovation (1988), not one of the otherwise excellent 48 articles had an explicit definition or discussion on what technology is.
3. The terms "high technology" and "low technology" are probably better replaced with the terms "technologically complex" or "technologically simple", as they more accurately reflect the process element of technological systems.

11

SPECIFICATION OF A THEORY OF VALUE CREATION, SKILL-BASED STRATEGY AND ENTREPRENEURIAL LEADERSHIP: PROVIDING A TESTABLE THEORY

INTRODUCTION

In this chapter, a grounded theory that discusses how and why potential sustainable competitive advantage is related to value creation, entrepreneurial leadership and skill-based strategy is developed. The theory developed in this chapter extends the discussion outlined in the last chapter and centres on the practitioner-derived core categories of "value creation" and "solving the customers' problems"—and explores how the development of long-run opportunity for the firm, which is essentially a resource-based issue, is central to the value creation process.

This focus on value creation and competitive advantage is a deviation from the original intention of the book, which was to develop a grounded working theory of skill-based strategy. However, since the emergent core category from the field focused on value creation, and since the processes of skill-based strategic thinking and entrepreneurial leadership are integral parts of the value creation process, which itself is an integral part of achieving sustainable competitive advantages (as will be shown in more detail shortly), it is appropriate at this point to change the focus to building a theory of value creation—which encompasses the original substantive area of skill-based strategy.

The chapter opens with a discussion on the domain of the working theory that will be developed, and includes remarks on what is meant by "value" in the context of this work. Following these opening comments, a simple model that depicts how value is created through firm activity is introduced. This general model identifies the primary elements by which basic resources are transformed by organizations into productive services and competitive stocks, which can serve as the foundation for the firm's long-run productive opportunity—which reflects the firm's ability to deliver value to customers effectively over long periods of time. The processes of entrepreneurial leadership, skill-based strategy and management are discussed in light of this simple value creation model.

Since a primary objective of the book was to examine how skill-based strategy and strategic thinking is related to long-run firm performance, the simple model of the value creation process is then extended to incorporate performance issues. This more complex analysis is the core of the chapter, and a theory of firm performance is developed that examines how potential sustainable advantages (which are the performance constructs) are related to the value creation process—and particularly the phenomena of entrepreneurial leadership and skill-based strategic thinking. The theory is presented in four complementary forms: in diagrams (1), with accompanying textual explanation (2) and related propositions (3); and in a summary chart (4), which is in the form of a hierarchically-ordered spreadsheet.

EFFECTIVE THEORY: FALSIFIABILITY, UTILITY AND DOMAIN

In the previous two chapters, various idiographic-level models were reviewed, and a substantive-level grounded model of value creation and skill-based strategy was developed. The models that have been presented up to this point have been descriptive and typological, or have been in the form of analogies—they addressed questions of *what*, but not *how*, *why* and *when* – which are the domain of theories.

A theory, according to Bacharach (1989), "is a statement of relationships between units observed or approximated in the empirical world . . . it is a system of constructs and variables in which the constructs are related to each other by propositions and the variables by hypotheses" (p. 498).

Two of the important elements in evaluating a theory are that the theory be falsifiable and that it have utility (Bacharach, 1989). In terms of falsifiability, the theory must be presented in such a way that empirical refutation is possible—the theory must be testable. In terms of utility, good theory must provide a bridge between theory and research, and that it must be able to explain and predict phenomena within its domain.

In the remainder of this chapter, the domain of the theory developed in this book is discussed, and the theory is presented in both a propositional and diagrammatic form. The propositions and diagrams are the base from which researchers can begin to test the theory and evaluate it in terms of utility.

The Domain and Levels of Analysis of the Theory of Value Creation and Sustainable Competitive Advantage

One of the important elements of a theory is that it specifies the domain and units of analysis in which the theory is held to be relevant. The domain of a theory specifies the boundaries in which the theory is supposed to apply. The critical assumptions that bound any theory include the implicit values of the researcher and the explicit restrictions regarding space and time (Bacharach, 1989: p. 498). With respect to the implicit assumptions of the primary researcher, every attempt has been made to specify potential theoretical biases and to design the research so as to minimize their impact on the final product.[1]

Domain of the Theory with Respect to Space

With respect to space, this theory is inherently resource-based and it focuses on the activities and attributes of individual firms, as defined below. It is acknowledged that this resource-based theory is not a general theory of firm performance, and that it is complementary to externally or market-structure oriented theories of performance (Porter, 1980). The primary issue addressed in the theory is one of identifying the degree to which activities and factors internal to the firm affect its ability to compete over the long run. The theory is applicable to what might generically be called the "competitive firm" or "business unit", which is defined as:

> A cooperative system that constitutes and utilizes a unique and cohesive set of resources and capabilities, both in production and administration, to create and add value to customers in the pursuit of the making of more money now and in the future (Barnard, 1938; Penrose, 1959; Goldratt, 1990).

This definition includes all independent single-unit business organizations and any unique production and administration unit of a larger "corporate organization" that directly adds value in the creation of products and services (Andrews, 1987). It is assumed that the firm in question operates in an environment of competition—where the competition centres on the ability of a firm to meet the needs of customers better than other firms—or to be the preferred supplier of value.

Note that "value" in the context above is defined relative to the firm's ability to meet the needs of customers—customers determine whether, and what, value is actually created when they buy a product or service[2]. This conceptualization of value is not the same as "shareholder value". It is assumed that creating value for customers, through the leveraging of uniquely held and managed resources and capabilities, is the *most effective means* of pursuing the goal of "making more money now and in the future", and hence ultimately providing shareholders with part of the wealth that is generated through the actions of the firm.

Analysis of Firm Activity Requires Analysis of its Parts

The focus on the firm as the level of analysis that is addressed by this theory corresponds to the business strategy level of analysis as discussed in Hofer and Schendel (1978). "At the business level, strategy focuses on "*how to compete in a particular industry or product/market segment*" (p. 28). The emphasis of business unit analysis is two-fold: (1) on the one hand it concentrates on product/market evolution and segmentation issues so that choices can be made concerning what products a firm should offer, when, to whom, and at what price–value mix from a market perspective; (2) on the other hand, it also centres on the internal issues of firm capability, resource mix and resource position. Analysis of these "resource-based" issues speaks to the point that the firm can only offer products and services that it is *capable* of producing at a delivery speed, quality, and price that will satisfy potential customers.

Firm capability is assessed in terms of functional skills and financial and material capacities, which implies that any complete business-level analysis must also examine the underlying "functional-level" attributes of the firm. At the functional level the principal focus is on the acquisition, coordination and utilization of human and material resources, and the development of competences which reflect the firm's ability to blend the appropriate resources in pursuit of specific firm objectives.

The "business strategy" perspective of the firm, briefly reviewed above, is an integrative one in that: (1) it examines in detail *both internal and external* factors that are relevant to the decisions a firm's general managers must make with regards to present and future goals and actions of the firm; (2) it recognizes the integrated nature of the firm as an aggregate or "hierarchy" of functional skills and financial and material capacities.

These points indicate that another appropriate "level of analysis" from which to examine "the firm" includes both the discrete functional operations and activities of the firm and the managerial processes throughout the firm which link these functional skills and capacities to external demands.

Domain of the Theory with Respect to Time

With respect to the element of time, the theory that is developed is causal and recursive in nature and addresses the relationships among processes over the long term (which is assumed to be three to five years at a minimum). It is assumed that the starting conditions, or genesis conditions, of a firm are a critical aspect of understanding the subsequent development path of that firm—and its future productive opportunity path. It is imperative, therefore, that the historical context of a firm be examined fully in order to assess the efficacy of the theory.

Domain of the Theory in Substantive Terms

Figure 11.1 identifies the focus and domain of the theory that will be developed in relationship to Porter's (1980) Five-Forces model of competitive behaviour, and shows how it is complementary to Porter's work. The focus of the theory in this book is on the value delivery process between an individual firm and its customers. The framing question is: given a particular local test to supply value to a set of customers, among competitive firms (those potentially able to deliver value to a common customer) who is more likely to win the contest to supply value and solve the customer's problems, and why?

The area in Fig. 11.1 in dashed lines highlights the domain of the theory in terms of its substantive focus. The focus is on understanding the process of how firms establish and maintain productive opportunity (Penrose, 1959), and on explaining why such productive opportunity is a necessary base for establishing sustainable competitive advantage.

AN OVERVIEW OF THE VALUE CREATION PROCESS: VALUE AT THE FRONTIER OF PRODUCTIVE OPPORTUNITY AND CUSTOMER NEED

An individual firm is a complex human system that exists both to add value to customers and society at large and to appropriate rents for itself

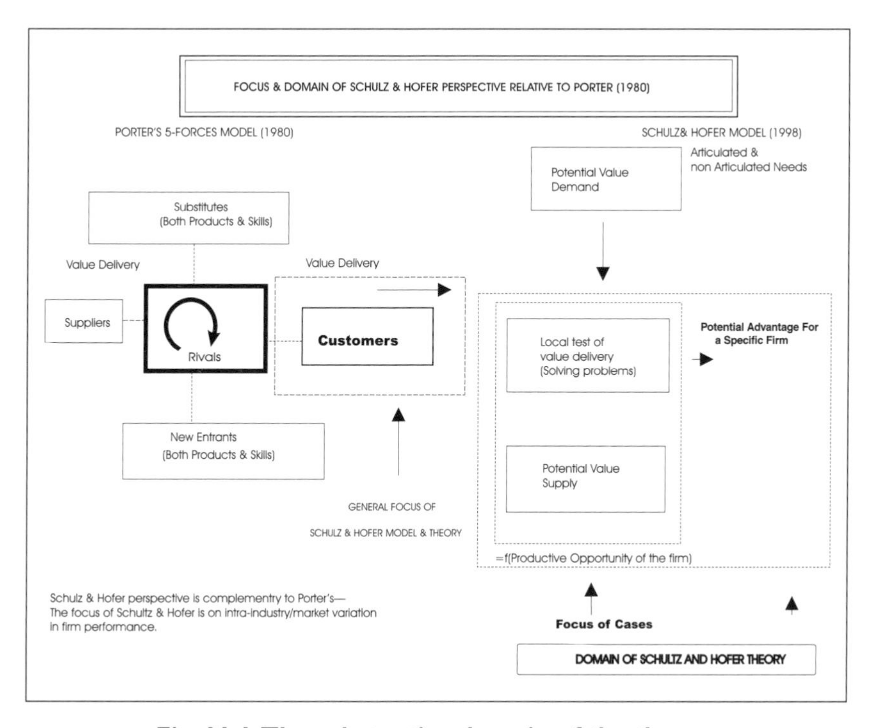

Fig. 11.1. The substantive domain of the theory.

in the process. The creation of value through the assembly and deployment of resources and the exercise of skills is the essence of enterprise in the economy. The direct appropriation, by the firm and its owners, of some of the value that is generated in the process is part of the benefit of engaging in enterprise. A very simple framework that illustrates the basic value creation process is illustrated in Fig. 11.2.

The value creation process is a recursive and ongoing process of change and transformation: given the perception, by entrepreneurial leaders, of extant and latent needs in the environment, action is taken to assemble basic resource stocks of human and material capital. These basic resources are the building blocks for the systems and productive service activities that the firm will generate through various transformation processes, guided by managers. The result of the application of productive services is a set of potential competitive stocks, with which the firm now competes to meet the needs of targeted customers. The overall mission of the organization and the organization's capacity to deal with change throughout the process are functions of leadership.

Entrepreneurial, managerial, and leadership processes are *all critical* to the firm's ability to recognize and establish the strategic links between its specific resources and skills and potentially attractive rent-generating

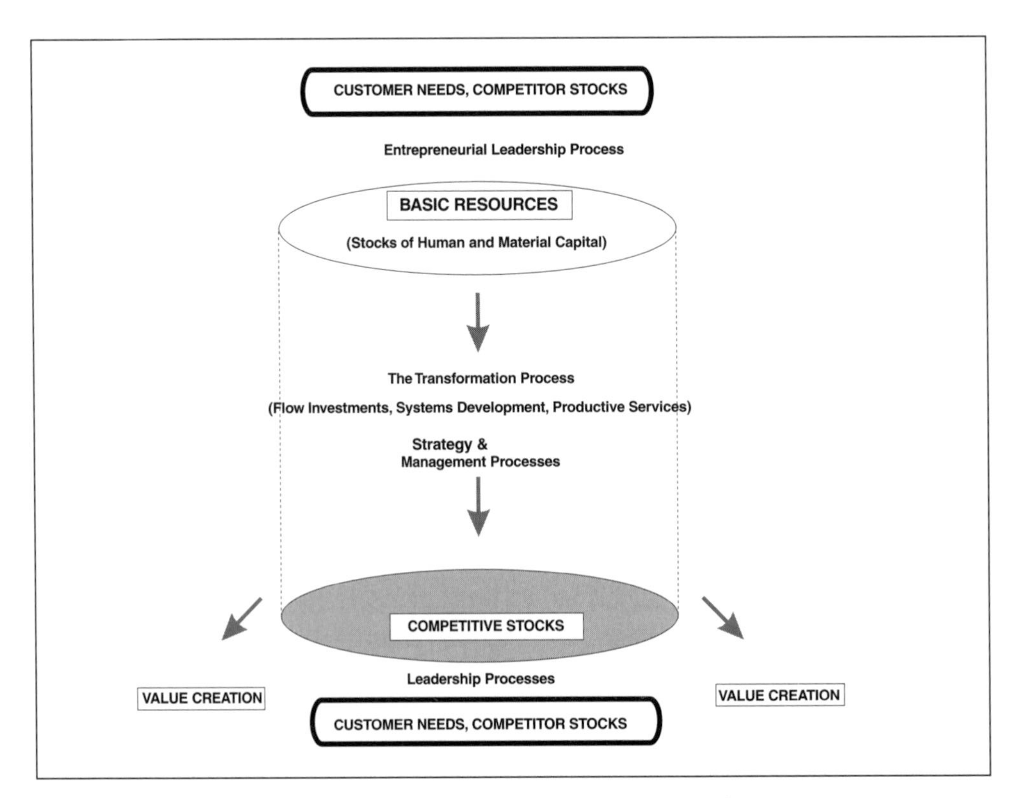

Fig. 11.2. A simple illustration of the value creation process.

opportunities (Schendel and Hofer, 1979; Kotter, 1990), as emphasized in Fig. 11.3 and discussed in the previous chapter. If a firm is to be able to produce value for customers over time, or to establish long-run productive opportunity (LRPO), which is a necessary condition for establishing sustainable competitive advantage, it must be adept at: (1) entrepreneurial processes, which are continuous processes used to identify and interpret opportunity in the environment; (2) managerial processes, which are continuous processes used to transform basic resources into dynamic capability; and (3) leadership processes, which are continuous processes used to integrate the external and internal aspects of the firm, and to provide a foundation for self-identification and meaning for the firm's members.

The Firm as an Engine of Knowledge, Skills and Competence that Creates Value

Viewed from this resource transformation perspective, the firm might be conceptualized as an *engine of knowledge, skills and competence* that is an instrument in producing creative and consumptive value in the market-place through the exercise of entrepreneurial choice (Schendel and Hofer,

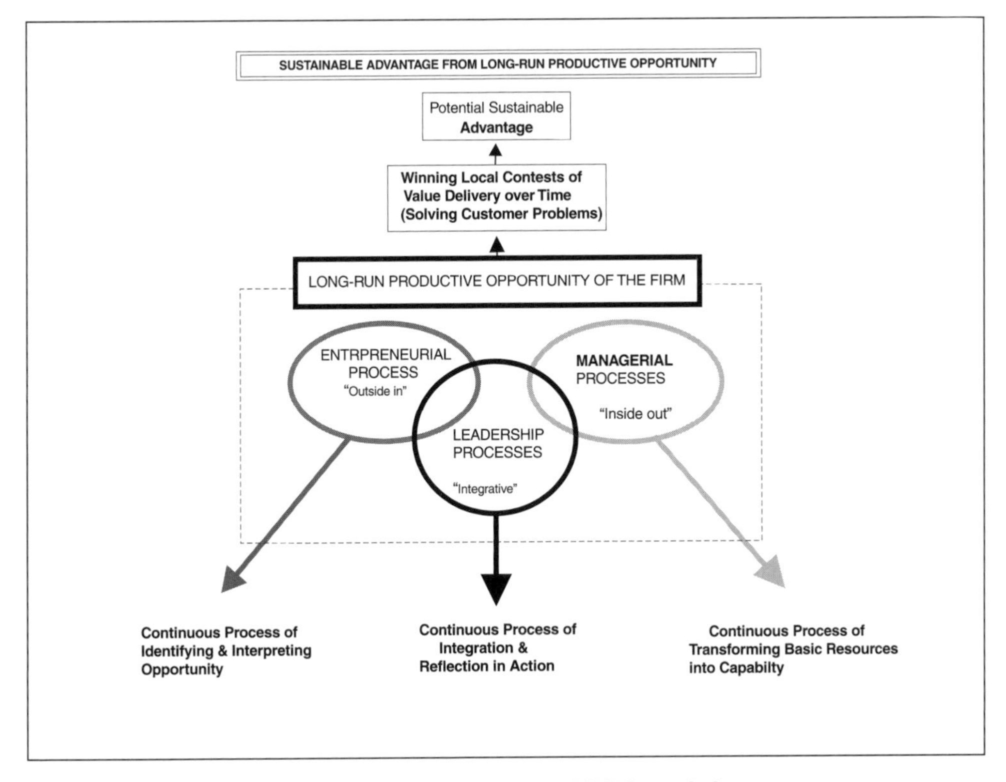

Fig. 11.3. Sustainable advantage from LRPO and three processes.

1979). Entrepreneurial choices are those which reflect the decisions of general managers to seek value-added endowments.

Three aspects of this perspective are important to emphasize:

(1) The role of leaders and general managers in this conceptualization is decidedly entrepreneurial and "line oriented". That is, the people responsible for guiding the firm must be keenly aware of the functional and managerial capabilities of their firms *and* "must consider the fact that new functional skills may provide the spark for the formation of many new businesses, none of which can be clearly defined" (Naugle and Davies, p. 37).

   Functional-skill definition and development and the crafting of an organization and its strategy (Mintzberg, 1987b) are critical aspects of general management—perhaps more important in the long run than external "competitive" environmental analysis (Porter, 1980). This "crafting of organization" involves the processes by which firms create value and advantage in the economy—by exercising their capabilities in those areas best suited to their talents and character (Selznick, 1957).

(2) For the vast majority of businesses, the heart of this "engine" is the creative and trained skills and competences of the technical and managerial people of the firm. All "material-based" resources ultimately serve supporting and complementary roles to the skill-based assets, which transform raw materials into service activities, which can then be used to create potentially valuable stock services and products for clients. The basics of this process argument have been laid out as part of the development of an organizational resources typology.

(3) Competition in the market-place is as much a *contest of skills and capabilities as it is products and services*, and hence skill-based perspectives of strategy are needed (Wernerfelt, 1984; Prahalad and Hamel, 1990; Hamel, 1991). If a firm is to function and sustain its operations, is must depend on and channel both cooperative and contentious behaviour in ways that allow it to continually learn and unlearn and improve its capabilities in both an evolutionary and revolutionary manner (Hedberg, 1981; Goldratt, 1990; Pascale, 1991).

Competitive Advantage: a Potential Reward in the Value Creation Process

An important possibility that arises during a firm's pursuit of value creation, through the exercise of its resources, is that a well-led and managed firm may be able to establish an advantage over competitors that allows it to appropriate both entrepreneurial and efficiency rents over time, which are profit levels above those needed to stay in business (Rumelt, 1987). Most entrepreneurs, leaders and managers seek such rents and advantages.

The interesting question, of course, is how can firms establish such advantages?

The theory that will be developed shortly argues that sustainable competitive advantages, and the rents they help generate, are rooted: (1) in the very fabric of a firm's resources and, *more importantly*, (2) in the abilities of the firm's leadership and administration to manage the process of acquiring, assembling and reassembling, nurturing, improving, and changing the nature of its resource-base, given their perception and interpretation of opportunities to create value for customers.

Resource Deployment and Competitive Advantage

Hofer and Schendel (1978) contend that, at the business level, the "level and patterns of [an] organization's past and present resource and skill deployments" (resource deployments), may be more important than the "extent of the organization's present and planned interactions with its environment (scope—including market/product domain) in determining the "unique position [the firm] develops *vis-à-vis* its competitors" (p. 25). This "unique position" is called competitive advantage. As Tom Peters (1984) cites, *value creation for customers is at the heart of competitive advantage*:

> Winners almost always compete by delivering a product that supplies superior value to customers, rather than one which costs less. (American Business Conference, 1983; Peters, 1984, p. 117)

Michael Porter (1985), among others (Barney, 1991, p. 102), expands on this theme:

> Competitive advantage grows fundamentally out of the value a firm is able to create for its buyers. It may take the form of prices lower than competitors' for equivalent benefits or the provision of unique benefits that more than offset a premium price. (1985, p. xvi)

Competitive advantage for a firm can arise only if customers (buyers) (1) perceive a value distinction among the products and services they are considering, (2) the firm has the ability to know what the customer values (Stoner, 1987), and (3) the firm can do something about it (Ulrich and Lake, 1991). Points 2 and 3 address the firm's productive opportunity, which is a function of the firm's facility in pursuing appropriate ends (formulating strategy and making entrepreneurial choices) and its ability to develop, coordinate and utilize its skills and resources in pursuit of those ends (implementing strategy and building dynamic capability).

Resources and Competitive Advantage: Summary

The perspective of firm dynamics in this book is decidedly resource- and process-oriented. According to this resource-based process strategy approach, a firm can establish sustainable competitive advantage by being competent at utilizing entrepreneurial, managerial and leadership processes.

Through the development of entrepreneurial competence, in which the firm develops an ability to innovate and either find new applications for its current skills, or build new skills to develop new applications, the firm is often able to realize entrepreneurial rents (Rumelt, 1987).

Through the development of dynamic capability, which is a function of managerial competence, the firm is able to exercise skill-based strategies, which hone a firm's ability to assemble and coordinate its resources and create value for its customers, on a day-to-day level, more effectively than its competitors, thus generating potential and *sustainable* efficiency rents over time (Rumelt, 1987).

Through effective leadership, the firm develops an institutional environment (Selznick, 1957) which enhances its ability to continually integrate its dynamic capabilities with its changing opportunity set—making the firm a true "learning organization".

ENTREPRENEURIAL LEADERSHIP AND SKILL-BASED STRATEGY AS KEYS TO COMPETITIVE ADVANTAGE: A THEORY OF VALUE CREATION THROUGH FIRM ACTIVITY

It has been argued to this point that the long-run productive opportunity (LRPO) of the firm, or its ability to effectively deliver value to customers over long periods of time, is determined by the firm's ability to be competent with respect to entrepreneurial, managerial and leadership processes. The LRPO of a firm is critical in determining whether a firm is able to win local contests of value delivery over time—which is a necessary condition for establishing potential sustainable competitive advantage.

In the remainder of this chapter, a more detailed exposition of a theory of value creation, through the establishment of long-run productive opportunity, is developed that explains how and why sustainable competitive advantages are likely to arise from effective entrepreneurial leadership and the utilization of skill-based strategies. Definitions of key constructs are provided, as are integrative diagrams, text narration and propositions, and a summary figure in spreadsheet form.

Primary Resource-Based Determinants of The Long-Run Productive Opportunity of a Firm

Figure 11.4 illustrates the causal chain that ties potential sustainable advantage to the three basic processes which determine the LRPO of a firm. The sustainability of advantage is a function of the number of contests a firm engages in at any time and the firm's LRPO. The firm's LRPO is a function of both deliberate and emergent (or autonomous) strategic processes that overlap with one another (Mintzberg & Waters, 1985; Burgleman and

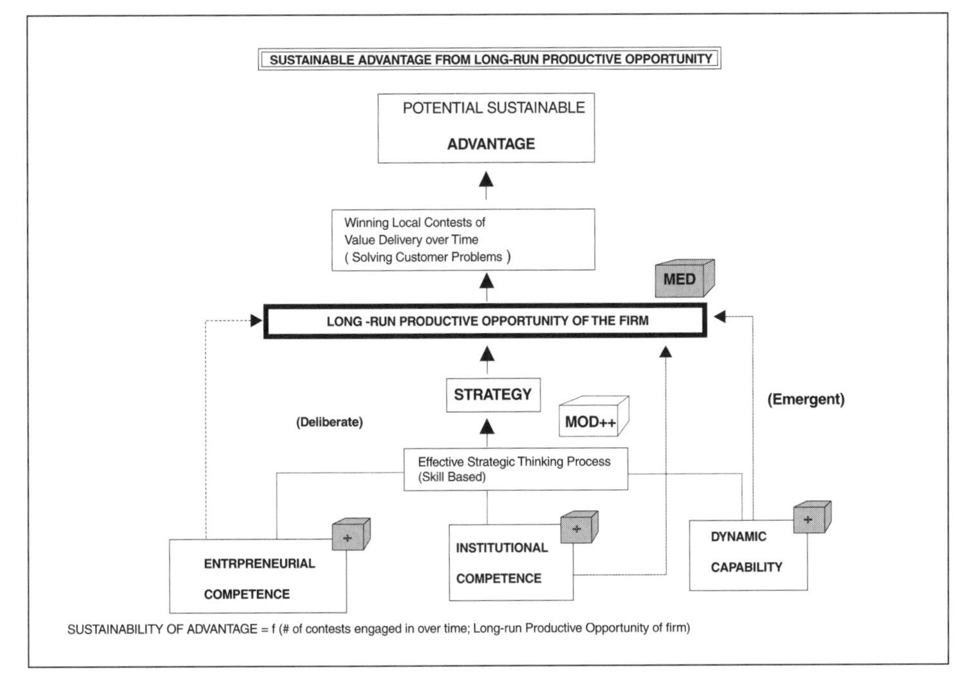

Fig. 11.4. The resource-based determinants of long-run productive opportunity.

Sayles, 1986). The first strategic process is a deliberate strategy process, whereby the strategists of the firm utilize effective strategic thinking (Rumelt, 1979) in an attempt to craft a strategy that integrates its entrepreneurial and institutional competences and its dynamic capabilities. The second is an emergent process, whereby a firm's entrepreneurial and institutional competencies and its dynamic capabilities affect its LRPO in unanticipated and emergent ways (Mintzberg & Waters, 1985). Both strategic processes encompass the three basic organizational processes discussed earlier, which are now defined in terms of discrete constructs:

(1) *Intrepreneurial competence* reflects the ability of the firm and its leadership to absorb relevant information about opportunities in the external environment, to interpret them, and then to mobilize resources in the pursuit of these opportunities. The determinants of entrepreneurial competence will be addressed shortly.
(2) *Institutional competence* reflects a firm's ability to adequately build a unique and viable organizational personality that is not easily imitated nor destroyed; and to establish a credible institution that can add value to both its internal and external constituents in the face of continuous pressures (Selznick, 1957). It is the construct that embodies the integrative leadership processes discussed earlier. The determinants of institutional competence will be addressed shortly.
(3) *Dynamic capability* reflects the firm's ability to acquire, structure, and

transform its basic resources into a set of productive services that can be used in a variety of ways and which can be changed flexibly and quickly. It is the construct that embodies the managerial processes discussed earlier, and which skill-based strategies are intended to develop and exercise. The determinants of dynamic capability will be addressed shortly.

Construct Relationships

Figure 11.4 suggests that LRPO mediates the relationship between the competences and capabilities of the firm and sustainable competitive advantage. Without LRPO there can be no sustainable competitive advantage for a firm. Likewise, without competencies in all three of the areas of entrepreneurship, leadership and management the firm cannot establish long-run productive opportunity.

With respect to strategic processes, it is reasonable to think that deliberate strategies moderate the relationship between the firm's competencies and capabilities, and LRPO, as indicated in the diagram. That is, deliberate strategies that are based upon effective strategic thinking (Rumelt, 1979) should increase the probability that a firm will develop and maintain long-run productive opportunity, relative to more emergent strategies.

Propositions with Respect to LRPO

LRPO-1: If a firm cannot sustain competency in all three of the primary process areas over time, then it will not be able to sustain its long-run productive opportunity frontier and will not be as effective in winning value delivery contests. It will therefore shrink in terms of customers it can serve and resources it can control.

Lrpo 1-A: Firms that grow faster than their capability to maintain competence in all three areas will be more prone to suffer negative business cycle effects than firms that control their growth according to their ability to sustain competencies in all three areas.

LRPO-2: Firms which attempt to integrate their entrepreneurial and institutional competences and their dynamic capability through the crafting of deliberate strategies will achieve a greater long-run productive opportunity set than those that rely on primarily emergent strategy processes.

The Resource-Based Determinants of Entrepreneurial Competence

Entrepreneurial competence reflects the ability of the firm and its leadership to absorb relevant information about opportunities in the external environment, to interpret them and to mobilize resources in pursuit of these opportunities. It is a measure of the firm's ability to manage the "outside-in" process discussed in the previous chapter.

Figure 11.5 identifies the primary elements, and the causal relationships among these elements, that affect a firm's ability to establish entrepreneurial competence. Three basic constructs affect entrepreneurial competence at the firm level: absorptive capacity, interpretation systems and leadership.

(1) *Absorptive capacity* reflects the ability of the firm to absorb and exploit external knowledge: to recognize the value of new information, assimilate it and apply it to commercial ends (Cohen and Levinthal, 1990).[3]

The absorptive capacity of a firm is a function of the conceptual and technical skills and deep knowledge of its leaders and other boundary-spanning personnel, and is also a function of the firm's ability to learn and to establish effective systems—especially with respect to communication.

(1-a) *Deep knowledge* reflects the tacit element of raw information about a topic or process; it reflects relationship information, and creative information that is gained by an individual through a synthesis of aptitude, study, experience and practice that usually takes time to develop.

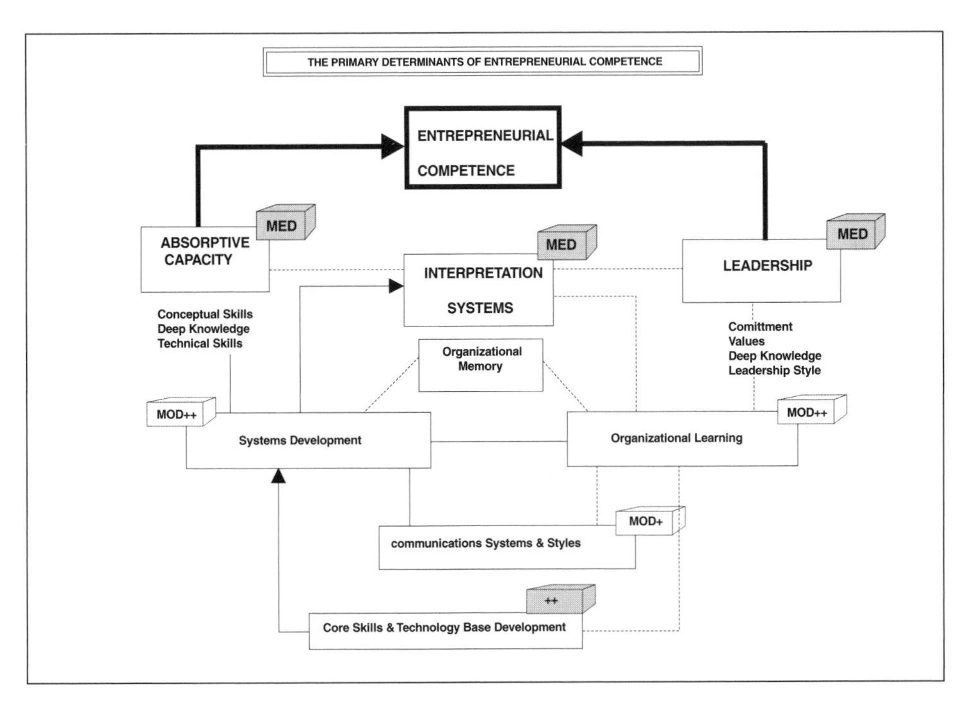

Fig. 11.5. The primary resource-based determinants of entrepreneurial competence.

(2) *Interpretation systems* are those formal and informal systems in the organization that are designed to facilitate "the process of translating events and developing shared understandings and conceptual schemes among members of upper management. Interpretation gives meaning to data, but it occurs before organizational learning and action" (Daft and Weick, 1984, p. 286).

(3) *Leadership* "is about coping with change . . . the function of leadership is to produce change" (Kotter, 1990, p. 18). Leaders challenge the *status quo* and articulate vision.

Leadership in a firm is necessary to establish entrepreneurial competence because it is one thing to recognize a set of opportunities and another to be able to both interpret which particular opportunities the firm should pursue, and also to mobilize the resources of a firm in pursuit of the chosen opportunity. The process that leads to entrepreneurial competence is inherently dynamic and requires vision.

Leadership in an organization is a function of the deep knowledge, spiritual skills, commitment, and leadership or managerial style of the individuals that are responsible for the long-term health of the organization (the entrepreneurs, the strategists etc.). Leadership also depends on the organizational context with respect to organizational memory, learning and systems development.

(3-a) *Spiritual skills* ("to spirit" means to "breathe life into") are a type of gnostic skill (gnostic means knowledge-related), and include such skills as mentoring, creating, advising and synthesizing.

(3-b) *Leadership or managerial style* refers to the approach that general managers or leaders take to conducting their activities with respect to governing their firm. Eastlack and McDonald (1970), for example, identified eight general approaches to leadership and management styles, including growth entrepreneur and R&D planner (which represent entrepreneurial styles), growth director, product manager, aloof strategist, remote controller, acquirer and activist.

Construct Relationships

Absorptive capacity, interpretation systems and leadership are all related to one another. Each of these constructs is a function of the organization's ability to learn and develop systems, which in turn are related to the core skill and technology base of the firm.

The existence of interpretation systems and leadership mediates the relationship between general systems development and organizational learning, and entrepreneurial competence, as does absorptive capacity. This suggests that one cannot have entrepreneurial competence without being able to absorb relevant external data, interpret it in such a way as to provide informational meaning in the organization, and then mobilize resources to act on this information.

The development of organizational systems and organizational learning routines (Nelson and Winter, 1982) strongly moderates the relationship between core skills and technology in the firm, and the ultimate levels of absorptive capacity in the firm, and the effectiveness of interpretation systems and leadership to mobilize resources.

Propositions Related to Entrepreneurial Competence, Absorptive Capacity, Deep Knowledge and Interpretation Systems

Entrepreneurial Competence

ENTCO-1: A firm cannot exhibit entrepreneurial competence without the ability to identify, interpret and choose which opportunities to pursue in the external environment, and also the ability to mobilize its resources with respect to the chosen opportunities.

ENTCO-2: Entrepreneurial competence affects the ability of the firm to renew itself over time through the pursuit of opportunity. If the firm loses the ability to identify, interpret and choose opportunities, it has important implications for the firm's ability to control key resources.

Ent 2-A: The skill-base of firms that lack entrepreneurial competence in their leadership or general management functions will, over time, erode more quickly than firms who have a higher degree of entrepreneurial competence.

Ent 2-B: The technology-base of firms that lack entrepreneurial competence will be more static, over time, than firms who have a higher degree of entrepreneurial competence.

Ent 2-C: Note: both 2-A and 2-B imply that the firm will lose its dynamic capability in time as it loses its entrepreneurial competence. The lose of entrepreneurial competence affects the knowledge–technique base—which affects the potential for dynamic capability (see dynamic capability section).

Absorptive Capacity

ABSCA-1: Careful attention to building absorptive capacity is a necessary condition for a business to grow and survive. Products can be changed; profits flow from having the right focus on the value creation process.

Abcap 1-A: Entrepreneurs that get into the field and "ethnographically" observe and analyze the culture of their customers will obtain greater insights into customer needs and how they can potentially manage the value creation process to meet those needs than those entrepreneurs that concentrate more heavily on product or profit.

ABSCA-2: The absorptive capacity of a firm is limited by its ability to learn and remember. Firms with organic structures and integrated knowledge systems will have greater absorptive capacities than firms that are more mechanistic and have strong control systems.

ABSCA-3: The absorptive capacity of a firm is dependent on the level of deep knowledge that resides in the members of the firm.

Deep Knowledge

DEEP-1: Entrepreneurs and general managers that build businesses in areas that they hold deep knowledge in, particularly in terms of value creation, will be more able to build a sustainable advantage than others.

Deep 1-A: Entrepreneurs or general managers that pursue opportunities without reflection on what deep knowledge, skills and competences they possess will be less effective in building a sustainable organization than those who understand their own skill-base.

Deep 1-B: An entrepreneur's or general manager's propensity to learn, improve and adapt will be greater when they pursue opportunities that allow them to exploit their key interests and skills (which lead to deep knowledge) than if they pursue other opportunities.

Deep 1-C: Intuitive leaps and insights are possible and more probable with the development of deeper knowledge and absorptive capacity. These leaps are the basis for successful Schumpeterian innovation.

DEEP-2: The deeper the knowledge, the more tacit it is. The more tacit the knowledge–technique base is in the firm, the less other firms will be able to imitate a skill-based strategy.

Interpretation Systems

INTERP-1: Organizations are more than transformation processes or control systems. To survive, organizations must have mechanisms to interpret ambiguous events and to provide meaning and direction for participants. Organizations are meaning systems, and this distinguishes them from lower level systems (Daft and Weick, 1984, p. 293)

The Resource-Based Determinants of Institutional Competence

Institutional competence reflects a firm's ability to adequately build a unique and viable organizational personality that is not easily imitated nor destroyed; and to establish a credible institution that can add value to both its internal and external constituents in the face of continuous pressures (Selznick, 1957). It is the construct that embodies the integrative leadership processes discussed earlier. Figure 11.6 illustrates the primary resource-based determinants of institutional competence.

Institutional competence is a basic function of leadership, which affects organizational identification and cohesion as represented by the establishment of mission, the infusion of value in the firm and the integrity of purpose. Leadership also affects, and is affected by, the firm's ability to become a learning organization that has memory. Organizational memory and learning processes are the vehicles by which the vision and values of leadership are instilled in the entire firm, and which shape the institutional nature of the firm (Hedberg & Starbuck, 1981; Walsh and Ungson, 1991).

The ability of a firm to learn and develop an institutional competence is a function of the boundary-spanning skills both the leaders and other personnel in the firm have—can they communicate across knowledge and social barriers within the firm, and can they communicate with the outside world with respect to their institutional mission and shared values?

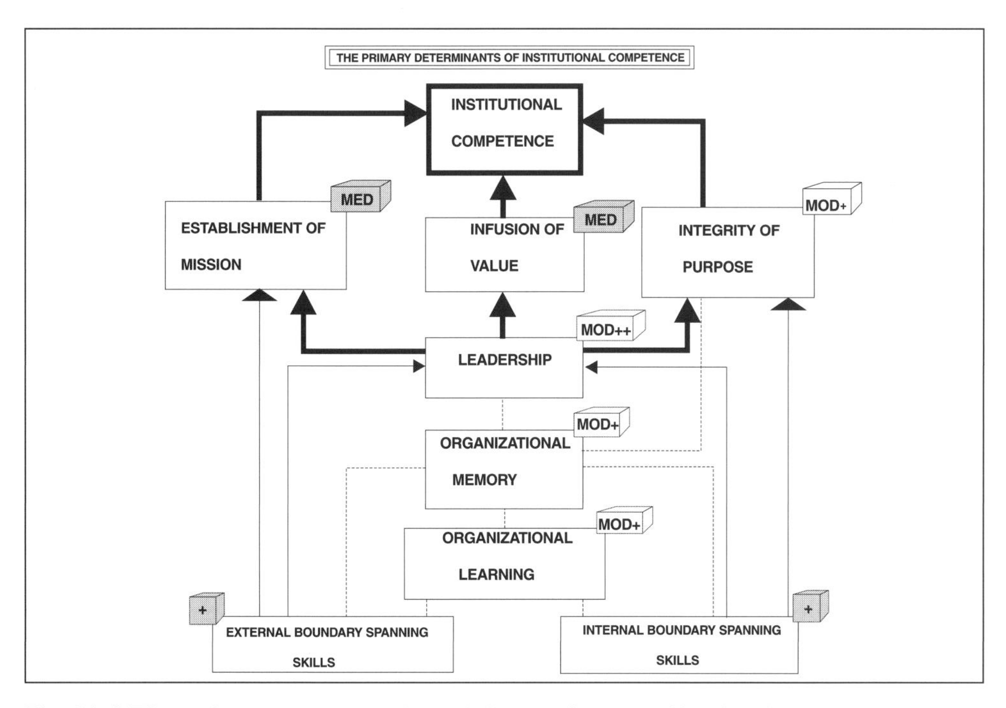

Fig. 11.6. The primary resource-based determinants of institutional competence.

Construct Relationships

Institutional competence is a function of the ability of the firm and its leaders to establish a mission and infuse the organization with shared values. Both the mission and values mediate the relationship between institutional competence and leadership processes. The building of integrity of purpose moderates the same relationship. Leadership in an organization is a function of the deep knowledge, spiritual skills, commitment and leadership style of the individuals that are responsible for the long-term health of the organization (the entrepreneurs, the strategists etc.). Leadership also depends on the organizational context with respect to organizational memory, learning and systems development. Without such organizational mechanisms, the abilities of the leaders to build an institution are limited, as there is no way to establish shared values and build organizational memories and myths (Hedberg, 1981; Walsh and Ungson, 1991). The existence of such systems moderates the relationship between leadership and basic boundary-spanning skills in the organization—which represent the skills of individuals to communicate both across internal boundaries and external boundaries.

Propositions Related to Institutional Competence

INSTU-1: Without the infusion of value and character into an organization, and their direct tie to an institutional embodiment of purpose, the technical functions and activities of a firm are imitable and will not provide sustainable competitive advantage (Lippman and Rumelt, 1982; Peters, 1984; Barney, 1986a; Fiol, 1991; Ulrich and Lake, 1991).

INSTU-2 Human capital growth may be the most distinctive feature of the economic systems of western societies (Schultz, 1990). One of the primary functions of the institutionalization of a firm is to foster human capital development.

Ins 2-A: Organizations that exhibit institutional competence will be able to attract better qualified members to join their core skill areas; and will retain the members of their core skill areas longer than other firms—providing potential learning benefits to the firm.

Ins 2-B: Firms that invest in and develop their human capital base over time will outperform those that primarily invest in physical capital or financial capital.

-1 When human capital is treated as a fixed, rather than a variable, asset, there is more pressure on general managers to be entrepreneurial leaders.

-2 Over time, firms that treat human capital as a fixed asset will outperform those that treat it as variable in both non-financial and financial measures of operations performance.

-3 Firms that treat human capital as fixed are more likely to exploit skill-based strategies that leverage economies of scope than those firms that treat human capital as variable.

INSTU-3: The establishment of an acceptable vision that provides direction for meeting and creating value for both internal and external stakeholders is the highest measure of institutional competence.

The Primary Resource-Based Determinants of Dynamic Capability

Dynamic capability reflects the firm's ability to acquire, structure and transform its basic resources into a set of productive services that can be used in a variety of ways and which can be changed flexibly and quickly. It is the construct that embodies the managerial processes discussed earlier, and which skill-based strategies are intended to develop and exercise.

Figure 11.7 identifies the primary elements and causal relationships among these elements that affect a firm's ability to establish dynamic capability. Dynamic capability is a function of a firm's ability to: (1) *develop* new organizational systems and organic structures that support organizational learning (as represented by the architectural competence construct); (2) to *maintain* and improve on these systems and processes

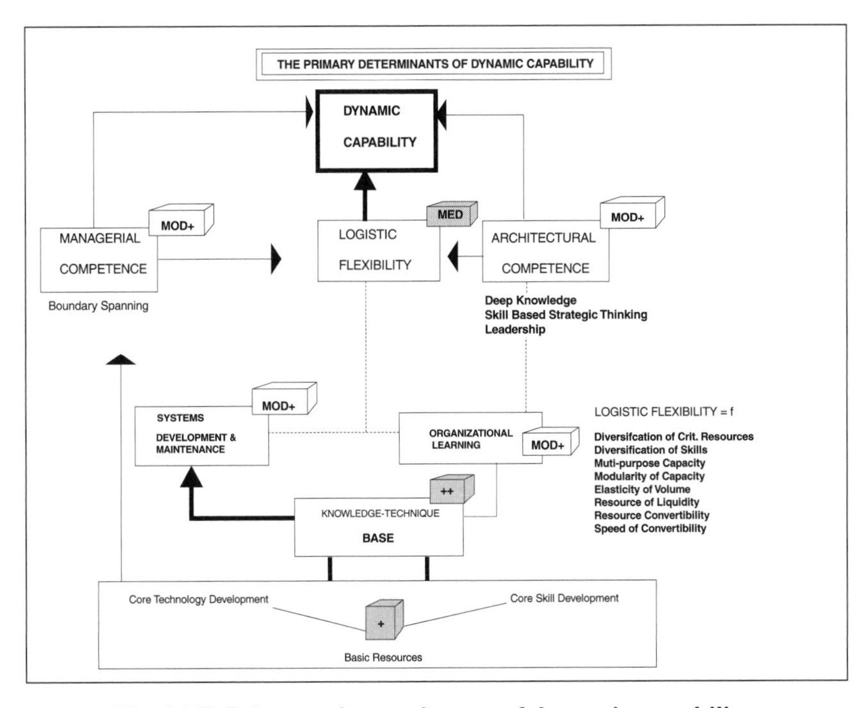

Fig. 11.7. Primary determinants of dynamic capability.

through boundary spanning and coordination of resources (as represented by the managerial competence construct); and (3) to utilize its basic and systemic resources in as *flexible a configuration* as possible (as represented by the logistic flexibility construct).

(1) *Architectural competence* reflects the ability of general managers and organizational leaders to design an organization, in terms of systems and structure, that builds upon the firm's core skills and provides an effective context for value creation.
(2) *Managerial competence* reflects the ability of managers to maintain effective organizational systems and to blend and coordinate the transformation of the organization's basic and systemic resources into valuable productive services.
(3) *Logistic flexibility* is a measure of the firm's capacity to change its resource configuration to meet changing needs (Ansoff, 1978). It is a function of both architectural and managerial competence.

Logistic flexibility is a function of the diversification of skills in the firm, the diversification of critical resources in the firm, the modularity and multipurpose of the firm's capacity, the liquidity and convertibility of the firm's resources and the elasticity of the firm's volume. Each of these elements is in some way related to the architectural competence of the firm's leadership, which must be cognizant of what resources to acquire, develop and grow, as well as be capable of building an organizational structure that can learn to adapt and change its resource configurations to meet the changing needs of its customers.

Construct Relationships

Dynamic capability is primarily a function of the firm's ability to achieve logistic flexibility, which mediates the relationship between dynamic capability and systems development, maintenance and organizational learning. Managerial and architectural competencies moderate the relationship between logistic flexibility and dynamic capability.

Organizational learning strongly moderates the relationship between the knowledge–technique base of the firm and logistic flexibility. Organizational learning is a vital component of resource management because it represents an important feedback (control) and feed-forward mechanism, which links implementation-oriented processes of skill development and utilization, with the more formulation-oriented processes of environmental scanning and entrepreneurial choice (Walsh and Ungson, 1991). Hedberg & Starbuck(1981) refers to organizational learning as the link between individual skills and organizational capability.

Propositions Related to Dynamic Capability, Logistic Flexibility, Managerial and Architectural Competence

Dynamic Capability

DYCAP-1: The process of establishing dynamic capability is inherently resource-oriented. Skill-based strategies are required to ensure that a firm has the dynamic capabilities it requires to create value for a target customer segment.

DYCAP-2: Firms with dynamic capability will be able to serve more product market segments, with a higher order of quality, speed and flexibility, than firms without it.

DYCAP-3: Firms with dynamic capability will be able to exploit economies of scope more readily than other firms, and may also be able to achieve scale economies in certain core functional skill areas. Such a combination of scope and scale economies should lead to both entrepreneurial and efficiency rents.

DYCAP-4: The establishment of dynamic capability is an ongoing process that is closely tied to the organizational systems and learning processes of the firm—as such, it is not easily imitable by competitors.

Logistic Flexibility

LOGFLX-1:Logistic flexibility is the primary determinant of dynamic capability. To achieve logistic flexibility, the firm must have competent organizational architects to develop organizational systems, and competent managers to maintain them.

Architectural Competence

ARCH-1: Organizational systems, and eventually structure, develop through the routinization of the skills of the earliest entrepreneurial, managerial and technical personnel. The development of such systems and structure is a function of architectural competence.

ARCH-2: The deeper the knowledge-base of the general managers or leaders of firm, with respect to the value creation process, the more likely it will be that they will competently structure their firms.

Summary Diagram and Spreadsheet of the Theory: Value Creation from LRPO and the Determinants of LRPO

Figure 11.8 provides an integrative summary of the previous arguments. On the left side of the diagram one can see the “outside-in” entrepreneurial

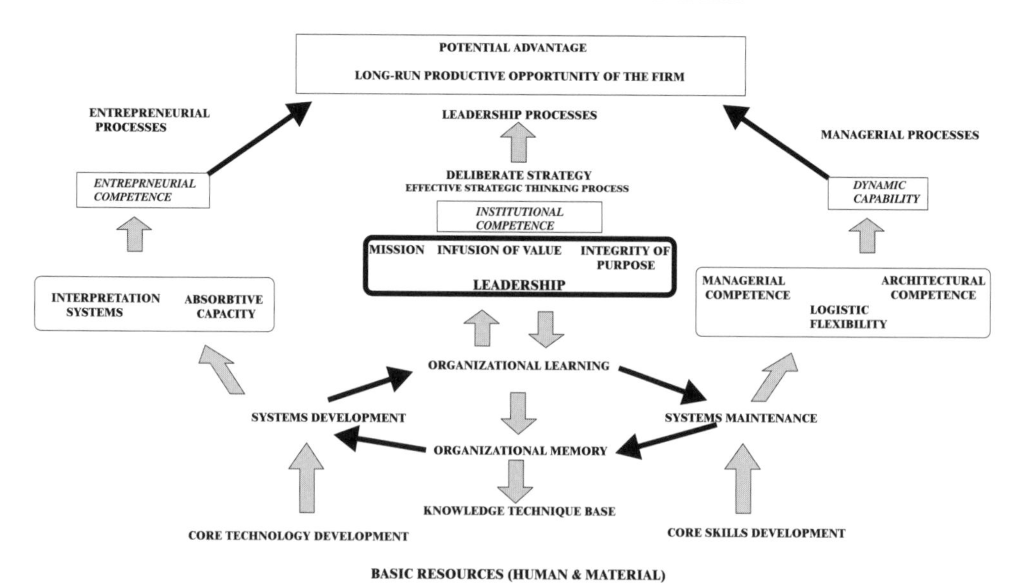

Fig. 11.8. Summary diagram of value creation and LRPO.

process of opportunity discovery, while on the right side one can see the "inside-out" process of resource generation and management. In the centre of the diagram is the "integrative" leadership and institutionalization process.

Note that the development and maintenance of organizational systems, learning routines and organizational memory are common to all three major processes, and are the bridge between the organization's resources and its potential for action.

Note too, that at the bottom of the diagram, underlying all the processes, is the basic resource foundation of the firm, and the cumulative knowledge–technique base (Lenz, 1980), which is affected by organizational learning, and which provides the ultimate limits of action of the firm.

Figure 11.9 is a hierarchical representation of the theory in spreadsheet form. The initial constructs are exogenous to the theory. For each of the constructs that are endogenous to the theory, a list of the major determinants is identified. The spreadsheet generally moves from a dependent to independent construct direction (from beginning to end).

CONCEPTUAL REVIEW: ENTREPRENEURIAL LEADERSHIP AND SKILL-BASED STRATEGY AS INTEGRAL PARTS OF THE VALUE CREATION PROCESS

It has been argued in the previous two chapters that a firm can establish sustainable competitive advantage through the mastery of three basic

PRIMARY VARIABLE NAME		SECONDARY VARIABLE NAME		EXOGE ENDOG	TYPE	L1	L2
BASIC MATERIAL FACTOR MKTS	= f(	DEMAND		EXOG		E	
BASIC MATERIAL FACTOR MKTS		SUPPLY	)	EXOG		E	
COMPETITORS	= f(	DEMAND FOR LIKE CAPABILITIES		EXOG		E	
COMPETITORS		DEMAND FOR LIKE RESOURCES		EXOG		E	
COMPETITORS		SUBSTITUABILITY OF PROD-SERV	)	EXOG		E	
CUSTOMER NEEDS & VALUES	= f(	SOCIETAL CONDITIONS		EXOG		E	
CUSTOMER NEEDS & VALUES		TASTE, ETC.	)	EXOG		E	
EXTERNAL TECHNOLOGY FACTOR M	= f(	DEMAND		EXOG		E	
EXTERNAL TECHNOLOGY FACTOR M		SUPPLY	)	EXOG		E	
INITIAL DEEP KNOWLEDGE	= f(	APTITUDE		EXOG		E	
INITIAL DEEP KNOWLEDGE		GNOSTIC SKILLS		EXOG		E	
INITIAL DEEP KNOWLEDGE		PASSION		EXOG		E	
INITIAL DEEP KNOWLEDGE		PRACTIC SKILLS		EXOG		E	
INITIAL DEEP KNOWLEDGE		PRIOR EXPERIENCE	)	EXOG		E	
MARKET DYNAMISM	= f(	COMPETITORS		EXOG		E	
MARKET DYNAMISM		REGULATION, ETC.		EXOG		E	
MARKET DYNAMISM		SUPPLIERS	)	EXOG		E	
SKILLED FACTOR MARKETS	= f(	DEMAND BY COMPETITORS		EXOG		E	
SKILLED FACTOR MARKETS		EDUCATION SYSTEM SUPPLY		EXOG		E	
SKILLED FACTOR MARKETS		POPULATION GROWTH	)	EXOG		E	
FIRM GROWTH	= f(	ABILITY TO SUSTAIN ADVANTAGE		ENDOG	DV	1	
FIRM GROWTH		COMPETITITIVE ADVANTAGE	)	ENDOG	DV	1	
SUSTAIN COMPETITIVE ADVANTAGE	= f(	PRODUCTIVE OPPOR-LONG-TERM	)	ENDOG	DV	1	
PRODUCTIVE OPPOR-LONG TERM	= f(	DYNAMIC CAPABILITY		ENDOG	MEDV	2	
PRODUCTIVE OPPOR-LONG TERM		ENTREPRENEURIAL COMPETENCE		ENDOG	MEDV	2	
PRODUCTIVE OPPOR-LONG TERM		INSTITUTIONAL COMPETENCE	)	ENDOG	MEDV	2	
COMPETITITVE ADVANTAGE	= f(	BEST LOCAL SUPPLY OF VALUE	)	ENDOG	DV	3	1
VALUE CREATED	= f(	POTENTIAL VALUE DEMAND		ENDOG	DV	3	2
VALUE CREATED		PRODUCTIVE OPPTY--SHORT-TER	)	ENDOG	DV	3	2
PRODUCTIVE OPPTY-SHORT-TERM	= f(	MANAGERIAL COMPETENCE		ENDOG	MODV	3	3
PRODUCTIVE OPPTY-SHORT-TERM		STOCK RESOURCES	)	ENDOG	MODV	3	3
SKILL-BASED STRATEGY	= f(	ENTREPRENEURIAL CHOICES		ENDOG	MODV	3.5	4
SKILL-BASED STRATEGY		RESOURCE INVESTMENT PATTERN		ENDOG	MODV	3.5	4
SKILL-BASED STRATEGY		SKILL-BASED STRATEGIC THINKIN	)	ENDOG	MODV	3.5	4
ENTREPRENEURIAL CHOICES	= f(	PERCIEVED OPPORTUNITY		ENDOG	MODV	4	1
ENTREPRENEURIAL CHOICES		SKILL-BASED STRATEGIC THINKIN	)	ENDOG	MODV	4	1
ENTREPRENEURIAL COMPETENCE	= f(	ABSORPTIVE CAPACITY		ENDOG	MEDV	4	2
ENTREPRENEURIAL COMPETENCE		INTERPRETATION SYSTEMS		ENDOG	MEDV	4	2
ENTREPRENEURIAL COMPETENCE		LEADERSHIP	)	ENDOG	MEDV	4	2
INSTITUTIONAL COMPETENCE	= f(	BOUNDARY SPANNING		ENDOG	MEDV	4	3
INSTITUTIONAL COMPETENCE		ESTABLISHMENT OF MISSION		ENDOG	MEDV	4	3
INSTITUTIONAL COMPETENCE		INFUSION OF VALUE		ENDOG	MEDV	4	3
INSTITUTIONAL COMPETENCE		INTEGRITY OF PURPOSE	)	ENDOG	MEDV	4	3
DYNAMIC CAPABILITY	= f(	ARCHITECTURAL COMPETENCE		ENDOG	MEDV	4	4
DYNAMIC CAPABILITY		KNOWLEDGE-TECHN BASE		ENDOG	MEDV	4	4
DYNAMIC CAPABILITY		LOGISTIC FLEXIBILITY		ENDOG	MEDV	4	4
DYNAMIC CAPABILITY		MANAGERIAL COMPETENCE		ENDOG	MEDV	4	4
DYNAMIC CAPABILITY		ORGANIZATIONAL LEARNING	)	ENDOG	MEDV	4	4
LOGISITIC FLEXIBILITY	= f(	SYSTEMS MAINTENANCE		ENDOG	MODV	4	5
LOGISTIC FLEXIBILITY		CRITICAL RESOURCE DIVERS		ENDOG	MODV	4	5
LOGISTIC FLEXIBILITY		ELASTICITY OF VOLUME		ENDOG	MODV	4	5
LOGISTIC FLEXIBILITY		MODULARITY OF CAPACITY		ENDOG	MODV	4	5
LOGISTIC FLEXIBILITY		MULTI-PURPOSE CAPACITY		ENDOG	MODV	4	5
LOGISTIC FLEXIBILITY		RESOURCE CONVERTIBILITY		ENDOG	MODV	4	5
LOGISTIC FLEXIBILITY		RESOURCE LIQUIDITY		ENDOG	MODV	4	5
LOGISTIC FLEXIBILITY		SKILL DIVERSIFICATION		ENDOG	MODV	4	5
LOGISTIC FLEXIBILITY		SPEED OF CONVERTIBILITY	)	ENDOG	MODV	4	5
LEADERSHIP	= f(	COMITTMENT		ENDOG	IV	5	2
LEADERSHIP		DEEP KNOWLEDGE		ENDOG	IV	5	2

Fig. 11.9. Hierarchical spreadsheet format of the theory.

PRIMARY VARIABLE NAME		SECONDARY VARIABLE NAME		EXOGE ENDOG	TYPE	L1	L2
LEADERSHIP		LEADERSHIP STYLE		ENDOG	IV	5	2
LEADERSHIP		SPIRITUAL SKILLS	)	ENDOG	IV	5	2
TRANSILIENCE	= f(	ABSORPTIVE CAPACITY		ENDOG	MODV	5	2
TRANSILIENCE		DEEP KNOWLEDGE		ENDOG	MODV	5	2
TRANSILIENCE		LEARNING SYSTEMS		ENDOG	MODV	5	2
TRANSILIENCE		LOGISITIC FLEXIBILITY	)	ENDOG	MODV	5	2
SKILL-BASED STRATEGIC THINKING	= f(	ABSORBTIVE CAPACITY		ENDOG	IV	5	3
SKILL-BASED STRATEGIC THINKING		COMITTMENT		ENDOG	IV	5	3
SKILL-BASED STRATEGIC THINKING		DEEP KNOWLEDGE		ENDOG	IV	5	3
SKILL-BASED STRATEGIC THINKING		LEARNING		ENDOG	IV	5	3
SKILL-BASED STRATEGIC THINKING		TRANSILIENCE	)	ENDOG	IV	5	3
MANAGERIAL COMPETENCE	= f(	BOUNDARY SPANNING		ENDOG	MODV	5	4
MANAGERIAL COMPETENCE		COMMUNICATIONS		ENDOG	MODV	5	4
MANAGERIAL COMPETENCE		CORE SKILL ID & BUILDING		ENDOG	MODV	5	4
MANAGERIAL COMPETENCE		SYSTEMS DEVEL & MAINT	)	ENDOG	MODV	5	4
ARCHITECHTURAL COMPETENCE	= f(	DEEP KNOWLEDGE		ENDOG	MODV	5	5
ARCHITECHTURAL COMPETENCE		SKILL-BASED STRATEGIC THINKING		ENDOG	MODV	5	5
ARCHITECHTURAL COMPETENCE		LOGISTIC FLEXIBILITY		ENDOG	MODV	5	5
ARCHITECHTURAL COMPETENCE		VALUES	)	ENDOG	MODV	5	5
ORGANIZATIONAL LEARNING	= f(	LEADERSHIP STYLE		ENDOG	MEDV	5	6
ORGANIZATIONAL LEARNING		ORGANIZATIONAL MEMORY		ENDOG	MEDV	5	6
ORGANIZATIONAL LEARNING		STRUCTURE		ENDOG	MEDV	5	6
ORGANIZATIONAL LEARNING		SYSTEMS	)	ENDOG	MEDV	5	6
ABSORPTIVE CAPACITY	= f(	CONCEPTUAL SKILLS		ENDOG	MODV	5	7
ABSORPTIVE CAPACITY		DEEP KNOWLEDGE		ENDOG	MODV	5	7
ABSORPTIVE CAPACITY		INTERPRETATION SYSTEMS		ENDOG	MODV	5	7
ABSORPTIVE CAPACITY		LEARNING SYSTEMS		ENDOG	MODV	5	7
ABSORPTIVE CAPACITY		TECHNICAL SKILLS	)	ENDOG	MODV	5	7
STRUCTURE	= f(	MARKET INTERFACES		ENDOG	MODV	5	8
STRUCTURE		RESOURCES		ENDOG	MODV	5	8
STRUCTURE		STRATEGY		ENDOG	MODV	5	8
STRUCTURE		STYLE		ENDOG	MODV	5	8
STRUCTURE		SYSTEMS DEVELOPMENT		ENDOG	MODV	5	8
STRUCTURE		TECHNOLOGY	)	ENDOG	MODV	5	8
LEADERSHIP STYLE		EXPERIENCE		ENDOG	IV	5	9
LEADERSHIP STYLE		LEARNING		ENDOG	IV	5	9
SYSTEMS DEVELOPMENT	= f(	ARCHITECTURAL COMPETENCE		ENDOG	MODV	6	1
SYSTEMS DEVELOPMENT		COMMUNICATION		ENDOG	MODV	6	1
SYSTEMS DEVELOPMENT		CORE SKILL DEVELOPMENT		ENDOG	MODV	6	1
SYSTEMS DEVELOPMENT		SKILL-BASE RESOURCES	)	ENDOG	MODV	6	1
KNOWLEDGE-TECHNIQUE BASE	= f(	CORE SKILL DEVELOPMENT		ENDOG	MEDV	6	2
KNOWLEDGE-TECHNIQUE BASE		TECHNOLOGY DEVELOPMENT	)	ENDOG	MEDV	6	2
TECHNOLOGY DEVELOPMENT	= f(	ACQUISITION		ENDOG	IV	6	3
TECHNOLOGY DEVELOPMENT		COMITTMENT		ENDOG	IV	6	3
TECHNOLOGY DEVELOPMENT		KNOWLEDGE-BASE		ENDOG	IV	6	3
TECHNOLOGY DEVELOPMENT		SKILL-BASE		ENDOG	IV	6	3
TECHNOLOGY DEVELOPMENT		STRATEGY		ENDOG	IV	6	3
TECHNOLOGY DEVELOPMENT		SYSTEMS DEVELOPMENT	)	ENDOG	IV	6	3
CORE SKILL DEVELOPMENT	= f(	ACQUISITION		ENDOG	IV	6	4
CORE SKILL DEVELOPMENT		COMITTMENT		ENDOG	IV	6	4
CORE SKILL DEVELOPMENT		KNOWLEDGE BASE		ENDOG	IV	6	4
CORE SKILL DEVELOPMENT		LEARNING PROCESSES		ENDOG	IV	6	4
CORE SKILL DEVELOPMENT		SKILL-BASE		ENDOG	IV	6	4
CORE SKILL DEVELOPMENT		STRATEGY	)	ENDOG	IV	6	4
FIRM RESOURCES		STOCK RESOURCES		ENDOG	IV	7	1
FIRM RESOURCES		SYSTEMIC RESOURCES		ENDOG	IV	7	1
FIRM RESOURCES		SERIVCE RESOURCES		ENDOG	IV	7	1
FIRM RESOURCES		BASIC RESOURCES		ENDOG	IV	7	1

Fig. 11.9. Continued

processes, which must work as an integrated system: (1) through the exercise of entrepreneurial competence, which reflects a firm's ability to innovate and either find new applications for its current skills or build new skills to develop new applications—in which case the firm is often able to realize entrepreneurial rents, and then possibly efficiency rents (Rumelt, 1987); (2) through the exercise of dynamic capability, which reflects a firm's ability to acquire, structure and transform its basic resources into a set of productive services that can be used in a variety of ways and which can be changed flexibly and quickly; and (3) through the exercise of institutional competence, which reflects the ability of the firm to adequately build a unique and viable organizational personality that is not easily imitated nor destroyed; and to establish a credible institution that can add value to both its internal and external constituents in the face of continuous pressures.

Entrepreneurial Leadership

Leadership is a fundamental requirement for the development of both entrepreneurial and institutional competence and is at the heart of the process that integrates all three processes. The primary task of the entrepreneurial leader is to ensure that the firm can create new value through the deployment and exercise of resources and skills at their potential employ. This requires that leaders must not only be competent at recognizing opportunities, but that they must also have a deep knowledge of the resources, skills and capabilities of their organization, and of the process of transforming those resources into services valued by the customer. This conception of entrepreneurial leadership is about the proactive management of change and opportunity, especially with respect to the creation of value through the development of an organizational ethos that encourages growth rooted in ever-improving capability. It is an "outside-in" process of moving outside opportunity into the firm's skill-based domain.

Skill-Based Strategy

Skill-based strategies are those designed to translate organizational skills into sustainable competitive advantages via the establishment of dynamic capability. Skill-based strategies focus on the continuous process of creating and adding value in the market-place through the making of entrepreneurial choices which exercise relevant and critical organizational skills, competences and capabilities. As such, skill-based strategies are inherently dynamic and require an approach to leadership and decision-making that is inherently entrepreneurial. The primary challenge to leadership, from a skill-based strategy perspective, is to find new applications for current skills and to build new skills for new applications—both of which involve creating new value and potentially new organizational forms.

What becomes most important in the skill-based strategic thinking process is knowing, as precisely as possible, what capabilities a firm has — and hence what set of opportunities it might feasibly pursue. The essence of skill-based strategic thinking, and what makes it different from opportunity-based strategic thinking (Porter, 1980), is that it focuses on developing a general proficiency in the strategist's ability to understand her firm in terms of generalizable capabilities rather than product markets. That is, strategy is less about product positioning in specific markets than it is about developing superior skills, material resources and capabilities that can be deployed in a number of market settings.

Institutionalization and Sustainability of Advantage

One of the more important issues with respect to the sustainability of an advantage that is generated through skill-based strategic thinking and entrepreneurial leadership revolves around the question of institutionalization. If a firm is going to be able to sustain advantage in the markets it operates in, it must have leaders who are able to protect their vital skill-based assets from depletion, obsolescence and imitation (Rumelt, 1987; Barney, 1986b, Fossum *et al.* 1986). One major element of this "protection of assets" comes in the form of creating and maintaining a viable, cooperative organizational ethos, personality or character, shared culture or institution (Barnard, 1938; Selznick, 1957; Pascale and Athos, 1981; Peters and Waterman, 1982; Ansoff and Baker, 1986; Barney, 1986b). Selznick (1957) comments:

> Institutionalization is a process. It is something that happens to an organization over time, reflecting the organization's own distinctive history, the people who have been in it, the groups it embodies and the vested interest they have created, and the way it has adapted to environments . . . in what is perhaps its most significant meaning, "to institutionalize" is to infuse with value beyond the technical requirements of the task at hand. (pp. 16–17)

This process is not easily imitable (Rumelt and Lippman, 1982; Barney, 1986b, 1991) and naturally leads one to focus on the question of what core a firm might "institutionalize around"? What role does general management play in shaping and changing the institution? Selznick himself suggests a test that might answer the question, "when can one tell if an organization or an institution exists?":

> The test of infusion with value is expendability. If an organization is merely an instrument, it will be readily altered or cast aside when a more efficient tool becomes available. (p. 18)

One of the implications of this book is that firms that institutionalize around both the concepts of "delivering value to the customer" *and* "taking care of the human resources of the firm" will in all likelihood outlive others that fail to do so.

It is a vital task of the leadership of a firm to recognize the importance of achieving institutional competence in order to avoid expendability. A company must possess skills and configure itself so that the customer chooses to stay with the firm both because it offers superior value and because it distinctly appeals to the values of its employees through its "integrity" and "personality", which are born of the process of building institutional competence.

Some Assumptions of the Theory

The skill-based perspective developed in this chapter is grounded in the evolutionary *neo-Schumpeterian* perspectives of the firm and its development. The skill-based model shares many of the basic assumptions of the *process oriented resource-based* theories of the firm with regards to the processes related to the generation of efficiency and entrepreneurial rents, and shares the resource-based view that the firm is a "dynamic creator of a positive", rather than the avoider of a negative as treated in transaction cost theories (Conner, 1991). As such, the skill-based perspective recognizes the vital aspect of the *entrepreneurial nature of firm development*, change and innovation; and acknowledges that firms and their decision-makers are boundedly rational and conflict-laden, and that they satisfice (*bounded-rationality theories*), interpret (*interpretavist theories*), learn and improve (*theory of constraints*) according to many of the principles identified by the managerial, behavioural and evolutionary theories of the firm.

Limitations of the Research and Theory

There are always limitations to the design and execution of a research project. The primary limitations of this book's research design and the subsequent theory and conclusions include:

(1) Reliability is threatened because the researcher is the primary research tool and has potential biases and weaknesses that may not be made explicit. Appendix Two contains a detailed discussion of the theoretical perspectives the primary researcher had going into the field work and examines potential sources of bias, and also shows how the research design attempts to control for such potential bias. In particular, the

primary researcher made explicit attempts to design a case-protocol that would reduce the potential negative effects of the researcher as instrument (Yin, 1989).

(2) Internal validity may be threatened by inaccurate or distorted information from the field which the researcher cannot detect; and by the fact that this is an uncontrolled design where many rival explanations may exist. The use of multiple data sources and analysis techniques was employed whenever possible to alleviate this potential limitation. The use of ethnographic interview techniques (Spradley, 1979), within the context of open and semi-structured interviews were employed to reduce the effects of potential researcher bias in collecting the practitioner-extracted theory.

(3) There are two issues that concern the external validity of the findings. The first is that one must recognize that the purposive sampling techniques that were used and the small sample size imply that the findings are not necessarily generalizable across other firms. One must be cautious in making generalizations about firms that do not meet the basic sampling criteria that were specified. The results of the analysis in this project are not statistically generalizable to a specific population or sample. Generalization has been made at the theoretical level through the use of theoretical sampling and constant comparative analysis techniques (Glaser and Strauss, 1967).

The second is that the interpretation of results and claims of discovery is more problematic with this project than some others because the standards for evidence and verification in this design are less stringent than for hypothesis-testing research designs. This exploratory research project has been designed to take advantage of cross-case analysis and an interdisciplinary perspective in an attempt to provide an adequate context for theoretical generalization.

(4) There is a sampling bias towards firms that have survived, so the attributes of failure have been under-represented. However, three of the four cases have a distinct turnaround aspect to them, so some failure analysis has been possible.

(5) The proof of predictive validity *per se* is low, as the development of theory is based on both contemporary and retrospective data, and focuses on genesis rather than prediction.

(6) The inductive theory specification process that has been used in Phase Two threatens the principle of theoretical parsimony, in that the data-intensive nature of the process may create a situation in which the analyst has been unable to assess which are the most important relationships and which are idiosyncratic to a particular case (Eisenhardt, 1989, p. 547). The careful use of a modified constant comparative analysis method, along with an attention to theoretical sensitivity (Glaser, 1978), was employed to minimize these risks.

NOTES

1. The use of the constant comparative analysis methods, within the context of *grounded theory-building*, is an example of how the research has been designed to minimize the effects of pre-project theoretical biases.
2. This concept of value parallels the concept of *throughput* as it is used in the Theory of Constraints (Goldratt, 1990, 1991).
3. Note: *Absorptive capability* is a similar construct that refers to an individual's ability to absorb and exploit external knowledge etc.

PART FOUR:

IMPLICATIONS FOR PRACTICE AND THEORY

12

OVERVIEW OF FINDINGS AND IMPLICATIONS OF THE RESEARCH

> If our findings of the relative importance can be generalized, it would suggest that the critical issue in firm success and development is not primarily the selection of growth industries or product niches, but it is the building of an effective, directed, human organization in the selected industries.
>
> Hansen and Wernerfelt (1989, p. 409)

INTRODUCTION

The work of Hansen and Wernerfelt (1989) suggests that in order to understand the more powerful attributes of firm success and failure, one must explore and understand the *strategic and evolutionary process* by which organizational leaders and managers build their firms' skills and capabilities, and examine how they integrate these skills and capabilities into a coherent organizational whole. The grounded theory-building approach taken in this book has provided a rigorous base from which to explore the evolutionary processes of skill-based strategy development which are part of the process of building effective, directed, human organizations that create value in the economy. In this chapter the overall findings of the book are discussed, and the implications of this research for the academic fields of entrepreneurship and strategic management are presented, as are the implications for effective management practice.

OVERVIEW OF FINDINGS

Some of the more general findings of the book, which have implications for both academic research and management practice, are that:

(1) *Leadership processes and style—especially the entrepreneurial leadership process—are critical aspects of creating value with skill-based strategies.* The leadership of an organization must take the responsibility for actively finding new applications for the skills of the people in the organization and for updating the skills of its workforce to ensure dynamic capability for the organization. Successful entrepreneurship is as much about developing and managing skills as it is about developing and managing products or services.

(2) *A detailed, process-oriented typology of organizational resources has been developed, and based on the understanding gained during the research about resource management, it is reasonable to argue that resource deployment patterns, over time, may in fact be more important than firm scope in determining competitive advantage. As part of this, it appears that an organizational* "skill life cycle" *exists*, and that it is related to the traditional product market life cycle in specific ways. More importantly, it appears that management of this skill life cycle is *more fundamental* to long-term firm performance than management of the product market life cycle.

(3) *Start-up conditions and the initial skill-sets of the entrepreneurs are critical* aspects of the firm's long-run skill-base development and capability.

(4) Successful and sustainable entrepreneurial strategies require that the *entrepreneurs be competent at managing the systems they develop*—they must be both Kirznerian and Schumpeterian!

(5) The organizational skill-development process is just as critical, and in many ways just as complex, in "low" technology environments as it is in "high" technology environments. There are reasons, based on the theory developed, to believe that there is *not as significant a difference between low- and high-technology resource-management processes as is commonly thought.*

Each of these findings is discussed in more detail below.

(1) *Leadership processes and style—especially the entrepreneurial leadership process—are critical aspects of creating value with skill-based strategies.*

Above all else, leadership matters! The role that an active entrepreneurial leader plays in guiding his or her firm through the tough choices of what skills to develop in-house and which to acquire and subcontract, all have a critical impact on what competitive productive services their firm can offer in the future—and on what opportunities they can pursue. The results of Eastlack and McDonald's 1970 survey of 211 Chief Executive Officers suggests that the general approach to leadership and management that a CEO takes *can affect the performance of his or her organization*. The theoretical perspective developed in this book would lead one to believe that the Eastlack and McDonald results are quite strong and have substantive theoretical underpinnings.

Active leadership plays a great deal of importance in the maintenance of a culture in the organization that encourages individual and organizational learning and failure, all of which deal with change. Leadership does indeed deal with change, as Kotter (1991) suggests, and the leader's role in developing institutional competence is one of the more important ones in the protection of a firms skill and general resource base.

Another important aspect of the skill-based and entrepreneurial leadership phenomena in this model is that the leaders and managers of a firm must have the requisite abilities to absorb (Cohen and Levinthal, 1990) and interpret (Daft and Weick, 1984) relevant information about consumer needs and values if they are to develop effective skill-based strategies to satisfy them. The recognition of need, which may present itself as opportunity, is one of the primary inputs to this value creation model.

(2) *A detailed, process-oriented typology of organizational resources has been developed, and based on the understanding gained during the research about resource management, it is reasonable to argue that resource deployment patterns, over time, may in fact be more important than firm scope in determining competitive advantage. As part of this, it appears that an organizational* "skill life cycle" *exists*.

The emerging set of structural and process model resource-based theories of firms have put a much needed focus on examining the characteristics of a firm's resources and internal management processes in relationship to the generation of various types of "rent". As Wernerfelt (1984) suggests, it is not clear that identifying resources is as easy as identifying products. This book has attempted to advance our understanding of resources by introducing a typology of organizational resources, as suggested by Yuchtman and Seashore in 1967, among others (Wernerfelt, 1984; Conner, 1991; Farjoun, 1990), that identifies resource levels, types and attributes. From a structural perspective the typology of resources provides a way to look at all of the firm's resources in terms of their basic attributes and potential for rent generation. From a process perspective the book shows why persistent efficiency rents will always exist in a changing environment—dynamic capability is a reflection of ever-changing and (hopefully) improving efficiency and effectiveness.

Conner (1991) argues that "explicit attention needs to be given to understanding the levels of resources that may exist within firms and to the potential contribution of each to performance differentials. Recognizing such levels appears especially important in preventing resource-based theory from becoming tautological: at some level, everything in a firm becomes a resource, and hence resources lose explanatory power" (pp. 144–145). This book has provided the details of an organizational resources typology that clearly makes progress in preventing resource-based theories from being tautological.

One of the more important findings of the book, with respect to institutional leadership issues and resource management, is that this research supports the contention that Hofer and Schendel made in 1978, that, at the business level, the "level and patterns of [an] organization's past and present resource and skill deployments" (resource deployments), may be more important than the "extent of the organization's present and planned interactions with its environment" (scope—including market/product domain) in determining the "unique position [the firm] develops vis-a-vis its competitors" (p. 25). In a dynamic, truly competitive environment, it seems reasonable to believe that the ability to deploy and redeploy the firm's assets, across traditional "industry boundaries" (leveraging dynamic capability), is far more important than its ability to defend turf within a particular product market-defined territory (leveraging position). Two elements support this line of reasoning: (1) skills and capabilities erode quickly if not exercised appropriately, and, (2) the ability to generate economies of scope, which are fast becoming a more important element to success than scale economies (Rothwell and Whiston, 1990), depend on a flexible and fast firm—which come with resource deployment mobility.

Figure 12.1 illustrates four general skill decay patterns, given various competitive market characteristics. Note that in competitive and unstable markets (like computers, Hauser, Precision), which is beginning to characterize more and more markets, the skill base erodes quickly and

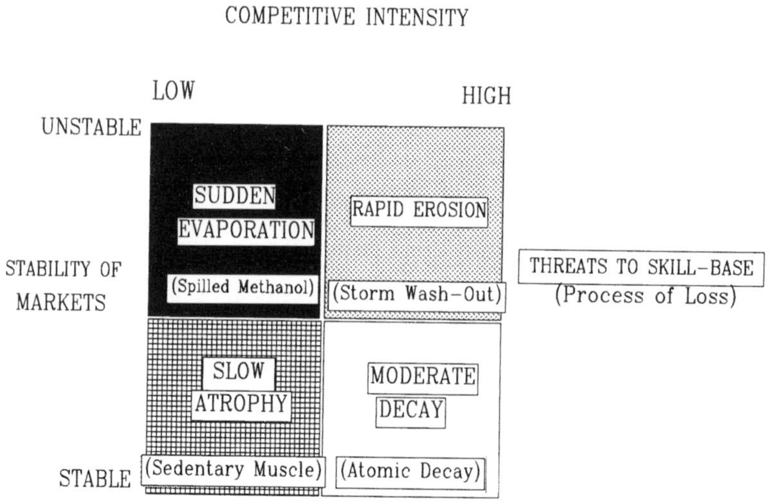

COMPETITIVE INTENSITY
LOW
HIGH
UNSTABLE
SUDDEN EVAPORATION
(Spilled Methanol)
RAPID EROSION
(Storm Wash-Out)
STABILITY OF MARKETS
THREATS TO SKILL-BASE
(Process of Loss)
SLOW ATROPHY
(Sedentary Muscle)
MODERATE DECAY
(Atomic Decay)
STABLE

Fig. 12.1. Skill erosion patterns.

constantly, like sand on a Pacific beach. Constant attention to resource deployment issues is vital to successful strategic management.

Figure 12.2 illustrates that the specific configuration of the skill-based resources of the firm, in terms of its breadth and depth, has implications for the potential of a firm to achieve scale or scope economies (Chandler, 1990). Again, a focus on external or opportunistic issues of strategy that prevent leadership from actively developing and investing in the human capital base will lead to a non-competitive position in the long run, for competition in the market-place is as much a contest of skills and capabilities as it is products and services (Wernerfelt, 1984; Prahalad and Hamel, 1990; Hamel, 1991). If a firm is to function and sustain its operations, is must depend on and channel both cooperative and contentious behaviour (Pascale, 1991) in ways that allow it to continually learn and unlearn (Hedberg & Starbuck, 1981) and improve its capabilities (Goldratt, 1990) in both an evolutionary and revolutionary manner.

Another element that supports the contention that, at the business level, the "level and patterns of [an] organization's past and present resource and skill deployments" (resource deployments) may be more important than the "extent of the organization's present and planned interactions with its environment (scope—including market/product domain)" (Hofer and Schendel, 1978, p. 25) in determining competitive advantage, is that it appears that an organizational "skill life cycle" exists, and that it is related to the traditional product market life cycle in specific ways. More importantly, it appears that management of this skill life cycle is *more fundamental* to long-term firm performance than management of the product market life cycle.

As Fig. 12.3 suggests, based on the substantive analysis of the cases,

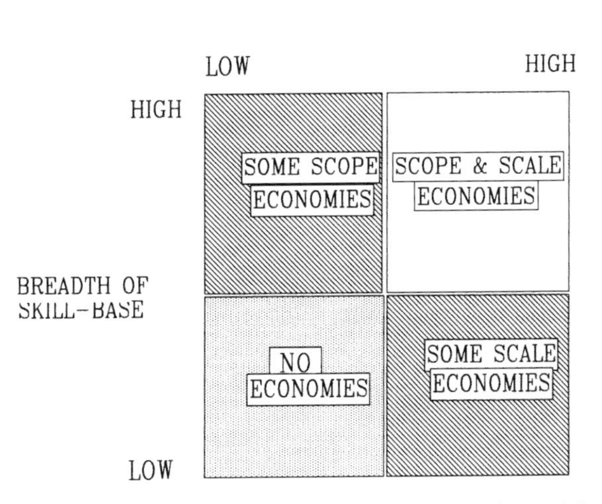

Fig. 12.2. Scale and scope implications of skill-base breadth and depth combinations.

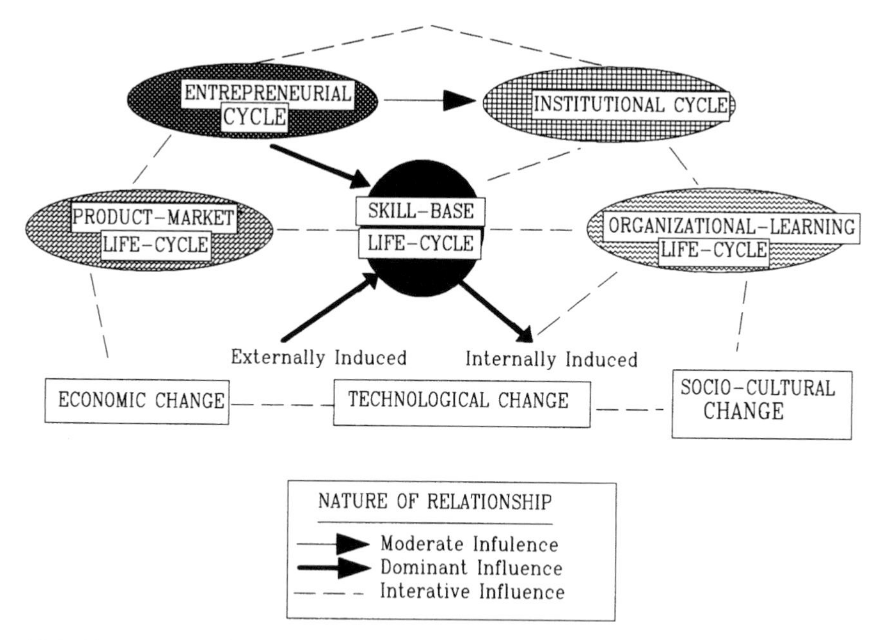

Fig. 12.3. Four general and interrelated life cycles that affect the skill life cycle.

there are at least four basic life cycles of change which leaders must be able to affect and which contribute to an organizational "skill life cycle" (Hofer, 1975). These major life cycles include: (1) an organizational learning cycle, (2) an institutional competence cycle, (3) an entrepreneurial cycle, and (4) a product market cycle. Each of these cycles, as well as the skill cycle, are driven by continuous and fundamental changes in society, technology and the economy. Figure 12.4 maps out how various skill cycles are related to the product market cycle, and shows what a generic total skill cycle might look like.

These skill life cycles are affected by skill erosion patterns and by the breadth and depth of skills an organization has. In all cases, however, it is the responsibility of general managers to be cognizant of the firm's skill-base development when attempting to formulate and implement competitive strategy.

(3) *Start-up conditions and the initial skill-sets of the entrepreneurs are critical aspects of the firm's long-run skill-base development and capability.*

In each of the cases analyzed in this book the initial startup or turnaround conditions played a profound role in shaping the eventual capabilities of the firm. It was argued earlier that skill-related issues were the primary element that led to the development of systems, and that systems were a large factor in determining the structure of the organization. The evolutionary and recursive process that is illustrated in Fig. 12.5 suggests that the

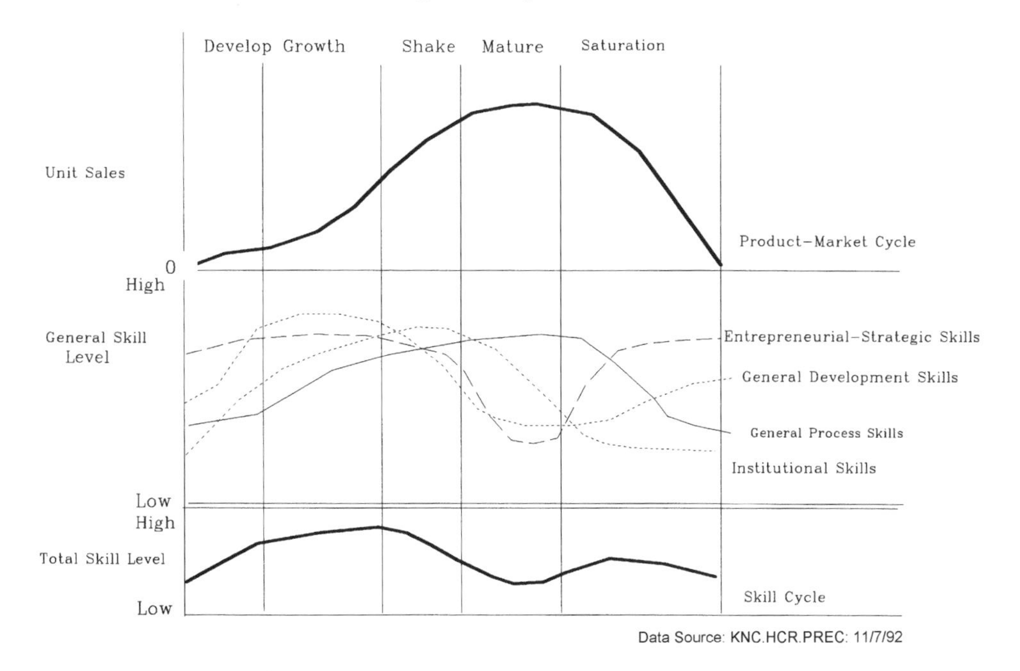

Fig. 12.4. Detailed skill components of a general skill life cycle.

evolution of organizations, like the evolution of humans and virtually every natural biological organism, follows the general principle of "ontogeny recapitulates phylogeny". Herbert Simon (1981) discusses this concept within the context of the architecture of complexity within self-reproducing systems and explains the concept:

> The DNA in the chromosomes of an organism contains some, and perhaps most, of the information needed to determine its development and activity. We have seen that, if current theories are even approximately correct, the information is recorded not as a state description [static], but as a series of "instructions" for the construction and maintenance of the organism from nutrient materials . . . (p. 225)
>
> The individual organism in its development stages goes through stages that resemble some of its ancestral forms [ontogeny recapitulates phylogeny] . . . . Biologists today like to emphasize the qualifications of that principle—that ontogeny recapitulates only in the grossest aspects of phylogeny, and these only crudely. These qualifications should not make us lose sight of the fact that the generalization does hold in rough approximation—it does summarize a very significant set of facts about the organism's development. (p. 226)

The analysis of the Mike Viera data and the analysis of Precision Instruments and Hauser data all point to this evolutionary pattern. The impact of the initial startup conditions, whether at original startup, restart (Dent, 1991) or at a crisis point, such as a turnaround (Hofer, 1980), *especially*

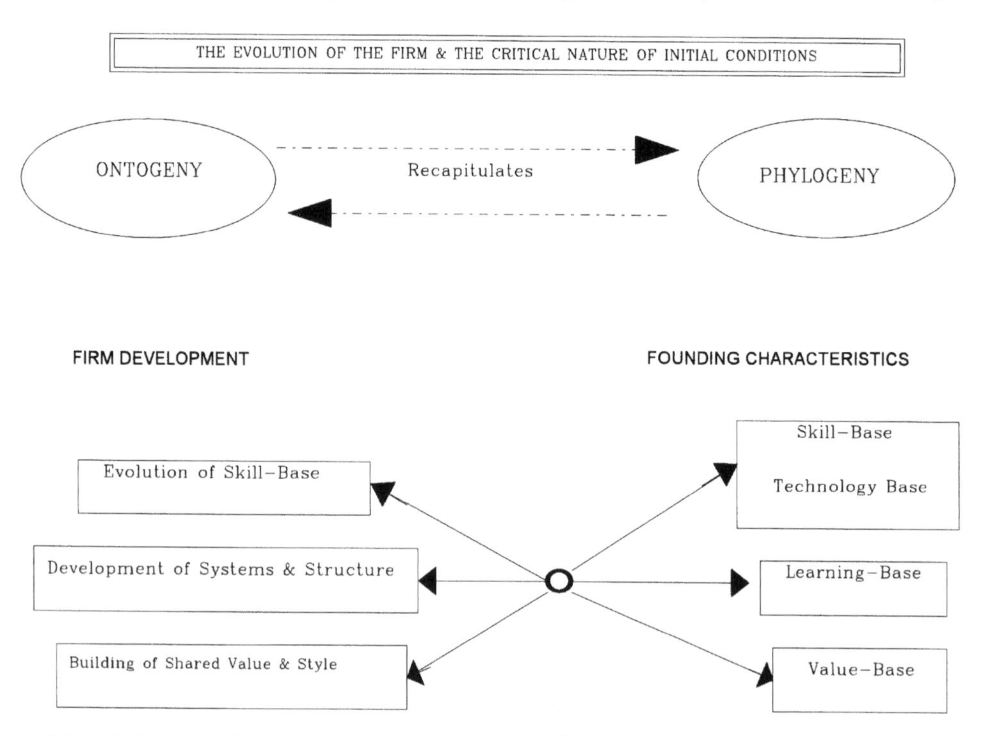

Fig. 12.5. The critical nature of startup conditions—ontogeny recapitulates phylogeny.

in terms of the entrepreneurial leadership and skill-base characteristics, is phylogenic (architectural) and persistent. While each of the companies studied in this book is relatively small and young, even giants like IBM, Ford, Hewlett-Packard and Wal-Mart, are still reflections of their early entrepreneurial leaders' visions and skills. The skills of the founders and the skill-base that evolves after many years in an organization are like the DNA of an individual (Nelson and Winter, 1982), which contains in it the seeds for maintenance and future evolution.

The initial skills of the entrepreneurs and general managers in the book sample were reflected in the long-run development of their organization's systems, structure and style, and certainly the firm's skills. The process of organizational development, from a skill perspective, followed a path where the skills and styles of the leaders were "routinized" in the systems and structure. This provides theoretically interesting evidence that Herron's (1990) empirical findings that the skills of entrepreneurs are indeed important to the ultimate success of the firm. An interesting follow-up to Herron's research would be to see if a more refined skill typology, such as the one developed in this book, would provide even more evidence that the entrepreneur, and the skills of leaders in general, do matter.

One of the more important aspects of the entrepreneurial leadership equation, which relates to initial starting conditions, is that the entrepreneurs must start with deep knowledge of the area in order to benefit from a skill-based strategy. Deep knowledge, and the inherent motivational boosts that come with such knowledge, are prerequisite conditions for grasping opportunity, as opposed to just perceiving it.

In terms of the renewal factor, it is critical that all top management teams have some degree of deep knowledge or competence in the technical aspects of the business they compete in so that they may understand what actually creates value in the business—what the core of competence is and should be. Top management teams that have members that were "staff oriented" throughout their careers, with no line orientation, will not be able to perceive and recognize when there are threats to the core value-creating assets.

(4) *Successful and sustainable entrepreneurial strategies require that the entrepreneurs be competent at managing the systems they develop—they must be both Kirznerian and Schumpeterian!*

If a firm wishes to remain effective over the long term, then it must maintain a pool of talented leaders on hand who will be able to span the external and internal worlds of opportunity and capability. However, once a choice about what markets to target with what skills is made it is critical that firms also have a pool of competent managers—managers who can cope with complexity and can successfully blend component skills, techniques and material resources together to create valuable productive services and dynamic capability.

There appears to be a close theoretical link between the process of competently blending resources, which is a management phenomenon, and the ultimate creation of value. Given the perspective outlined in this book, value is not created when an entrepreneur perceives an opportunity, but only when the opportunity is grasped and a venture has become an ongoing concern—when one can manage. Therefore an important finding is that effective entrepreneurs who wish to build sustainable organizations must have *both* Kirznerian skills at recognizing opportunity and Schumpeterian skills at being able to implement systems that allow for sustainable growth and continued innovation.

(5) *The organizational skill-development process is just as critical, and in many ways just as complex, in "low" technology environments as it is in "high" technology environments.*

There are reasons, based on the theory developed, to believe that there is *not as significant a difference between low- and high-technology resource-management processes as is commonly thought.* In each of the companies studied in this book, the interaction between "technology", "human resources" and "markets" was complex and iterative. Technology, and the technical knowledge and expertise that is embedded and related to technology, is a major and underlying factor in the resource mix available to entrepreneurial leaders and strategists. Technology is one of the fundamental ingredients in understanding organizations from a resource perspective; and based on the analysis performed on the data from the widely disparate firms in the book, it is an inseparable aspect of the skill-based development process.

The entrepreneurial leaders of all the companies studied in this book, including Mike Viera, who managed in a relatively technologically simple environment, managed their companies with the idea that the skills and knowledge of human beings are what drives technology and innovation, and are at the heart of the process of value creation.

This suggests that the distinction between "low" and "high" technology businesses may not be as important, relative to managing the process of skill development and resource management, as is believed. While technologically intensive businesses may have to deal with more complex streams of technology than technologically simple organizations, both types of firm must manage the basic processes of skill and resource development and enhancement if they are to sustain themselves. The process of managing these resources appears to be similar in both "low" and "high" technology environments.

IMPLICATIONS OF THE BOOK

Implications of the Research Methods for Theory and Research

The primary implication of the book's methods for theory and research in the fields of entrepreneurship and strategic management is that the research design and methods work for developing *both* practitioner-based idiographic models and cases and substantive-level, more formal, theory that integrates extant academic work in areas that were not adequately explained using previous research methods.

It does this by combining the best of both areas: it gathers rich, insightful, grounded data from practitioners, who have spent much more time living with the issues under study, but who do not necessarily have an explicit theoretical framework or language to discuss the issues; and it uses the academic's theory-building skills to develop more formalized, grounded theory from the data.

The general research design developed in this book should be used under these general conditions:

(1) Where there is an established need to explore and develop new theory in an area. Inconsistent research findings may indicate such a need, and a lack of precision in definitions in a field may indicate such a need, among others.
(2) When the issue under study demands a data-rich process-oriented

perspective that attempts to trace developmental patterns in organizations.

Overall, it is critical that the language of practitioners be analyzed and compared systematically with extant academic language, and that theory builders be diligent in specifying definitions of constructs clearly, and show how these definitions either converge or diverge with current thought. The research methods described in this book provide a means to do such comparisons.

Implications for Entrepreneurship Theory and Research

Theory Implications: a Focus on Both Individual and Organizational Skills is Relevant and Needed in Entrepreneurial Theory

Hofer and Bygrave (1992) identify nine particularly important aspects of the entrepreneurial process *vis-à-vis* the development of conceptual frameworks, models and theories in the field—the entrepreneurial process: (1) is *initiated* by an act of *human volition*; (2) occurs at the level of the *individual firm*; (3) involves a *state of change*; (4) involves a *discontinuity*; (5) is a *holistic process*; (6) is a *dynamic process*; (7) is *unique*; (8) involves *numerous antecedent variables*; and (9) generates outcomes that are *extremely sensitive to the initial conditions of those variables*.

Figure 12.6 presents these nine aspects of the entrepreneurial process in the form of a framework, based on the research and findings of the book. The research conducted in this project suggests that all nine aspects of the entrepreneurial process are also aspects of the skill-based strategy process – the two are inextricably linked. Indeed, one of the primary arguments has been that entrepreneurial processes, as they relate to skill-base development and nurturing, are a vital factor in ensuring that firms can renew themselves and survive and are a function of leadership.

There are two major implications for theory in the area of entrepreneurship, based on the results of this book and the Hofer/Bygrave (1992) elements:

(1) Effective theory must address the central role of the skills of the entrepreneur, as distinct from constructs such as family background and mental attitude. Herron (1990) has found empirical evidence, based on nascent theory, that the skills of the individual and teams of entrepreneurs are related to firm performance. More detailed theory of skill types and skill evolution must be developed. Given the importance of initial conditions in determining the potential long-run performance of a given firm, it is important that theory is developed that can identify and trace the relevant skill attributes of the entrepreneur, and can show how these skills are embodied in the organization as a whole.
(2) The holistic and dynamic nature of the entrepreneurial process requires

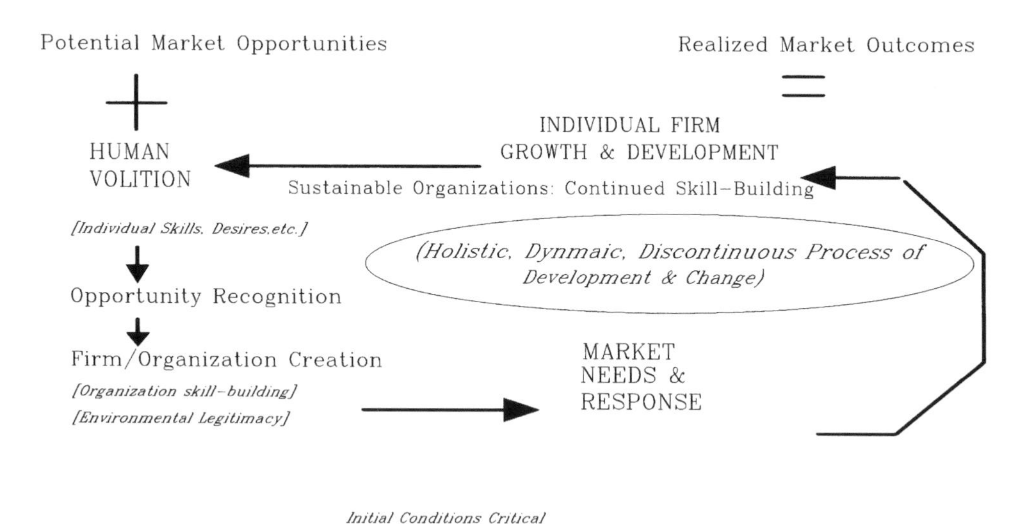

Fig. 12.6. Framework for examining the entrepreneurial process, based on findings and Hofer and Bygrave (1992).

that theorists focus on the entire organization over time, not just the lead entrepreneurs. The ability of entrepreneurs to develop the skill-base of the entire organization is a critical, and often overlooked, aspect of entrepreneurial theory.

Research Implications: Process-Oriented Test Methods are Needed

Despite having chosen a sample that is theoretically diverse, it is difficult to ascertain the generality of the basic dynamic processes that have been discussed in this book. The only way to discover whether these developmental patterns, and the effects of initial startup conditions, are indeed valid over a broad array of contexts, is to explore more case histories, from an analytical inductive perspective (Katz, 1987) and to employ longitudinal and historical techniques to look at the developmental characteristics of a wide variety of companies; looking for common threads and interesting divergences. New research techniques imported from the study of human personality and development and chaos theory statistics may prove to be quite useful for research in entrepreneurship.

It is important to note that entrepreneurship was studied in this book as an *evolving process*, and that appropriate tests of the theory must proceed with this basic supposition in mind. Testing techniques which rely on assumptions of linear independent relationships among constructs will, in all likelihood, fail to measure the substance of the theory appropriately. More appropriate would be the use of quasi-experimental techniques, including cohort analysis, time series impact analysis and simulation.

Implications for Strategic Management Theory and Research

Theory Implications: the Appropriate Level to Build Resource Theory is at the Functional and Individual Firm Level

Competence, which is one of several general outcome constructs that were examined in this research, reflects management's ability to build, blend and utilize *both* human and material resources to create services that will best enable it to offer value to customers and give it leverage in the competitive environment. Competence is a critical ingredient in the process of converting basic resources (individual skills and single-component material resources) into organizational services. Likewise, capability is a similarly important ingredient in the process of converting an organization's productive services into viable, value-added and relevant goods and services.

If one accepts the above perspective, competence is a concept that indicates how *effective* a firm's managers and leaders are in deploying specific skill-based and material assets in pursuit of a given opportunity. As such, the appropriate level at which to measure a firm's competence is at the "functional" or "operating unit" level, which is the level at which individual skills are deployed in pursuit of organizational ends (Cleveland *et al.*, 1989). Competence is a non-financial measurement of "operating unit" skill utilization or effectiveness. Similarly, McKelvey (1982) concludes that "competence" is an excellent term to convey the idea that [his organizational species or classification framework] is based on differentiating organizations in terms of how they arrange their affairs so as to compete and survive (p. 171).

Competition as a Construct Needs to Be Augmented to Take in Resource Elements

The issue of examining "competence" in the light of "competition" leads to an interesting etymological point of comparison. The word "compete" also comes from the Latin root *COMPETERE*, to strive together. It is closely associated with the word "competence" in that "competition" means, among other things, "a contest or similar test of skill or ability". This suggests that a firm's "competitors" are any outside forces that are striving to deploy similar resources or "to be competent" with regards to a shared opportunity-set. From a strategic management point of view, emphasizing the "competence" base of "competition" directs one's attention more to the resource side of the firm than the market side, when considering the attributes of the competitive environment (Day 1981).

More Work on Systems Theories of the Firm, Like the Theory of Constraints, that Help Identify Core Competence Activities and Their Evolution is Needed

One of the more surprising aspects of this research was the discovery that the theory of constraints (Goldratt, 1990) is such an effective tool for both strategic thinking and implementation. In general a firm has an incentive to use the most valuable and specialized of its resources as fully as possible (Penrose, 1959, p. 71) and to pace its growth in terms of its critical least-reducible resources (Thompson, 1967). This line of reasoning is interesting if we connect it to the driving force and theory of constraints concepts, and recognize that, from a strategic perspective, the firm should choose the location of the least reducible element or the primary internal constraint (Goldratt, 1990), which paces the growth and scope of the firm's activities, in the area it is most likely to be able to develop core competencies in.

The existence of a core competence implies that the firm will be able to get the most out of its critical strategic resource areas, and that protective capacity (Goldratt, 1990) exists in all other areas to allow smooth flow of throughput away from the core area. The core competencies of the firm also mirror the central skills and skill-sets that the firm has acquired and built, and reflects the strategic commitments (Ghemawat, 1991) general managers have made for their firms through time.

These commitments are dynamic constraints on strategy and are cumulative (Ghemawat, 1990, p. 14), but they can provide a firm with the potential for building isolating mechanisms (Rumelt, 1987) based on organizational skills. According to Prahalad and Hamel (1990), "since core competencies are built through a process of continuous improvement and enhancement that may span a decade or longer, a company that has failed to invest in core competence building will find it very difficult to enter an emerging market" (p. 85).

Implications for Strategic Management Research

There has been virtually no formal effort to measure any of the constructs identified in this book. There are a tremendous number of challenges along this path that must be met if the working theory posited here is to be part of an operational theory. On the content side: how can the resource attributes discussed in this paper be operationalized and measured? Much work has gone into measuring variables external to the firm that affect the firm's mobility and strategy (Porter, 1980). Relatively little work has gone into understanding and measuring variables that are internal to the firm, and which reflect the firm's resource position and organizational capability. Farjoun's (1990) work in specifying "Resource Industry Groups" and developing a resource-based alternative to the SIC code system is a step in the right direction, but more work needs to be done to get inside the "black-box" of business level strategy and to develop identifiers and measures of resources that go beyond outdated cost-accounting measures (Dierkes and Coppock, 1975; Goldratt, 1990).

Finally, what constructs and variables can be identified that best represent the degree to which a firm is effectively utilizing a particular resource? Given the wide variety of resource types and the different resource levels, how many constructs and variables that measure resource effectiveness are needed?

These are a few questions that suggest what directions work in future theory and research on organizational resources and firm performance might take. If, as Conner (1991) suggests, the resource-based approach to organization theory and strategic management is "reaching for a new theory of the firm" (p. 143), it is imperative that both inductive and deductive theory-building work be done in this area in a timely manner, and that pure "theory-testing" not begin prematurely, before the relevant constructs and variables are mapped out.

CONCLUSIONS AND IMPLICATIONS FOR MANAGEMENT PRACTICE

This book has been designed to be both rigorous in its academic approach and relevant. Nearly all of the conclusions and implications discussed above are relevant to practising leaders and managers. Most of the value added from this book comes from the information gleaned from very open and interesting practitioners.

Customers' Needs Drive the Show—Skills Provide the Dough

This book has focused on how the leaders of firms create value in the economy by assembling and leveraging the resources, especially the human resources, at their employ. Interestingly, the focus on the "value creation process" was not part of the researchers' original conceptualization of the skill-based strategy process. However, the concept of value creation emerged consistently across all four cases as one of the most important concepts to the practitioners, and was identified under the various *practitioner-derived* codes of "value-added", "solving the customer's problem", "meeting the customer's needs", "building relationships that would bring value for the customer", "we could bring things together and deliver them to the customer in a new way" and "I will do my best to make a customer happy". The leader's concern for meeting customer needs and expectations and solving their problems all reflected a value-added approach to managing their businesses. The selected code "meeting customer needs, expectations and solving problems" thus emerged as the core category around which others have been analyzed and subordinated.

What is suggested by the analysis in this book is that the leaders of these organizations have each recognized that they must take a resource-based approach to the business in order to meet customer needs and expectations, and to solve their problems. The challenge to the leadership is to know which customers to serve (this is a function of entrepreneurial leadership), which is a function of the firm's ability to serve them, based on the human and material resources at their employ. The leaders of a firm can tap two fundamental sources of knowledge which they can attempt to transform into value for the customer: (1) general knowledge, experience, know-how and creativity, borne in its human resources, and (2) technical knowledge and expertise, which is embedded in both human and material resources, and always in relation to some technology.

It is not enough to know what resources are at the employ of a firm. The leaders of a firm must know what types of knowledge and other resources they can mix together to create a composite that will be of value in the market-place, and which will thus generate returns. Since the customer is the ultimate filter in determining value, this implies that as leaders they must know the specific needs and problems of customers, and must also recognize, in detail, how they can mix their resources to create an appropriately valued composite.

The challenges to leadership are entrepreneurial, strategic and operational. The selection of customers to serve (which filters to understand) and which resources to blend is a function of entrepreneurial choice and strategic thinking. The mixing process, the integration of human and material resources, the formation of teams and shared values, is at the heart of the concept of core and institutional competence building, which define the limits of a firm's operational capability.

Skills and Systems are the Heart of the Value Creation Engine

If one looks at the research and conclusions of this book from the McKinsey 7-S perspective, an interesting relationship emerges. Unlike the original model, which shows how each of the seven elements is interdependent on the others, and in which no element seems to dominate, one can make a reasonable argument, based on this research, that skills drive systems and that skills and systems are the engine of value creation, and that they are the primary determinants of structure. This is consistent with what Waterman *et al.* (1982) observed about their work, although it is not reflected in the model: "if there is a variable in our model that threatens to dominate the others, it could well be systems. Do you want to understand how an organization really gets things done? Look at the systems" (in Quinn and Mintzberg, 1991, p. 311).

Figure 12.7 illustrates a revised 7-S model, based on the observations and analysis of this book. There are three basic contexts to the model: a soft, a hard and an action context. The soft context reflects the role of institutional leadership, which is to provide a sense of mission and shared

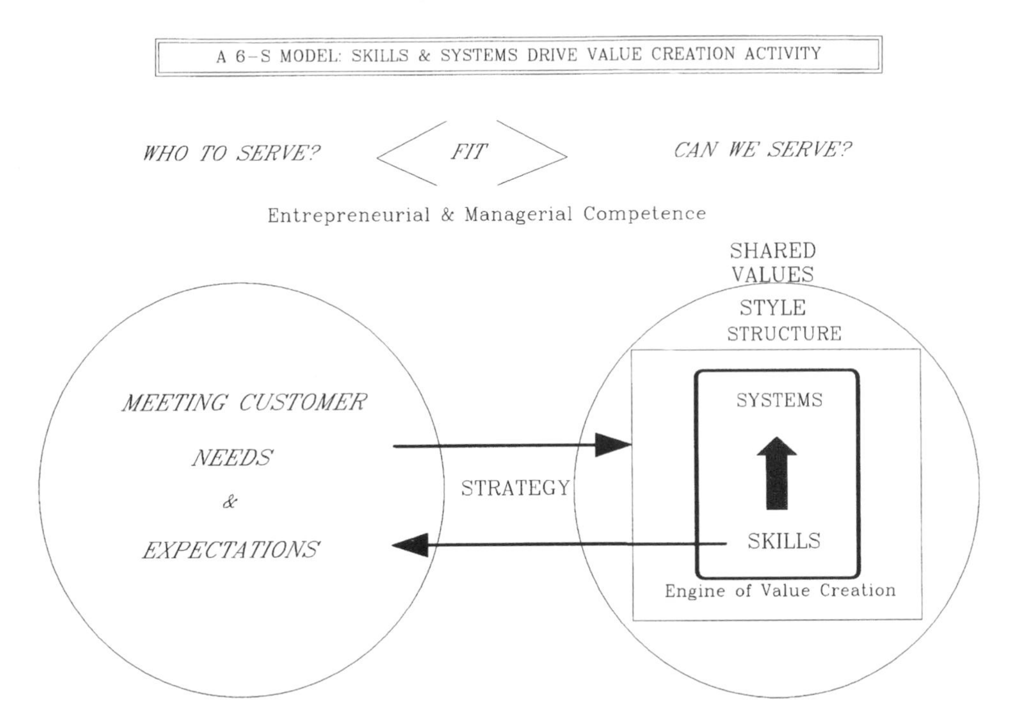

Fig. 12.7. McKinsey 7-S framework re-cast—skills drive value creation activity.

values for the organization, and a style that will encourage the growth of skills and capabilities. That is, shared values and style are important "soft" contextual variables that affect how skills and systems evolve.

The hard context reflects the structural elements and the strategy of the organization. Strategy is the sole bridge between the potential value-creation activities of the firm and the potential set of applications in the environment. And although strategy is really dominated by structure in this model (strategic choice is constrained by the specific attributes and structure of the firm's resources), it is the vital communications link to the world of opportunity. Strategy is the domain of the entrepreneurial leader, who is responsible for ensuring that the firm can pursue new opportunities as they are perceived. The action context, of course, is where the value creation occurs—in the development and exercise of organizational skills, staff and systems. Given a "soft" context of overall meaning and direction, and a "hard" context of current strategic choice and structural limits, the firm seeks to create value within the action context.

This perspective of the 7-S model suggests that the role of effective leadership in defining the institutional context of the firm and the entrepreneurial and strategic context cannot be overemphasized—as they are what both protect the engine of value creation (skills, staff and systems) from internal erosion and ensure that it actually produces valued services.

The Theory of Constraints and Focused Activities

As discussed above, one of the more powerful tools a manager can employ to understand and strategically manage the resource elements of their firm is the theory of constraints. A primary constraint, viewed as a strategic tool, is quite similar to the core competence concept discussed earlier (Prahalad and Hamel, 1990). A firm must thoroughly understand the nature of the primary constraint it has chosen to manage and be competent in the skills required to move products and services through the constraint (throughput per constraint minute becomes a key performance indicator). As Quinn *et al.* noted in a 1990 article in *Sloan Management Review*,

> the key point is that a few selected activities should drive strategy. Knowledge bases, skill sets, and service activities are the things that can generally create continuing value added and competitive advantage. Competitors must be defined as those with substitutable skill bases, not those with similar product lines (Quinn *et al.*, 1990).

Figure 12.8 illustrates how the Theory of Constraints fits into the more general theory of value creation discussed in the previous chapter. This figure illustrates how the primary measures of “throughput”, “inventory” and “operating expense” relate to the basic resource transformation process, and to the value constructs discussed in the last chapter.

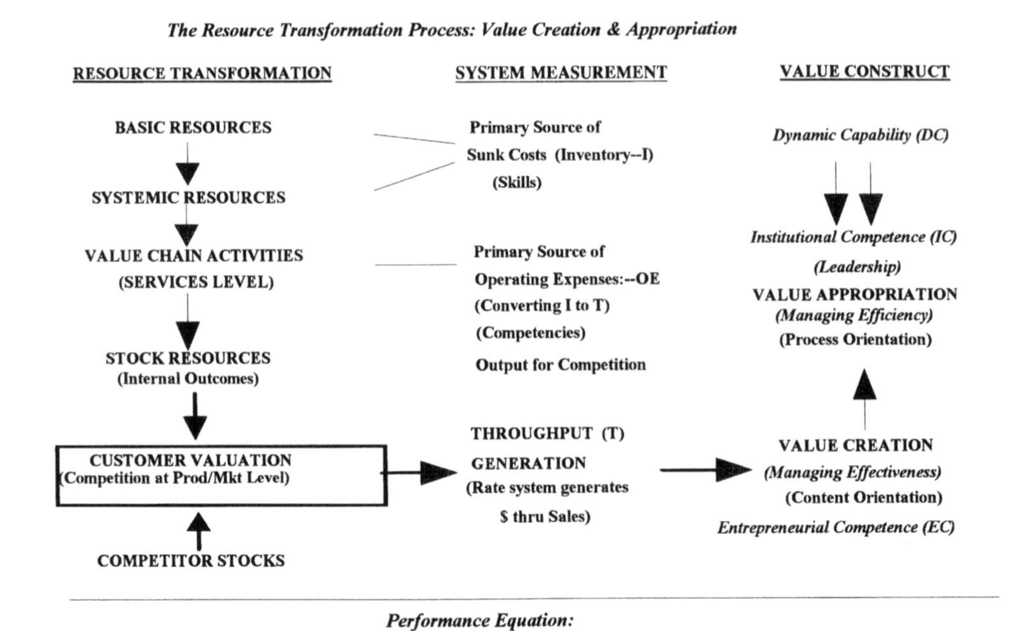

Fig. 12.8. TOC and the value creation process.

Know Your Technologies—They are Really in Your People

In each of the companies that was studied in the book, the interaction between "technology", "human resources" and "markets" was complex and iterative. Technology, and the technical knowledge and expertise that is embedded and related to technology, is a major and underlying factor in the resource mix available to entrepreneurial leaders and strategists. Technology is one of the fundamental ingredients to understanding organizations from a resource perspective; and based on the analysis performed on the data from the widely disparate firms in the book, it is an inseparable aspect of the skill-based development process.

The ability to acquire, generate and regenerate needed technologies and technical knowledge in a firm is closely tied to the nature and development of a firm's skill-base of human capital (Schultz, 1990). As the need for technological advancement increases, through either demand-pull or business-push mechanisms (Burgleman and Sayles, 1986), there is a concomitant need for a more sophisticated skill-base, in terms of both skill depth and breadth. The leaders of a firm must continually choose an investment path in either the acquisition of technology and skilled capital or in their internal development, or some combination, if they are to continue developing more broad or deep capabilities in their firms.

As a firm's systems evolve, they provide the base for the firm to exercise its "dynamic capabilities", which become a source of further development and *also of potential competitive advantage*. Here then is the real link between skill-base development and strategy—skills and technology are the building blocks of the value creation process, and if a firm can develop its skill-base more quickly and effectively than competitors, they will potentially benefit from opportunities that only they are ready to mine.

CONCLUDING COMMENTS

A Short Story About "Matters of Consequence"

In 1943 Antoine de Saint Exupéry wrote a remarkable fable about the journeys of a Petit Prince. In the story, the tiny hero leaves his asteroid planet and all that he cares about in search of the secret of what is important in life. He discovers many fascinating things about "matters of consequence". Take for example, the little prince's sojourn to the fourth planet (Saint Exupéry, 1943), which belonged to a businessman:

> This man was so much occupied that he did not even raise his head at the little prince's arrival. "Good morning," the little prince said to him. "Your cigarette has gone out."
>
> "Three and two make five. Five and seven make twelve . . . twenty-six and five makes thirty one. Phew! Then that makes five-hundred-and-one million, six-hundred-twenty-two thousand, seven-hundred-thirty-one."

"Five hundred million what?" asked the little prince . . . who never once let go a question once he had asked it.

" . . . I was saying then, five-hundred-and-one millions—"

"Millions of what?"

"Millions of those little objects," he said, "which one sometimes sees in the sky."

"Flies?"

"Oh, no. Little golden objects that set lazy men to idle dreaming . . . there is no time for dreaming in my life."

"Ah! You mean the stars?"

"Yes, that's it. The stars."

"And what do you do with five-hundred millions of stars?"

"What do I *do* with them?"

"Yes."

"Nothing. I own them . . . ."

"And what do you do with them?"

"I administer them," replied the man. "I count them and recount them. It is difficult. But I am a man who is naturally interested in matters of consequence . . . "

"And that is all?"

"That is enough," said the man . . .

"I myself own a flower," the little prince continued his conversation with the man, "which I water every day. I own three volcanoes, which I clean out every week (for I also clean out the one that is extinct; one never knows). It is of some use to my volcanoes, and it is of some use to my flower, that I own them. But you are of no use to the stars . . ."

The man opened his mouth, but he found nothing to say in answer. And the little prince went away. (St Exupéry, 1943, pp. 52–57)

The Little Prince had stumbled upon a man intent on owning assets, amassing wealth and counting his earnings, without regard to the underlying reality of what those assets were and what the responsibility of ownership of assets entailed. The man failed to recognize that his ownership of the stars added no sustaining value to them and that his "profits" were thus illusory.

This book has explored a central theme intrinsic to the story: it is not enough for owners of organizations and managers to count "assets" and amass "wealth", which can then be traded and redistributed in "efficient" financial markets. Genuine commitment by leaders and general managers to the development of firm-specific skills and distinctive competencies is required (Selznick, 1957; Peters and Waterman, 1982; Peters, 1984). If a firm is to *endure and thrive* in the turbulent international business environment, leaders and managers must focus on understanding the specific nature of their firms' knowledge, skills and capabilities, and must learn how to nurture them, invest in them, change them and utilize them strategically (Teece, 1982, 1987; Badarocco, 1990).

To the extent that this research can help those who study business management, and can help managers to understand the value of putting their own people first, not as a slogan, but as a viable approach to dynamic competition, we have accomplished one of our central objectives as scholars of business—to help make the practice of management more effective.

APPENDIX ONE

THE MACRO-RESEARCH DESIGN

INTRODUCTION

In this appendix, an overview of the research design and the methodological approach of the book is presented. Following the presentation of the general research design, the methodological approach taken in the book (theory building with cases) is introduced. The reasons for using the case method are presented, and three major extant approaches to building theory with cases are examined and compared. Following the comparison of the three extant approaches and a discussion of their overall strengths and weaknesses, an integrative model for building theory from cases is presented. This integrative model is the one that is then used for the fieldwork and analysis phases of the book.

Overview of the Research Design

Research design provides an analytic road map for, and tells how, a researcher should proceed in a study (Bogdan and Biklen, 1982). Given that the investigation of this topic is considered exploratory and developmental in nature, and that the emphasis is on the construction of a more complete and consistent theoretical perspective on skill-based strategy phenomenon than currently exists, a multi-phase research design has been utilized. This is to ensure that both practitioner and academic perspectives and theories are examined and that they are integrated.

The research has been conducted in four basic phases in order to accommodate both academic and field-based inquiry. In Phase One, an in-depth literature survey is conducted, with respect to organizational resources and theories of the firm, of the fields of economics, entrepreneurship, organization theory and strategic management; from which a typology of organizational resources is synthesized and an initial base for building an initial vocabulary about resource and skill-based management is developed.

In Phases Two and Three the book explores the processes by which entrepreneurial leaders and their firms develop and implement *skill-based strategies* that explicitly focus on and seek to convert the organization's skills and resources into sustainable competitive advantages. Four firms, which are theoretically diverse *vis-à-vis* best exemplars selection criteria, are chosen to be the sites for in-depth field research. Using interview and archival data, as well as secondary source data at each site, a multi-stage, modified constant-comparative data analysis process (Glaser and Strauss, 1967; Eisenhardt, 1989) is used to develop four idiographic-level models of skill-based strategy and entrepreneurial leadership. During the first stage of field data analysis, a focused case history (Strauss, 1987) is written that traces the entrepreneurial birth and competitive development of the skill-base and resource capabilities of each of the three exemplary single-business firms, and also the "skill-based" turnaround effort by the leadership of a large high-technology multi-business corporation. These case histories serve as the foundation for the development of the four idiographic-level models presented in the book (Yin, 1989). During the second stage of field data analysis, the idiographic-level histories and models are analyzed for similarities and differences across the four sites (Eisenhardt, 1989), and a more general substantive-level model of skill-based strategy and entrepreneurial leadership is developed.

In Phase Four of the research, the idiographic-level models and the substantive-level grounded model developed in Phase Two are systematically examined and enfolded with relevant extant academic literatures (Eisenhardt, 1989), and an integrated theory of skill-based strategy and entrepreneurial leadership is developed. An inventory of propositions that describes how the important concepts that emerged in the grounded theory are systematically related to one another is developed, as is a graphical and spreadsheet summary of the theory.

This general design has been used to ensure that the resulting grounded theory is strong with respect to issues of clarity, validity, interest, logical adequacy and explanatory potential. Use of this general theory-building design also helps to ensure that the book meets all of the criteria for "good strategic management process research", as identified by Huff and Reger (1987). Specifically, they suggest that good process research in strategic management should:

> (1) Build on existing theory and research, (2) it should import concepts and research from related areas, (3) it should consider the organizational and environmental context (4) it should reflect the content of the strategic decision being studied, (5) it should vary research methods, and (6) it

should aim for non-intuitive, but supportable [propositions and] hypotheses and should value surprising conclusions (p. 226).

The four phases of the research project are summarized in Fig. A.1 and discussed further in the next section.

PHASE ONE: EXPLORATION AND ANALYSIS OF EXISTING ACADEMIC LITERATURE

Phase One of the research design involves an examination of academic literature and an analysis that attempts to refine current theoretical understandings of what resources are and how they are related to skill-based strategy resource deployment concepts. This integrative work builds on existing theory and research in strategic management and imports concepts from other relevant areas. Phase One serves as a platform from which to develop a clearer understanding of the issues and to develop a conceptual base from which to conduct the field research.

Data in Phase One consist of the theories, concepts, models etc. that were thought to have potentially been relevant to aspects of skill-based strategy, drawn from the literatures in strategic management, entrepreneurship, human resources management, production and operations management, economics, organizational theory, and organizational

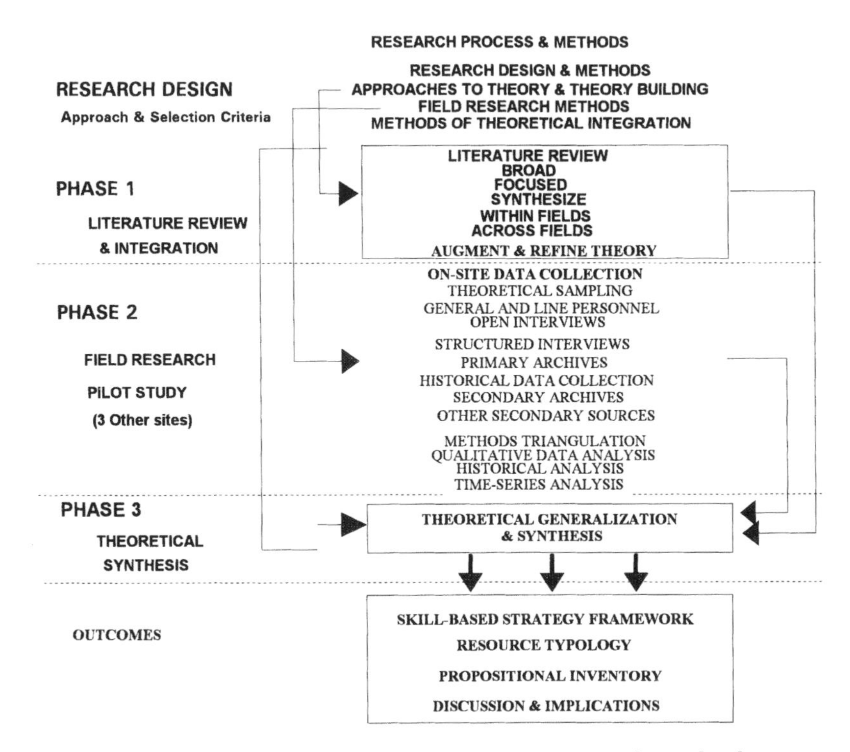

Fig. A.1. Overview of the research process and methods.

psychology. Table A.1 identifies some of the literature on theory construction and evaluation that was used to help in both the analysis of Phase One and also the development and presentation of theory in Phase Four.

PHASE TWO: THE PILOT STUDY

The purpose of conducting the pilot study, which is the primary focus of Phase Two of the research, is to establish a best exemplars, accessible field site where the process of collecting empirical data and practitioner knowledge about skill-base development and strategy can begin. The pilot study site is used to refine the case study protocol (Yin, 1989), including refinement of site selection criteria, data collection and data analysis methods, and case-writing procedures (Strauss, 1987; Eisenhardt, 1989). It is also one of the four sources for an idiographic-level model of skill-based strategy and entrepreneurial leadership.

The firm chosen to be the pilot study, which was highly accessible to the researcher, was thought to be representative of many of the diverse aspects of the skill-based strategy phenomenon, and a broad array of data was collected about this company. The initial pilot site data was collected using a general protocol based on relevant extant literature. Preliminary analysis of the pilot study data led to a narrowing of the focus of data collection for the subsequent cases and a second data collection effort at the pilot site itself.

Table A.1. Primary literature used to inform theory construction and evaluation.

	Research Methodology Sources and Uses	
Primary topic Phase 1	Citation (alphabetical)	Relevant concepts and uses in book
Theory construction and evaluation	Bacharach (1989)	Criteria for evaluation of organizational theories; components of a theory; utility and falsifiability
	Davis (1971)	What makes a theory interesting; denying the assumption base of your audience; theory building as a phenomenological process
	Daft (1982)	Research as storytelling; theory as poetry
	Osigweh (1989)	Concept Formation processes; concept fallibility
	Weick (1989)	Theory construction as disciplined imagination; plausibility; disconfirming assumptions

PHASE THREE: FOLLOW-ON CASE STUDIES AND AUGMENTATION OF PILOT STUDY

Three other appropriate sites were chosen to expand the field-based inquiry, based on preliminary analysis of the pilot site data. The purpose of conducting three additional case studies in Phase Three is to collect additional, more focused, data on the topic of organizational skill-development from practitioners in theoretically diverse settings. By analyzing data from firms with diverse skill-bases and operating environments, one should be able to gain a more complete understanding of the issues related to skill-based strategy and entrepreneurship, and can begin to employ the constant comparison analytic process of inductive and grounded theory-building (Glaser and Strauss, 1967), extending the data and theory beyond the pilot study.

The data collection at these sites was more focused than the collection effort at the pilot site. The data analysis methods employed at all the sites studied in the book include: (1) modified constant comparative qualitative research techniques (Glaser and Strauss, 1967; Glaser, 1978), which depend on an continuing recursive-interactive process between data collection and analysis; (2) case study techniques (Eisenhardt, 1989; Yin, 1989); and (3) historical analysis techniques (Lawrence, 1984; Goodman and Kruger, 1988; Brundage, 1989; Menard, 1991). This multi-tool analysis process was constructed to ensure that the emerging theory has content validity, primarily through subjecting construct development to the tests of convergent and discriminant validity. Within-method triangulation (Jick, 1979) has also been employed for each analysis technique.

The analyses of interview data, historical archive information, general policies, organization charts, HRM procedures and policies, and other organizational information have been used to try to generate a picture of the different organizational and intellectual processes that are linked to skill development and strategic action in exemplary companies. Table A.2 identifies the primary literature used to guide Phase Two research.

PHASE FOUR: INTEGRATION OF PRACTICE AND THEORY

Phase Four is an integrated analysis effort aimed at the development of a more general and formal theoretical synthesis that integrates and extends the perspectives of both practitioners and academicians identified and developed in Phases One to Three. Using comparative analysis methods (Glaser and Strauss, 1967; Strauss, 1987; Eisenhardt, 1989), the concepts and idiographic-level models, derived from practitioner data at both the Phase Two pilot site and Phase Three supplemental sites, have been systematically compared with each other and enfolded with concepts and substantive-level theory from relevant extant literatures.

Table A.2. Primary literature used to guide Phase Two research.

Research methodology sources and uses		
Primary topic Phase Two	Citation (alphabetical)	Relevant concepts and uses in book
Grounded or inductive theory building	Christensen (1976)	Validity through language assessment; development of formalized language
	Eisenhardt (1989)	Theory-building process from cases; embedded designs; enfolding literature; saturation
	Hambrick and Lei (1985)	Empirical evidence on contingency variables
	Hofer (1975)	Theory on Contingency variables for sample
	Glaser (1978)	Theoretical sensitivity; generating substantive theory
	Glaser and Strauss (1967)	Generating theory by comparative analysis; theoretical sampling; constant comparative method
	Yin (1989)	Pattern matching; minimizing researcher bias; evidence; generalization of theory
Methods and data triangulation	Harrigan (1983)	Fine and medium grains methods; hybrid designs
	Jick (1979)	Variance of trait, not method; unique variance; within-method (reliability); between-method (validity)
Qualitative analysis	Kirk and Miller (1986)	Field research as a validity check; theoretical validity; field notes as reliability
	Patton (1990)	Triangulation; researcher bias; theoretical generalization
	Spradley (1979)	Ethnographic interview process; asking descriptive, structural, and contrast questions
	Strauss (1987)	Analysis: open, axial, selective coding; memo writing; diagrams and integrative mechanisms
Historical and longitudinal analysis	Goodman and Kruger (88)	Building theory based on evidence; generalizability; pattern recognition; hypothesis generation
	Lawrence (1984)	Historical perspective to sharpen vision of present; complexity into higher relief
	Menard (1991)	Retrospective panel design; genesis vs. prediction
Time series analysis	Cook and Campbell (79)	Interrupted time series designs (ARIMA)

METHODOLOGICAL APPROACH: THE CASE METHOD AND BUILDING THEORIES FROM CASES

The theory-building approach in this book is decidedly substantive – that is focused on the issue of strategic resource management and skill-based strategy – and is based on a grounded or inductive analysis process. The research design has been constructed so that a middle-range working theory of skill-based strategy may be produced (Eisenhardt, 1989). The heart of the theory-building process in this book is the analysis and presentation of four new cases. The approach focuses primarily on the substantive level of theory development, based on the presentation and cross-case analysis of the four idiographic level theories, or stories.[1] An overview of the reasons for using cases in this project is presented in the next section, as is a discussion on the case study methods that have been chosen.

The Case Method

Given the discussion in Chapter 1 about the state of prior research in the topic area of skill-based strategy and the goals of the book, in terms of theory-building the use of grounded cases to tell the story of skill-based strategy has been chosen because they provide a rich and dynamic context from which to look at the role that entrepreneurs and strategists play in the actual processes involved in the development of the firm's skill-base and the creation of value, each of which is central to the character and essence of an organization (Schendel and Hofer, 1979; Hofer and Bygrave, 1993).

Philip Selznick argues in his 1957 book, *Leadership in Administration*, that it is critical for organizational scholars to recognize the dynamic and evolutionary processes that affect the "character" of an organization over time. Selznick suggests that, in order to understand the development of an organization, one must "see [the character of an organization] as the product of self-preserving efforts to deal with inner impulses and external demands" (1957, p. 141). According to Selznick:

> The study of institutions is in some ways comparable to the clinical study of personality. It requires a genetic and developmental approach, an emphasis on historical origins and growth stages. There is a need to see the enterprise as a whole and to see how it is transformed as new ways of dealing with a changing environment evolve. (p. 141)

Selznick continues, and argues that cases are a means for developing general knowledge about organizational phenomenon:

> In approaching [the problems of character], there is necessarily a close connection between clinical diagnosis of particular cases and the development of sound general knowledge. Our problem [as organizational scholars] is

> to discover the characteristic ways in which *types* of institutions respond to *types* of circumstances. (p. 142)

The problems of organizational character, as Selznick discusses it, are relevant to the specific issues addressed in this book (strategic resource management and skill-based strategy), so the general developmental case approach to studying organizations that Selznick advocates appears to be reasonable to take here.

In this book, which is a theory-building effort aimed at developing a working theory about skill-based strategy, the primary methodological approach taken is to use new business cases as a base from which to build theory. As Yin (1989) argues,

> A case study is an empirical inquiry that investigates a contemporary phenomenon within its real-life context, when: (1) the boundaries between phenomenon and context are not clearly evident, and in which, (2) multiple sources of evidence are used. (1989, p. 23)

Yin argues that the case study has a distinctive advantage over other research strategies when "a *how* or *why* question is being asked about a contemporary set of events, over which the investigator has little or no control" (p. 20). A review of the research objectives and questions in Chapter 1 indicates that the primary focuses of this book are: (1) to explore what primary concepts and phenomena are involved in the practice of skill-based strategic thinking and the execution of skill-based strategies, and (2) to seek to understand why these particular concepts are important, and how they relate to one another. As such, using a case study approach in this work is viable and appropriate.

THREE EXTANT APPROACHES TO BUILDING THEORY WITH CASES

In this section, three extant approaches to building theory with qualitative evidence and cases are briefly reviewed: (1) Glaser and Strauss's (1967) work, *The Discovery of Grounded Theory*, (2) Robert Yin's (1989) *Case Study Research: Design and Methods*, and Kathleen Eisenhardt's (1989) *Building Theory from Case Study Research*. Based on an analysis of the strengths and weaknesses of each approach, an integrated process for building theory from cases is presented in the next section.

The Glaser and Strauss (1967) Approach to Theory Building

In 1967 Glaser and Strauss wrote what Charmaz (1983, p. 1) calls "a pioneering book that provides a strong intellectual rationale for using qualitative research to develop theoretical analyses". Originally aimed at theorists in sociology, *The Discovery of Grounded Theory: Strategies for Qualitative Research* has been used as a basis for building theory from qualitative evidence in a wide range of academic disciplines, including business management.[2]

Glaser and Strauss identify five general approaches to qualitative research, particularly as they affect data analysis, which they assert is the heart of strong qualitative research designs, and argue that the *constant comparative approach* is most effective for the generation of grounded theory. The five general approaches are identified in Table A.3.

In the first general design for qualitative analysis, *Ethnographic description*, which Glaser and Strauss identify but do not discuss, the researcher is primarily interested in describing the attributes and context of a phenomenon as richly as is possible, without any formal concern for analysis to develop or test theory.

In the second general analysis approach, *inspection for hypotheses*, the researcher does not formally code data, but "merely inspects his data for new properties of his theoretical categories, and writes memos on these properties" (p. 101), which become the basis for the specification of hypotheses and theory generation.

In the third general approach, *coding for test, then analyzing data*, the analyst codes the data first and then analyzes it. He makes an effort to code "all relevant data that can be brought to bear on a point", and then systematically assembles, assesses and analyzes these data in a fashion that will "constitute proof for a given proposition" (Glaser and Strauss, 1967, p. 101). The focus of this approach is on testing narrow hypotheses from extant theory without the intention of developing additional theory.

In the fourth approach, the *constant comparative method of analysis*, the goal of the analyst is to generate new, grounded theory. This approach is an integration of the *inspection for hypotheses* and the *coding for test approaches*. It addresses the development of theory and the discovery of hypotheses through a process of constant comparison of data with emerging theoretical insights, which then guide the theoretical sampling of additional data (Glaser and Strauss, 1967, p. 102).

The final and fifth approach to qualitative research analysis, *analytical*

Table A.3. Use of approaches to qualitative analysis (modified table from Glaser and Strauss, 1967, p. 105).

	Provisional testing of theory	
Generation of theory	No	Yes
No	Ethnographic description (1)	Coding for test, then analyzing data (3)
Yes	Inspection for hypotheses (2) Constant comparative method (4)	Analytic induction (5)

induction, combines the *coding for test then analysis of data* design with the *inspection for hypotheses* design in a different manner than the constant comparison approach. The primary thrust of analytical induction is to test extant theory rather than generate new theory. Theory can be extended using this approach, however. When a current theory appears to fail a given test (i.e. cannot not explain a negative case), the researcher can modify the theory and show how the negative case can now be explained.

The Constant Comparative Approach

Of the five general qualitative research analysis approaches discussed above, Glaser and Strauss (1967) contend that the constant comparative approach is the most effective for generating new grounded theory. Charmaz gives an excellent summary of the elements and procedures of the Glaser and Strauss approach to building grounded theory using the constant comparative method of analysis. She notes, in Emerson's (1983) collection of readings, *Contemporary Field Research*, that:

> The grounded theory method [of Glaser and Strauss] stresses discovery and theory development rather than logical deductive reasoning which relies on prior theoretical frameworks. These aspects of the method lead the grounded theorist to certain distinctive strategies. First, data collection and analysis proceed simultaneously. Since grounded theorists intend to construct theory from the data itself, they need to work with solid, rich data... grounded theorists shape their data collection from their analytic interpretations and discoveries, and therefore, sharpen their observations. Additionally, they check and fill out emerging ideas by collecting further data. These strategies serve to strengthen both the quality of the data and the ideas developed from it.
>
> Second, both the processes and products of research are shaped from the data rather than from preconceived logically deduced theoretical frameworks. Grounded theorists rely heavily on studying their data and reading in other fields during the initial stages of the research. They do not rely directly on the literature to shape their ideas, since they believe they should develop their own analyses independently . . . .
>
> In keeping with their foundations in pragmatism, grounded theorists aim to develop fresh theoretical interpretations of the data rather than explicitly aim for any final or complete interpretation of it.

Figure A.2 illustrates the generic research approach of Glaser and Strauss (1967). Their method includes: (1) given a phenomenon of interest, the researcher selects an appropriate site to begin initial observation, data collection and analysis; (2) the researcher begins to construct theory from the data itself and does so in an iterative manner, constantly comparing theoretical insights with gathered data and using these insights to inform

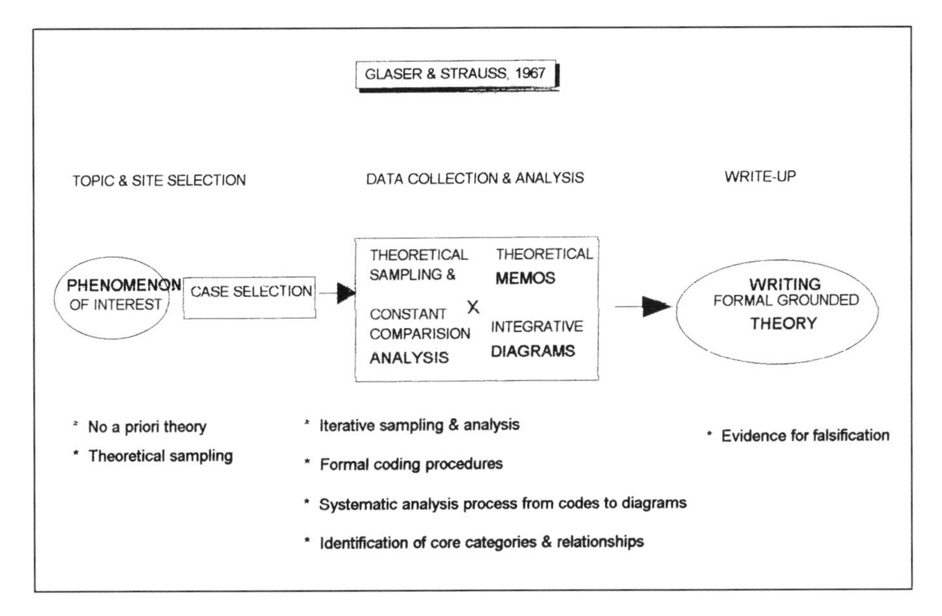

Fig. A.2. The Glaser and Strauss theory building approach.

the decision as to what new data to collect and analyze; (3) the researcher studies the data systematically, generating open, axial and selective *codes* (Strauss, 1987), and making extensive theory and methodological *memos*; (4) the researcher generates *integrative diagrams* (Strauss, 1987) that reflect how various categories in the data may be related; and finally (5) the researcher writes up the formal grounded theory in the form of a case study or history.

Yin's (1989) Approach to Case Studies

A second approach, which examines the use of cases as a research method, is outlined in Robert Yin's (1989) book *Case Study Research: Design and Methods*. From Yin's perspective, case study research should seek to establish, "analytic generalization, in which a *previously developed theory* is used as a template with which to compare the empirical results from the case studies" (1989, p. 38) [our emphasis]. Yin presents a figure, reproduced in Fig. A.3, to illustrate how the "replication logic approach to case studies" is structured to achieve analytic generalization. In Yin's approach, the first step in the process is to *develop theory*, then to select and conduct the case studies, and then finally to modify theory, based on the empirical "results".

For Yin, the case study research process follows the following chain of events:

(1) Theory emerges from the examination and synthesis of previous research and fieldwork in a given area.

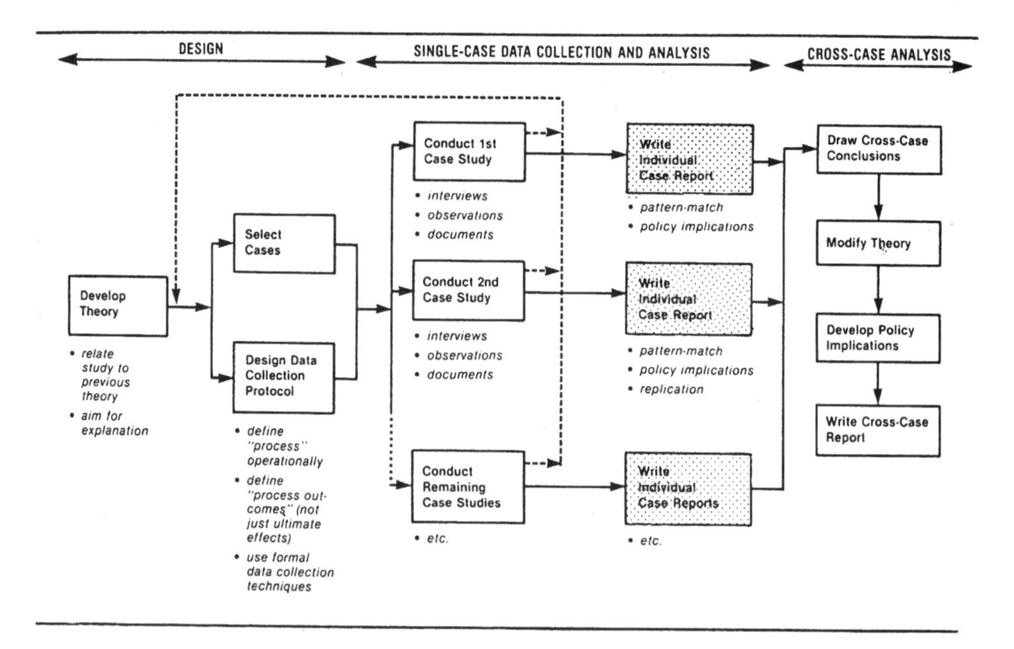

Fig. A.3. Yin's case study replication logic (1989, p. 56).

(2) Based on a careful selection of cases and a data collection protocol, evidence is collected that bears on the theoretical or analytic questions of the research. At this point in the process the researcher analyzes individual case data, and examines theory at an *idiographic level*. Theory that is described at an idiographic level attempts to understand and explain the phenomenon of a particular case study or unit separately from others.[3]

(3) The idiographic level evidence is analyzed, with one of two "analytic strategies" guiding the analysis effort. The two analysis strategies include "relying on theoretical propositions", which are developed prior to entering the field, to guide the analysis; and, "developing a case description", where the analyst is to develop a descriptive framework for organizing the case study. In either case Yin suggests that one of three "dominant modes of analysis" should be considered:

(a) *Pattern matching* of non-equivalent dependent variables, or dependent/independent variable relationships. In this context Yin suggests that the researcher should attempt to compare "an empirically based pattern (from the case) with a predicted one (from pre-specified theory)" (p. 109). In general the patterns he is talking about are with regards to the relationships among "constructs" and are hence "cross-sectional" in nature.

(b) *Explanation building*. In this context the goal is to "analyze the case study by building an explanation about the case" (p. 113). Yin states that the process of building an explanation may be "iterative", where initial propositions are revised as data is analyzed, and that the ideal is to be able to make causal links transparent. He briefly refers to Glaser and Strauss (1967) in this section and maintains that theirs is a similar approach for "exploratory" studies.

(c) *Time series analysis and chronologies*. Here the patterns of relationships among variables are analyzed through time and causal "postulates" generated. The idea is to examine the "how" and "why" about the relationship of events over time. The specific analysis techniques can involve simple observations of time series data or sophisticated statistical analysis of data.

(4) After the individual cases are written up, Yin suggests that the researcher analyze the data across each of the cases and that theory be addressed at a more general *substantive level*. Substantive level theorizing begins when a theorist moves away from explaining the specifics of a single case and focuses on a more general understanding of a substantive area, like skill-based strategy. As Glaser and Strauss (1967, p. 33) note, "the generation of [a more general] theory can be achieved by a comparative analysis between or among groups within the same substantive area".[4]

Eisenhardt's Approach to Theory Building with Cases

A third approach to using case studies as a tool for building theory is that of Kathleen Eisenhardt. In the October 1989 "Special Forum on Theory Building" issue of the *Academy of Management Review*, Eisenhardt presents a methodological approach, or "roadmap", for building theories from case study research (p. 532). She identifies eight major elements in this theory-building research design approach:

(1) There should be a *research focus*. There should be a definition of a research question(s), and possible development of some *a priori* constructs to focus efforts and better ground the constructs.

(2) The researcher should choose cases to examine that will be theoretically useful: that is, cases should be chosen for, "theoretical rather than statistical reasons" (1989, p. 536). The goal of such *theoretical sampling* is to, "choose cases which are likely to replicate or extend emergent theory" (p. 537).

(3) The researcher should *craft a data collection protocol* and other instruments that enable multiple sources of data to be collected for analysis. Both qualitative and quantitative data should be collected.

(4) There should be an *overlap of data collection and analysis*, and the researcher should keep field notes. The researcher should take advantage of flexible data collection procedures and make adjustments in the collection and analysis of data as the study proceeds.

(5) *Analyzing data* is considered the heart of building theory with cases

and the researcher should attempt to analyze "*within cases*" (at the idiographic level) and "*across cases*" (at the substantive level).

(6) The researcher should tabulate evidence on constructs in order to sharpen her definitions and measures. The *shaping of hypotheses* in theory-building research involves measuring constructs and verifying relationships.

(7) The researcher should seek to *compare emergent theory concepts with extant literatures*, looking for conflicts that might lead to a "frame breaking mode of thinking" (p. 544).

(8) Finally, the researcher should bring the project to a close when *theoretical saturation* has been reached – that is, "when they have reached the point where incremental learning is minimal because researchers are observing phenomena seen before" (p. 545).

Eisenhardt's process of building theory from cases is summarized in Table A.4.

SIMILARITIES AND DIFFERENCES AMONG THE THREE EXTANT APPROACHES

The three case-oriented theory-building approaches discussed above share some important similarities and differences which need to be addressed before we move on to identifying the actual approach that was used in the book.

Similarities of the Approaches

In terms of similarities, the most obvious is that all three approaches are designed to help researchers build theory with qualitative case evidence. Each provides a systematic process that an analyst can use to move from specific inductively gathered data about a phenomenon to a more general substantive-level of theorizing.

Eisenhardt's approach is the most general, and incorporates some of the elements of both Glaser and Strauss's and Yin's work. Like Glaser and Strauss, she supports the idea of theoretical sampling and iterative data collection and analysis, and considers data analysis to be the heart of the theory building process. Like Yin, she argues that it is important to craft a data collection protocol and to analyze both within and across case sites.

Differences in the Approaches

Despite some of the general similarities, there are also important differences among the approaches. Glaser and Strauss's approach is most different from the others in that they are primarily concerned with generating new grounded theory, and are not concerned with the elements of testing

Table A.4. Process of Building Theory from Case Study Research

Step	Activity	Reason
Getting Started	Definition of research question	Focuses efforts
	Possibly a priori constructs	Provides better grounding of construct measures
	Neither theory nor hypotheses	Retains theoretical flexibility
Selecting Cases	Specified population	Constrains extraneous variation and sharpens external validity
	Theoretical, not random, sampling	Focuses efforts on theoretically useful cases—i.e., those that replicate or extend theory by filling conceptual categories
Crafting Instruments and Protocols	Multiple data collection methods	Strengthens grounding of theory by triangulation of evidence
	Qualitative and quantitative data combined	Synergistic view of evidence
	Multiple investigators	Fosters divergent perspectives and strengthens grounding
Entering the field	Overlap data collection and analysis, including field notes	Speeds analyses and reveals helpful adjustments to data collection
	Flexible and opportunistic data collection methods	Allows investigators to take advantage of emergent themes and unique case features
Analyzing Data	Within-case analysis	Gains familiarity with data and preliminary theory generation
	Cross-case pattern search using divergent techniques	Forces investigators to look beyond initial impressions and see evidence thru multiple lenses
Shaping Hypotheses	Iterative tabulation of evidence for each construct	Sharpens construct definition, validity, and measurablility
	Replication, not sampling, logic across cases	Confirms, extends, and sharpens theory
	Search evidence for "why" behind relationships	Builds internal validity
Enfolding Literature	Comparison with conflicting literature	Builds internal validity, raises theoretical level, and sharpens construct definitions
	Comparison with similar literature	Sharpens generalizability, improves construct definition, and raises theoretical level
Reaching Closure	Theoretical saturation when possible	Ends process when marginal improvement becomes small

theory or shaping formal hypotheses. Yin's work is markedly different from Glaser and Strauss's in that it focuses mainly on the testing of theory, within the context of a general analytic induction design – theory is augmented within the context of a replication logic test. Eisenhardt's approach can accommodate either the generation or testing of theory.

STRENGTHS AND WEAKNESSES OF EACH APPROACH

Each of the three approaches has its strengths and weaknesses in terms of theory-building, which are discussed below.

Strengths and Weaknesses of the Glaser and Strauss Approach

The Glaser and Strauss approach provides a strong base from which to systematically discover theory from qualitative data, particularly in terms of understanding processes (Charmaz, 1983, p. 111). Their emphasis on firmly grounding the theory in data is a strength, as it requires the analyst to pay attention to issues of validity. Their flexible and iterative method of constant comparative analysis procedures is a strength because it allows the researcher to follow important theory trails that may not have been anticipated before the field research – again enhancing validity. The insistence by Glaser and Strauss that the researcher develop and follow specific coding procedures enhances reliability.

Weaknesses of the Glaser and Strauss approach include that the research process is quite complex and many of the techniques that Glaser and Strauss discuss are not fully explained in their work: much of the process seems tacit (Charmaz, 1983, p. 109). There is little guidance in knowing exactly what to do with extant theory examined before and during the fieldwork (how *tabula rasa* should the theorist be?), and the procedures for the initial collection of data are quite unclear (i.e. where should one start the inquiry?). Finally, since their book is not really about case writing, it does not provide enough direction in terms of designing a case protocol and actually writing up the results.[5]

Strengths and Weaknesses of Yin's Approach

A strength of Yin's approach to theory building is that he provides a comprehensive approach to defining a case study protocol and encourages the researcher to collect both qualitative and quantitative data (Eisenhardt, 1989). His insistence that both within-case and cross-case analysis be done enhances validity. In fact, Table A.5 is one of the more important strengths of Yin's work, as in it he offers various tactics the researcher can use to increase construct, internal and external validity, and reliability.

The primary weaknesses in Yin, from the perspective of theory discovery, is that his methods are really aimed at analytic induction designs, where

Table A.5. Yin's Case Study Tatics for Four Design Tests.

Tests	Case-Study Tactic	Phase of Research in Which Tactic Occurs	Further Description in this Book
Construct validity	use multiple sources of evidence	data collection	Chapter 4
	establish chain of evidence	data collection	Chapter 4
	have key information review draft case study report	composition	Chapter 6
Internal validity	do pattern matching data analysis	Chapter 5	
	do explanation-building	data analysis	Chapter 5
	do time-series analysis	data analysis	Chapter 5
External validity	use replication logic in multiple-case studies	research design	Chapter 2
Reliability	use case study protocol	data collection	Chapter 3
	develop case study data base	data collection	Chapter 4

extant theory is being tested using case evidence. Most of the theory development occurs at the beginning of Yin's process rather than throughout the process.

Strengths and Weaknesses in Eisenhardt's Approach

Eisenhardt's article is very useful in providing an overall picture of the process of building theory with cases. Her work builds on the work of both Glaser and Strauss and Yin, and attempts to show researchers how to do theory building with cases. One important element of her integrative work is that she helps the researcher to ensure that he or she can benefit from one of the primary strengths of theory-building from cases that she identifies, namely that:

> although a myth surrounding theory building from cases is that the process is limited by investigator's preconceptions, in fact, just the opposite is true. The constant juxtaposition of conflicting realities [that comes with case research] tends to "unfreeze" thinking, and so the process has the potential to generate theory with less researcher bias than theory built from incremental studies or armchair, axiomatic deduction. (1989, p. 546)

Eisenhardt is careful to identify how her eight elements of theory building from cases can improve the theory that is ultimately developed from a case study approach.

The primary weakness in Eisenhardt is that, despite her statement that, "analyzing data is the heart of building theory" (p. 539), she does not adequately explain the analysis methods that should be followed in building theory with cases. Eisenhardt provides little help in identifying a systematic process for analyzing the data. Eisenhardt's work, which is valuable due to its integrative nature as a whole, has not been used with regard to data analysis because it does not provide an adequate description of analysis procedures, nor does it reflect the Glaser and Strauss methods accurately.

THE INTEGRATIVE CASE STUDY APPROACH USED IN THIS BOOK

The three approaches examined above provide important and useful information on how one might build good theory from case evidence. After careful study of each method, the researcher felt that no one approach would be sufficient to meet the objectives of the book. Since many of the strengths of the different approaches appeared to be complementary, an integrative case study approach, building on the best from each of the extant approaches, was chosen.

Figure A.4 presents an overview of the process that was used in designing the case study approach for this book. This integrated approach attempts

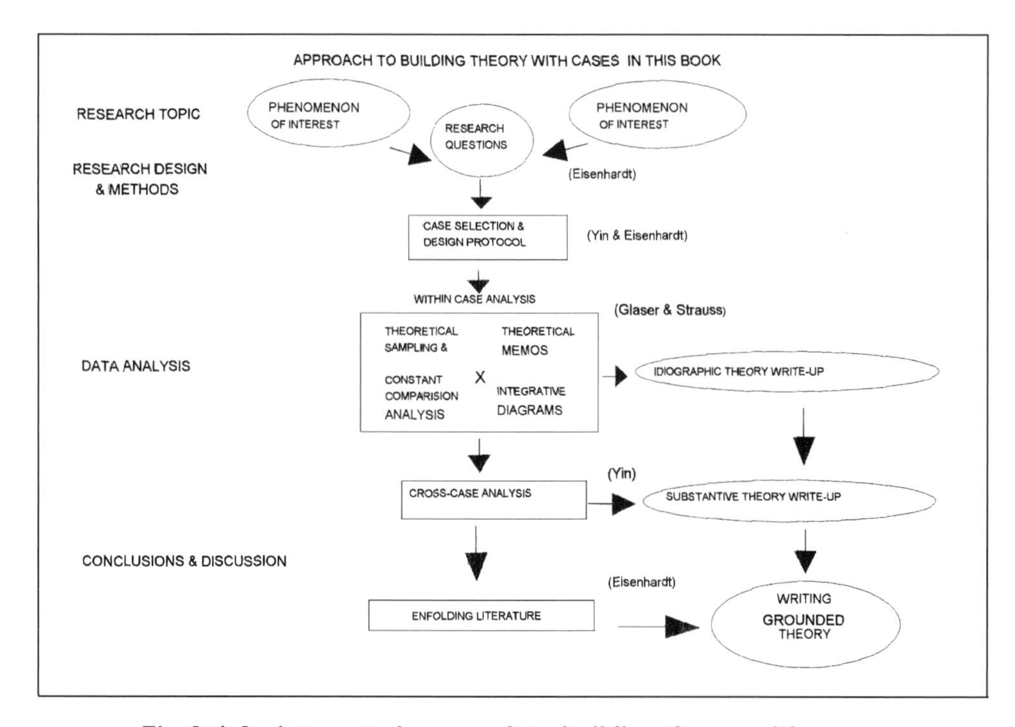

Fig. A.4. An integrated approach to building theory with cases.

to build on the complementary strengths of the ideas of Glaser and Strauss (1967); Yin (1989); and Eisenhardt (1989). The text below provides a summary of the case development approach used.

Specification of the Research Topic and Questions (Eisenhardt and Yin)

Eisenhardt and Yin both suggest that the researcher go into the field with some *a priori* definition of the objectives of the research project and the research questions. The initial inquiry in Phase One was quite broad and attempted to identify the substantive fields that may be relevant to skill-based strategy phenomenon. After a careful analysis of extant literatures, the inquiry was narrowed somewhat through the specification of the research objectives and accompanying research questions presented in Chapter 1. This narrowing of perspective followed Eisenhardt's advice to establish a focus that is sufficiently narrow to guide the data collection process, but which is also broad enough to allow for flexibility in the development of theory.

Development of a Case Study Protocol (Yin)

Given the research objectives and questions, a detailed case study protocol was developed, following Yin's advice. According to Yin (1989) a case study protocol, if designed properly, "is a major tactic in increasing the reliability of the case study research and is intended to guide the investigator in carrying out the case study" (p. 70). The protocol (which is reproduced in its entirety in Appendix Two) provides an overview of the case study project, specifies the field procedures used for identifying sites, including the pilot study site, and collecting data, and includes case study questions and other relevant tools for ensuring the systematic collection of evidence.

Data Analysis (Glaser and Strauss)

Once the protocol was developed, field research commenced on a pilot study and then the other cases. The analysis of the pilot and all other case data, which is the heart of the theory-building method used in the book, follows the Glaser and Strauss (1967), Glaser (1978), and Strauss (1987) constant comparative analysis method as closely as possible. This method is a recursive analysis–theory–analysis process that involves the creation of various analytic and theoretical memos and integrative diagrams, all of which are used in an attempt to make sense out of the data using systematic procedures.

Idiographic and Substantive Level Theory Write-Ups (Eisenhardt)

Based on the within-case analysis, an idiographic-level theory was written about each field site. Part Three of the book presents these idiographic theories, which provide the base for the more substantive-oriented analysis. This emerging substantive-level theory is then integrated with extant literatures, following Eisenhardt's example, which provides a base for the tentative development of a more formal theory, about the conditions under which value creation is more likely to occur than not.

APPENDIX SUMMARY

This appendix has provided an overview of the research design and research methodologies that have been used in the book.

The multi-phase research design for this book has been designed to ensure that both practitioner and academic perspectives and theories are examined and that they are integrated using a process that attempts to ensure that validity and reliability issues are addressed at each stage of the research.

The methodological approach introduced in this appendix is integrative and builds on the complementary strengths of three extant approaches to building theory from case evidence. Again, the specific approach has been selected in an effort to ensure that both issues of validity and reliability are addressed at all stages of the research.

NOTES

1. The terms "substantive" and "idiographic" theory are defined and discussed in more depth in the following sections.
2. Sutton's (1987) study of organizational death is a good example of business research based on the Glaser and Strauss methods.
3. The cases in this book represent four separate idiographic theories with respect to the components and attributes of skill-based strategy. Each case is a story about particular aspects of the concept of skill-based strategy and is the raw material for more substantive and formal theorizing.
4. All cross-case data analysis in the book has been conducted in an effort to address more substantive and general issues of skill-based strategy phenomena.
5. Strauss (1987) does provide important information as to how to write historical narratives.

APPENDIX TWO

SUMMARY OF RESEARCH METHODS BY PHASE OF RESEARCH

PHASE ONE DESIGN AND METHODS

Phase One: Conducting Literature Reviews to Establishing Focus for the Field Research

In Phase One, an in-depth survey of academic theories of the firm and of organizational resources was conducted in the fields of economics, entrepreneurship, organization theory and strategic management. These literature reviews served as a base for building an initial vocabulary about resource-management and its role in theories of the firm, and helped the researcher focus on the skill-base aspect of the topic. The information gathered and analyzed in Phase One of the research was intended to provide a base from which to better explore and understand the information to be gathered in Phase Two from practitioners. An initial case-study protocol was also developed during Phase One in order to serve as a preliminary guide for the field research.

Specification of the Research Topic and Questions: Literature Reviews

Eisenhardt (1989) and Yin (1989) both suggest that the researcher go into the field with some *a priori* definition of the objectives of the research project, including the identification of research questions. The initial inquiry in Phase One was quite broad and attempted to identify the substantive fields that might be relevant to skill-based strategy phenomenon. After a careful analysis of extant literatures, the inquiry was narrowed somewhat through the specification of the main research objectives and accompanying research questions presented earlier.

Development of an Initial Case Study Protocol (Yin)

Given the research objectives and questions, a detailed case study protocol was developed, following Yin's advice. According to Yin (1989) a case study protocol, if designed properly, "is a major tactic in increasing the reliability of the case study research and is intended to guide the investigator in carrying out the case study" (p. 70). The protocol provides an overview of the case study project, specifies the field procedures used for identifying sites, including the pilot study site, and collecting data, and includes case study questions and other relevant tools for ensuring the systematic collection of evidence.

Results Generated in Phase One

The results of Phase One included an 80-page computerized, annotated bibliographic database in AskSam® (a relational text database program), and a typology of organizational resources. The typology served as a synthesized framework for understanding resource-based issues, particularly in terms of understanding organizational skills as a potential resource. Both the bibliographic data and the typology were used, as needed, in both Phase Two and Phase Four of the research, when the academic and practitioner data was enfolded in a two-stage process (Eisenhardt, 1989).

PHASE TWO: THE PILOT STUDY

The purpose of conducting the pilot study, which was the primary focus of Phase Two of the research, was to establish a best exemplars accessible field-site where the process of collecting empirical data and practitioner knowledge about skill-base development and strategy could begin. The pilot study site was used to refine the case study protocol (Yin, 1989), including refinement of site selection criteria, data collection and data analysis methods, and case-writing procedures (Strauss, 1987; Eisenhardt, 1989). It was also one of the four sources for an idiographic-level model of skill-based strategy and entrepreneurial leadership.

Site Selection Procedures for the Pilot Site

The basic research design used in the book was a *purposive study of best exemplars*. Thus the firms studied in this research have been chosen purposively because they were thought to illustrate exemplary aspects of resource-based strategic management practice in a variety of competitive conditions. The purposes for doing a single case pilot study followed by several additional sites were two-fold. First, a pilot site could provide a base from which to look at the issues relevant to the book in depth. Given the critical nature of the fieldwork in the overall research design of the book, it was important to identify a firm that was considered a best exemplar of the phenomena under study, *and* that was readily accessible to the researcher so that a sufficient depth and breadth of analysis could occur. Second, due to the iterative nature of the constant comparison analysis techniques that were used in the field (Glaser and Strauss, 1967), at least one site needed to be chosen that could be visited repeatedly so that new questions and data collection techniques might be tested and relevant theoretical insights evaluated. The criteria that were used to select the pilot site are identified in Exhibit 1.

Exhibit 1. Criteria used for pilot site selection.

General criteria

1. *The firm should be readily accessible to the researcher.*
2. *The firm should be independent of a larger corporate organization.*
3. *The firm should not be a startup or a restart.*
4. *The firm should be small enough so that key managers and personnel can be identified and interviewed, and so that a relatively holistic understanding of the firm is feasible.*
5. *There should be information available about the genesis of the firm, through either primary or secondary sources.*
6. *There should be a willingness by the firm's personnel to talk about their company as openly as possible.*
7. *The firm should exhibit strong performance characteristics in terms of either financial or other non-financial competitive measures, such as market share and share growth, or time-based indicators of product delivery time, or manufacturing cycle time, or return on value added, or return on skill, or quality etc.*

Resource and skill-based criteria

1. *The firm should be large enough so that there is a clear division of resources among management.*
2. *The firm should possess some set of unique skills and competences that can be identified and analyzed.*
3. *The firm should have an identifiable technology-base that can be analyzed.*

Product market criteria

1. *The firm should have had a history that encompasses at least one market segment that has reached a mature stage of development. The more varied the life cycle stages of the product markets the firm competes in the better.*
2. *The firm should compete in at least one product market segment that is dynamic and is growing quickly.*
3. *The firm should compete in at least one product market segment where it makes a relatively high value-added contribution.*

The Pilot Site Selected: Precision Instruments Company

The Precision Instruments Company (hereafter, Precision Instruments; note that this and the Custom Cabinets Company cases were disguised at the request of the firms) was chosen to be the pilot site, as it was highly accessible to the researcher and met the other established selection criteria. Precision Instruments is one of the world's leading designers and manufacturers of hand-crafted surgical instruments. As of 1992, the company competed in five distinct product market segments and was well positioned in each of the five product markets. Two were dynamic and emerging markets that posed great technological and skill-base challenges to Precision, while two others were rather mature and posed a different set of challenges to the firm's skill-base. The company relied heavily on a highly skilled, craftsmanship-oriented workforce, and was in the process of developing its own sales force to augment its distribution capabilities. Given the nature of the skills the company appeared to have, the markets the company operated in, and the open accessibility to the general managers, Precision Instruments was an ideal pilot study site.

Data Collection Procedures at the Pilot Site

Exhibit 2 identifies the primary data sources for the pilot site.

Collection of data at the pilot study differed from that in the three subsequent cases in two primary ways: First, the *scope* of data collected at the pilot site was the most purposively broad of any of the cases, and the data collected at the pilot site was initially organized in terms of a data collection protocol that was developed using categories developed from several extant frameworks as guides. Second, the pilot site was *visited on multiple occasions* over a year and a half time period and served as a platform to evaluate the field techniques and explore emerging concepts more fully.

The initial pilot-site data was collected using a general data-collection

Exhibit 2. Primary data sources for the pilot site.

Company	Primary archives: type	Secondary archives: type	Interview information source (interviews) [total taped hours] location, date
Precision Instruments Company	Marketing manuals; plant tour; selected video screening of marketing material	*Inc. Magazine* Entrepreneur of the Year application	Len Petrush (4) [8] Dan Demaria (1) [0.0] Zeszinger (1) [1.5] Line workers (1) [0.5] All on-site, 1/92; 3/92; 4/92; 7/92.

protocol. In this protocol, four general extant frameworks and models that describe activities and processes of organizations were used to provide an initial broad data collection procedure at the pilot study site. They were: (1) the McKinsey Seven-S framework (Pascale and Athos, 1981; Peters and Waterman, 1982); (2) the Hofer and Schendel (1978) strategic management model; (3) the Tregoe and Zimmerman (1980, 1988) driving force framework; and (4) a generic, traditional, functional activities model of management. Each of these frameworks complemented the others in certain areas. A synthesis of activity categories was developed in order to guide initial data collection at the pilot site *vis-à-vis* extant strategic management concepts.

Pilot Study: Data Analysis Techniques

After systematically collecting data at the pilot site, across as many of the categories identified in the extant guide as was possible, preliminary analysis of the data was begun. Using primarily historical analysis techniques (Harrigan, 1983; Goodman and Kruger, 1988), the researcher sought to: (1) identify the important themes, concepts and language that was collected in the data from the participants; and (2) began the process of developing a formalized language about these important themes (Christensen, 1976).

These historical analysis techniques, which are designed to help in the analysis of retrospective data, enabled the researcher to put together the outlines of a comprehensive case history of the pilot site. Since a primary goal of the research was to explore how the processes of skill-based strategy unfolded and evolved over time, the *genesis* of firm behaviour and action has been examined through the use of analysis of historical evidence (Lawrence, 1984; Barzun and Graff, 1985; Goodman and Kruger, 1988; Brundage, 1989; Menard, 1991). Historical analysis techniques have also been employed in an effort to sharpen the focus of the idiographic-level models and to aid in maintaining reliability and validity. Historical analysis techniques were also employed to help identify and explore the patterns in relationships among constructs through time and have helped both to clarify the generalizability of the cases (Lawrence, 1984; Goodman and Kruger, 1988) and to understand the "why" behind relationships (Selznick, 1957). These writers suggest that "historiography does not limit Generalizability; it clarifies it". Harrigan (1983) also notes that, "research methodologies that focus on company histories (rather than a limited time horizon) provide greater insights concerning the antecedents of the strategies currently observed" (p. 403).

Pilot Study Results: Sequenced, Categorized Data Summary

The results of the data collection and analysis procedures employed at the pilot site consist of a data chapter in the book that presents the initial data from the pilot site in a time-sequenced and categorized format. Using the extant guide, data was placed into its appropriate historical chronology, and was further sorted by extant code category. Examination of the initial data across categories informed the researcher about areas that were potentially under-represented as theory categories, and steps were taken either to augment the data or to confirm why it was not central. This data was supplemented, using theoretical sampling techniques that will be discussed shortly, so that it could be reanalyzed and compared systematically with the data collected at the three follow-on sites.

PHASE THREE: FOLLOW-ON CASE STUDIES AND AUGMENTATION OF PILOT STUDY

The purpose of conducting three additional case studies in Phase Three was to collect additional more focused data on the topic of organizational skill development from practitioners in theoretically diverse settings. By analyzing data from firms with diverse skill-bases and operating environments, one could gain a more complete understanding of the issues related to skill-based strategy and entrepreneurship, and could begin to employ the constant comparison analytic process of inductive and grounded theory-building (Glaser and Strauss, 1967), extending the data and theory beyond the pilot study.

Site Selection Criteria and the Sites Selected

One of the goals in Phase Three was to select additional firms that operated in a variety of competitive conditions, and that utilized a potentially diverse set of organizational skills to meet the challenges of competition. The specific criteria for choosing these sites were similar to those used to select the pilot site, with the exception of the need for convenience criteria, which were disregarded. However, each firm was selected in a sequential process. That is, after the pilot site was chosen, a second site was chosen that would have a purposively different skill-base and/or operating environment (i.e. the intensity of technology development, the type of skills in the workforce, the stage of product market life cycle etc., played an important role in selecting the overall sample). If a firm that the researcher identified fit the selection criteria, then it was approached as a target site. If the practitioners accepted being studied, then another firm, complementary to the others, would be chosen in a similar fashion until a sufficiently diverse sample was established. These additional cases did not serve as "tests" of the pilot case *per se*, but rather as independent representative exemplars in various contexts. Exhibit 3 provides information on each of the company sites.

Exhibit 3. Description of sites chosen.

The first additional site: Custom Cabinets Company
The first additional site was a small cabinet manufacturing company. This firm was a custom cabinet manufacturer that possessed unique craft skills and that had recently adopted "Synchronous Manufacturing" methods, based on the Theory of Constraints (Goldratt, 1990). These methods were skill- and resource-based, and the firm had made a strong commitment to training its entire workforce in the methods. The firm had an exemplary growth record since the late 1980s. During this period the company turned itself around from bankruptcy and developed the ability to deliver custom cabinet sets in 10 days in a business where the industry delivery average is between six and eight weeks. Contact with the Founder and President of Custom Cabinets was made in October 1991 and a three-day site visit was authorized. It was conducted in February 1992.

The second additional site: Hauser Chemical Research
The second additional site was Hauser Chemical Research. This firm was particularly important to the sample because its base of expertise and skills was much more knowledge and service-oriented than craft-type manufacturing firms such as Precision and Custom Cabinets. Hauser Chemical Research, Inc. is one of the world's leading manufacturer's of extracted natural products and is the world's only current (1992) commercial-scale manufacturer of the anti-cancer agent taxol. Hauser Chemical research was included in this book because the company relies heavily on its base of highly trained scientists and technicians. Hauser has begun to learn how to tap into the value it creates for a particular customer and generalize its solutions. The case thus provides an interesting context from which to explore the issues of organizational learning and entrepreneurial leadership and development. Contact with Hauser was initially made in March 1992, and a three-day site visit was arranged for July 1992.

The third additional site: Combustion Engineering
The third additional site came from a personal contact with the former President of Combustion Engineering (C-E), George Kimmel. The researcher had a conversation in February 1992, with Kimmel, who was the President of Combustion Engineering during the time they attempted to turn round the company. During the conversation, Kimmel said that he would be happy to talk about the "skill-based strategy" (his words) that his company used to reorganize the company, and that he could introduce the researcher to other key members of his past management team. Despite the fact that C-E did not meet all of my sampling criteria (particularly with reference to size and independence), it was impossible to pass up such a unique opportunity to study skill-based strategy concepts. Though 1990 the various business units of the Combustion Engineering Corporation (C-E) had been leading providers of engineered products, systems and services to the worldwide power generation, process controls, and public sector and environmental waste and energy markets. The case traces the reinvention of Combustion Engineering since 1982. It is a special case in the book in that it is a corporate strategy case. The interviews with Kimmel and members of his management team occurred in July 1992, and were supplemented by secondary source information.

Data Collection Procedures Used at the Three Additional Sites

Within each of the three additional firms, data was collected using multiple data collection methods. The choice to use ethnographic interview techniques, as discussed by Spradley (1979), was made in an attempt to control for potential biases that the researcher might introduce by asking leading questions, defining terms etc. The data collection procedure at the additional sites was different from that used at the pilot site. Data collection at the additional sites *started with focused theoretical sampling techniques*, not with the extant collection guide. This approach was part of the constant comparative method of data collection and analysis that was begun after the initial analysis of pilot data was completed. A *critical facet* of the constant comparative analysis method is that a flexible living data collection procedure is employed in order to meet the needs of inductive theory-building. This dynamic collection procedure is called *theoretical sampling* (Glaser and Strauss, 1967). In order to augment the data initially collected at the pilot site, Glaser and Strauss (1967) suggest that a *theoretical sampling procedure* be used in an inductive theory-building environment. They define and discuss theoretical sampling as follows:

> Theoretical sampling is the process of data collection for generating theory whereby the analyst jointly collects, codes and analyzes his data and decides what data to collect next, and where to find them, in order to develop his theory as it emerges. This process of data collection is controlled by the emerging theory, whether substantive or formal . . . . The initial decisions for the theoretical collection of data are based only on a general subject or problem area... the [researcher] may begin the research with a partial framework of "local" concepts, designating a few principal or gross features of the structure and processes in the situations that he will study... .of course, he does not know the relevancy of these concepts to his problem – this problem must emerge – nor or they likely to become part of the core explanatory categories of his theory . . . . The basic question in theoretical sampling is: *what* groups or subgroups does one turn to *next* in the data collection? And *for what* theoretical purpose? (pp. 45–47)

In terms of this book, the *groups* that Glaser and Strauss discuss are two-fold. First, within a particular case, the theoretical sampling procedure guides the researcher as to what individuals in the company and/or what document information needs to be addressed next. Second, as information is analyzed about the entire case, the theoretical sampling procedure guides the researcher in terms of what other companies to possibly select for further analysis.

Exhibit 4. Primary data sources for additional sites.

Company	Primary archives: type	Secondary Archives: type	Interview sources (interviews) [total taped hours] location, date
Combustion Engineering	Annual Reports Contact: Schulz	Wall Street analyst reports, newspaper accounts	George Kimmel (2) [3] Gulf Breeze, FL 2/92; New Canaan, CT 8/92 Dale Smith (1) [4] Wilmington, NC, 8/92
Hauser Chemical Research, Inc.	Annual Reports; *Inc. Magazine* Entrepreneur of the Year application; "Into the Wilderness and Beyond", by Ray Hauser and Dean Stull; Internal organization charts; plant tour. Contact: Dr Dale Meyer, University of Colorado, Boulder	Wall Street analyst reports, newspaper accounts	Dean Stull (1) [2] Ray Hauser (1) [2] Randy Daughenbaugh (1) [0.5] Tim Ziebarth (1) [1] Tom Scales (1) [2] Sue Maynard (1) [1] Pat Roberts (2) [1] Greg Huckabee (1) [0.5] Ken Ammond (1) [0.5] All on-site, Boulder, CO 7/92
Custom Cabinets, Inc.	Company education books; company policy manual; company TOC implementation manual; company effect–cause–effect tree diagrams; company newsletters; plant tour Contact: Professor	Newspaper and magazine accounts	Mike Viera (2) [4] Scott Smith (1) [1.5] Joe Banks (1) [1] Stasi Bara (1) [0.7] John Zayac (2) [1] Coleen Russo (2) [0.5] David Hone (1) [0.5] P. R. Director (1) [1] Systems Analyst (1) [1] Computer Programmer (1) [1] All on-site, South, 2/92

Data Analysis at the Additional Sites and Reanalysis of the Pilot Data

The data analysis methods employed at all the sites during Phase Three included: (1) modified constant comparative qualitative research techniques (Glaser and Strauss, 1967; Glaser, 1978), which depend on an continuing recursive interactive process between site data collection and analysis; (2) case study techniques (Christensen, 1976; Eisenhardt, 1989; Yin, 1989); and (3) historical analysis techniques (Lawrence, 1984; Goodman and Kruger, 1986; Brundage, 1989; Menard, 1991). This multi-tool analysis process was constructed to ensure that the emerging theory had content validity, primarily through subjecting construct development to the tests of convergent and discriminant validity. Within-method triangulation (Jick, 1979) has also been employed for each analysis technique. (Note: data at the pilot site was reanalyzed with these tools so that a topical, focused case history could be developed and compared systematically with the additional case sites.) See Exhibit 4 for information on primary data resources.

Idiographic-Level Analysis: Constant Comparative Data Analysis Techniques

The primary data analysis approach used in this book for generating focused idiographic-level models was a modified *constant comparative method of analysis* (Glaser and Strauss, 1967, p. 101), which involved the following process: (1) given a phenomenon of interest, and some preliminary understanding of that phenomenon (from the phase one research), the researcher selected an appropriate site to begin initial observation, data collection and analysis (the pilot site, then each of the other sites separately); (2) then the researcher began to construct theory from the data at each site itself and did so in an iterative manner, constantly comparing theoretical insights with gathered data; (3) the researcher studied the data systematically, generating open, axial and selective *codes* (Strauss, 1987) that identified the language of the practitioner (Christensen, 1976); and (4) made extensive theoretical and methodological *memos*, that began to formalize the language of the practitioner (Christensen, 1976); (5) next, the researcher generated *integrative diagrams* (Strauss, 1987) that reflected how various categories in the data might be related; and finally (6) the researcher wrote up the formal grounded theory in the form of a case study or history (Yin, 1989).

The specific qualitative data analysis procedures used included:

(1) The generation of open, axial and selective *codes* from the raw data, which become the material for theory specification (Strauss, 1987). During the initial phase of the qualitative analysis of the case data, the starting, or "open" codes for the concepts that were identified were based on the principle of "in vivo coding" (Strauss, 1987, p. 30), which reflects a preference for identifying concepts in the terms used by the people being studied, in this case the managers and employees in each firm (Christensen, 1976). These participant-derived codes were the roots from which all subsequent analytical and selective coding emanated. Strauss (1987) and Glaser (1978) note that "selective coding pertains to coding *systematically* and concertedly for the core category [of the phenomenon under study]" (Strauss, p. 33). The other codes become subservient to the key code under focus. To code selectively, then, means that the analyst delimits coding only to those codes that relate to the

core codes in sufficiently significant ways as to be used in a parsimonious theory (Glaser, p. 61).

(2) Extensive *memo writing* in an analysis and theory journal (as described below) for each site. Memos were written to remind the researcher of questions to ask, people to interview, concepts to explore or relate to one another etc.

(3) The generation of *integrative diagrams* (Strauss, 1987) that reflect how various practitioner categories in the data are related. An example of such a diagram will be presented in the next section.

(4) The generation of three *journals* that were created in an attempt to minimize threats to reliability (Spradley, 1979, p. 76; Kirk and Miller, 1986). These journals included: (a) a condensed account journal which contains basic interview and other notes taken while at each site; (b) an expanded account journal, which contains the actual, full text of taped interviews, more complete accounts of non-taped interviews, and photocopied or original written materials from each site. This journal also contains the inductively derived theoretical codes and technical memos required by the qualitative data analysis tools being employed; and (c) a fieldwork journal, which contains the thoughts, potential biases, ideas, frustrations and inspirations of the researcher.

Taken together, the journals comprise the "qualitative field database" that has been used to conduct data analysis and to generate grounded theory. They also represent a full log of the fieldwork done during the project, much of which can be used by other researchers who might have questions about the reliability of the work or data. Exhibit 5 provides more information on the journals that were compiled for the book, including the master fieldwork journal, which includes all the expanded account journals combined, and the academic database that was compiled throughout the book.

PHASE FOUR: INTEGRATED ANALYSIS PROCESS AND RESULTS

Phase Four was an integrated analysis effort aimed at the development of a more general and formal theoretical synthesis that integrates and extends the perspectives of both practitioners and academicians identified and developed in Phases One to Three. Using comparative analysis methods (Glaser and Strauss, 1967; Strauss, 1987; Eisenhardt, 1989), the concepts and idiographic-level models derived from practitioner data at both the Phase Two pilot site and Phase Three supplemental sites have been systematically compared with each other and enfolded with concepts and substantive-level theory from relevant extant literatures.

Exhibit 5. Information on journals.

Field data journals				
Journal//(type), # pages	Precision Instruments	Custom Cabinets	Hauser Chemical	Combustion Engineering
Condensed account	Hand notes, 25	Hand notes, 12	Hand notes, 15	Hand notes, 15
Expanded account	Wordperfect®, 60	Wordperfect®, 100	Wordperfect®, 65	Wordperfect®, 20

Integrated database journal files		
Journal	Analysis tool, pages	Note: the fieldwork journal was a handwritten journal that was brought to all four sites. Notes on the researchers' personal thoughts and feelings during the research were recorded in this journal.
Master fieldwork journal	WordPerfect®, 237 AskSam®, 300	
Academic database	WordPerfect®, 80 AskSam®, 130	

REFERENCES

Aaker, D. A. (1989). Managing Skills and Assets. *California Management Review*, **31**, 91–105.

Abell, D. (1980). *Defining the Business: The Starting Point of Strategic Planning*. New York, NY: Prentice Hall.

Abernathy, W.J., Clark, K.B., (1985), “Innovation: Mapping the Winds of Creative Destruction”, *Research Policy*, v14n1, 3–22.

Amdendola, M. and Bruno, S. (1990). The Behavior of the Innovative Firm: Relations to the Environment. *Research Policy*, **19**(5), 419–433.

Ansoff, H. I. (1965). *Corporate Strategy: An Analytic Approach to Business Policy for Growth and Expansion*. New York, NY: McGraw-Hill.

Ansoff, H.I., (1979), *Strategic Management*, Wiley, New York.

Ansoff, H. I. (1980). Managing surprise and discontinuity: strategic response to weak signals, in *Strategy + Structure = Performance* (ed. H. B. Thorelli). Bloomington, IN: Indiana University Press, 53–82.

Ansoff, H. I. (1984). Corporate capability for managing change, in *Advances in Strategic Management* (ed. Lamb), Vol. 2, 1–30. Jai Press

Ansoff, H. I. (1987). The emerging paradigm of strategic behavior. *Strategic Management Journal*, **8**, 501–515.

Ansoff, H. I. and Baker, T. E. (1986). Is corporate culture really the answer? *Advances in Strategic Management*, **4**, 81–93.

Bacharach, S. B. (1989). Organizational theories: some criteria for evaluation. *Academy of Management Review*, **14**(4), 496–515.

Badaracco, J. L. Jr (1990). *The Knowledge Link: Competing Through Strategic Alliances*. Cambridge, MA: Harvard Business School Press.

Bain, J., (1959), *Industrial Organization*, Wiley, N.Y.

Barnard, C. I. (1938). *The Functions of the Executive*. Boston, MA: Harvard University Press.

Barney, J. B. (1986). Organizational culture: can it be a source of sustained competitive advantage? *Academy of Management Review*, **11**(3), 656–665.

Barney, J. B. (1986). Strategic factor markets: expectations, luck and business strategy. *Management Science*, **32**(10), 1231–1241.

Barney, J. B. (1986). Types of competition and the theory of strategy: toward an integrative framework. *Academy of Management Review*, **11**(4), 791–800.

Barney, J. B. (1991). Firm resources and sustained competitive advantage. *Journal of Management*, **17**(1), 99–120.

Barrett, W. (1979). *The Illusion of Technique*. New York, NY: Anchor Books.

Barretto, H. (1989). *The Entrepreneur in Microeconomic Theory: Disappearance and Explanation*. London: Routledge.

Barzun, J. and Graff, H. F. (1985). *The Modern Researcher*, 4th edn. San Diego, CA: Harcourt Brace Jovanovich.

Beard, D. W. and Dess, G. G. (1981). Corporate-level strategy, business level strategy and firm performance. *Academy of Management Journal*, **24**(4), 661–688.

Becker, H.S., (1964), "Problems in the Publication of Field Studies", in *Reflections on Community Studies*, Vidich, et.al., eds, 267–284, Wiley.

Becker, S. W. and Gordon, G. (1966). An entrepreneurial theory of formal organization: Part 1. *Administrative Science Quarterly*, **11**, 315–344.

Bogdan, R. C. and Biklen, S. K. (1982). *Qualitative Research for Education: An Introduction to Theory and Methods*. Boston, MA: Allyn and Bacon.

Bourgeois, L. J. (1979). Toward a method of middle-range theorizing. *Academy of Management Review*, **4**(3), 443–447.

Bourgeois, L. J. (1980). On the measurement of organizational slack. *Academy of Management Review*, **5**(1), 29–39.

Bourgeois, L. J. (1980). Strategy and the environment: a conceptual integration. *Academy of Management Review*, **5**(1), 25–39.

Bower, J. (1970). *Managing the Resource Allocation Process*. Homewood, IL: Irwin.

Bowman, E. H. (1976). Strategy and the Weather. *Sloan Management Review*, **17**(2) 49–62.

Bryant, J. (1989). Assessing company strength using value added. *Long Range Planning*, **22**(3), 34–44.

Brundage, A., (1989), *Going to the Source: A Guide to Historical Research & Writing*, Harland Davidson, Inc.

Buchele, R. B. (1962). How to evaluate a firm. *California Management Review*, **5**, 5–17.

Burgelman, R. A. and Sayles, L. R. (1986). *Inside Corporate Innovation: Strategy, Structure, and Managerial Skills*. New York, NY: Free Press.

Butler, J. E. (1988). Human Resource Management as a Driving Force in Business Strategy. *Journal of General Management*, **13**(4), 88–102.

Buzzell, R.D., (1987), *The PIMS Principles: Linking Strategy to Performance*, Free Press.

Bygrave, W. D. and Hofer, C. W. (1991). *Theorizing About Entrepreneurship*. Working Paper, Babson College.

Cameron, K. S. (1986). Effectiveness as paradox: consensus and conflict in conceptions of organizational effectiveness. *Management Science*, **32**, 539–553.

Campbell, D. T. (1988). *Methodology and Epistemology for Social Science: Selected Papers*. Chicago, IL: University of Chicago Press.

Cangelosi, V. E. and Dill, W. R. (1965). Organizational learning: observations toward a theory. *Administrative Science Quarterly*, **10**, 175–203.

Carlzon, J. (1987). *Moments of Truth*. New York, NY: Harper & Row.

Caves, R.E., Ghemawat, P., "Identifying Mobility Barriers", *Strategic Management Journal*, v13n1, 1–12.

Chakravarthy, B. S. (1982). Adaptation: a promising metaphor for strategic management. *Academy of Management Review*, **7**(1), 35–44.

Chakravarthy, B. S. (1986). Measuring strategic performance. *Strategic Management Journal*, **7**(5), 437–458.

Chandler, A.D., (1990), *Scale and Scope: The Dynamics of Industrial Capitalism*, Belknap Press.

Charmaz, K., (1983), "The Grounded Theory Method: An Explication & Interpretation", in Emerson, R., ed., *Contemporary Field Research*, Waveland Press.

Chatterjee, S. and Wernerfelt, B. (1991). The link between resources and type of diversification: theory and evidence. *Strategic Management Journal*, **12**, 33–48.

Chrisman, J. J. (1986). *Strategy, Skills, and Success: An Exploratory Study*. Unpublished doctoral dissertation, University of Georgia, Athens, GA.

Chrisman, J. J., Hofer, C. W. and Boulton, W. R. (1988). Toward a system for classifying business strategies. *Academy of Management Review*, **13**(3), 413–428.

Christensen, C. R. (1976). Proposals for a Program of Research into the Properties of Triangles. *Decision Sciences*, **4**, 631–648.

Clarke, C. J. and Brennan, K. (1990). Building synergy in the diversified business. *Long Range Planning*, **23**(2), 9–16.

Cleveland, G., Schroeder, R. and Anderson, J. (1989). A theory of production competence. *Decision Sciences*, **20**(4), 655–668.

Cohen, W. M. and Levinthal, D. A. (1991). Absorbtive capacity: a new perspective on learning and innovation. *Administrative Science Quarterly*, **35**(1), 128–152.

Collis, D. J. (1991). A resource-based analysis of global competition: the case of the ball bearings industry. *Strategic Management Journal*, **12**, 49–68.

Conner, K. R. (1991). A historical comparison of resource-based theory and five schools of thought within industrial organization economics: do we have a new theory of the firm? *Journal of Management*, **17**(1), 121–154.

Cook, T. D. and Campbell, D. T. (1979). *Quasi-Experimentation: Design and Analysis Issues for Field Settings*. Boston, MA: Houghton Mifflin.

Coyne, K.P., et.al., (1997), "Is Your Core Competence a Mirage?", McKinsey Quarterly, n1, 40–54.

Coyne, K.P., (1986), "Sustainable Competitive Advantage–What It Is, What It Isn't", *Business Horizons*, v29n1, 54–61.

Crawford, R. (1991). *In the Era of Human Capital: The Emergence of Talent, Intelligence, and Knowledge as the Worldwide Economic Force and What it Means to Managers and Investors*. New York, NY: HarperBusiness.

Cryer, J. D. (1986). *Time Series Analysis*. Boston, MA: PWS.

Curran, J., Goodfellow, J.H., (1990), "Theoretical and Practical Issues in the Determination of Market Boundaries", *European Journal of Marketing*, v24n1, 16–28.

Cyert, R. M. and March, J. G. (1963). *A Behavioral Theory of the Firm*. Englewood Cliffs, NJ: Prentice Hall.

Daft, R. L. (1982). Learning the craft of organizational research. *Academy of Management Review* **8**(4), 539–546.

Daft, R. L. and Weick, K. E. (1984). Toward a Model of Organizations as Interpretation Systems. *Academy of Management Review*, **9**(2), 284–295.

Day, G. S. (1981). Strategic market analysis and definition: an integrated approach. *Strategic Management Journal*, **2**, 281–299.

Deming, W. E., (1986), *Out of the Crisis*, MIT Center for Advanced Engineering Study.

Dettmer, W., (1997), *Goldratt's Theory of Constraints*, ASQC Quality Press, Milwaukee.

Dierickx, I. and Cool, K. (1989). Asset stock accumulation and sustainability of competitive advantage. *Management Science*, **35**(12), 1504–1510.

Drucker, P. F. (1954). *The Practice of Management*. New York, NY: Harper & Row.

Drucker, P.F., (1985), *Innovation & Entrepreneurship: Practice & Principles*, Harper & Row.

Dutton, J. M. and Freedman, R. D. (1985). External environment and internal strategies: calculating, imitation and experimentation, in *Advances in Strategic Management* (ed. Lamb), 39–67, Jai Press.

Dutton, J. M. and Thomas, A. (1985). Relating technological change and learning by doing, in *Research on Technological Innovation: Management and Policy*, 187–224, Jai Press.

Eastlack & McDonald, (1970), "CEO's Role in Corporate Growth", Harvard Business Review, May-June, 150–163.

Eisenhardt, K. M. (1989). Building theories from case study research. *Academy of Management Review*, **14**(4), 532–550.

Ellul, J. (1964). *The Technological Society*. New York, NY: Vintage.

Fahey, L. (1989). *The Strategic Planning Management Reader*. Englewood Cliffs, NJ: Prentice Hall.

Fahey, L., Christensen, H.K., (1986), "Evaluating the Research on Strategy Content", *Journal of Management*, v12n2, 167–183.

Farjoun, M., (1990), *Beyond Industry Boundaries: Human Expertise, Diversification, and Resource-Related Industry Groups*, Ph.D. Dissertation, Northwestern University.

Fayol, H. C. (1930). *Industrial and General Administration*. Geneva: International Management Institute.

Feeser, H.R., (1987), *Incubators, Entrepreneurs, Strategy and Performance: A Comparison of High and Low Growth High Tech Firms*, Ph.D. Dissertation, Purdue University.

Fiol, C. M. (1991). Managing culture as a competitive resource: an identity-based view of sustainable competitive advantage. *Journal of Management*, **17**(1), 191–212.

Fiol, C. M. and Lyles, M. A. (1985). Organizational Learning. *Academy of Management Review*, **10**(4), 803–813.

Fleishman, E. A. and Quaintance, M. K. (1984). *Taxonomies of Human Performance: The Description of Human Tasks*. Orlando, FL: Academic Press, Harcourt Brace Jovanovich.

Fossum, J., et.al., (1986), "Modeling the Skills Obsolescence Process: A Psycological/Economic Integration", *Academy of Management Review*, v11n2, 362–374.

Fossum, J. A., Arvey, R. D., Paradise, C. A. and Robbins, N. E. (1986). Modelling the skills obsolescence process: a psychological/economic integration. *Academy of Management Review*, **11**(2), 362–374.

Fox, H. W. (1973). A framework for functional coordination. *Atlanta Economic Review*, **23**(6), 8–11.
Freedman, D. H. (1991). Invasion of the insect robots. *Discover*, March, 42–50.
Freeman, R. E. and Lorange, P. (1985). Theory building in strategic management. *Advances in Strategic Management* (ed. Lamb), Vol. 3, 9–38, Jai Press.
Gael, S. (1988). *The Job Analysis Handbook for Business, Industry and Government: Volumes I and II*. New York, NY: John Wiley & Sons.
Gagne, R. M. (1970). *The Conditions of Learning*, 2nd edn. New York, NY: Holt, Rinehart and Winston.
Galbraith, J. R. and Kazanjian, R. K. (1986). *Strategy Implementation: Structure, Systems and Process*. St Paul, MN: West Publishing Company.
Garratt, B. (1987). Learning is the core of organisational survival: action learning is the key integrating process. *Journal of Management Development*, **6**(2), 38–44.
Garvin, D., (1988), *Managing Quality*, Free Press, N.Y.
Ghemewat, P., (1990), "Sustainable Advantage", Harvard Business Review, v64n5, 53–58.
Ghemawat, P. (1991). *Commitment: The Dynamic of Strategy*. New York, NY: Free Press.
Ghemewat, P., Ricart i Costa, J., (1993), "The Organizational Tension Between Static & Dynamic Efficiency", *Strategic Management Journal*, v14, 59–73.
Ginsberg, A. (1984). Operationalizing organizational strategy: toward an integrative framework. *Academy of Management Review*, **9**(3), 548–557.
Ginsberg, A. (1988). Measuring and modelling changes in strategy: theoretical foundations and empirical foundations. *Strategic Management Journal*, **9**(6), 559–575.
Glaser, B. G. (1978). *Theoretical Sensitivity*. Mill Valley, CA: The Sociology Press.
Glaser, B. G. and Strauss, A. L. (1967). *The Discovery of Grounded Theory: Strategies for Grounded Research*. New York, NY: Aldine Publishing.
Goldratt, E. M. (1990). *What is this Thing Called Theory of Constraints and How Should it be Implemented?* Croton-on-Hudson, NY: North-River Press.
Goodman, R. S. and Kruger, E. J. (1988). Data dredging or legitimate research method? Historiography and its potential for management research. *Academy of Management Review*, **13**(2), 315–325.
Gorman, P. and Thomas, H. (1997). The theory and practice of competence-based competition. *Long Range Planning*. **30**(4), 615–620.
Grant, R.M., (1991), "The Resource-Based Theory of Competitive Advantage: Implications for Strategy Formulation", *California Management Review*, v33n3, 114–135.
Grant, R. M. (1996). Prospering in dynamically-competitive environments: organizational capability as knowledge integration. *Organization Science*, **7**(4), 375–387.
Greiner, L. E. (1972). Evolution and Revolution as Organizations Grow. *Harvard Business Review*, **50**(4), 37–46.
Hagenaars, J. A. (1991). *Categorical Longitudinal Data*. Beverly Hills, CA: Sage.
Halberstam, D. (1983). *The Reckoning*. New York, NY: Morrow.
Hall, R. (1992). The strategic analysis of intangible assets. *Strategic Management Journal*, **13**(2), 135–144.

Hambrick, D. C. (1983). An empirical typology of mature industrial-product environments. *Academy of Management Journal*, **26**, 213–230.
Hambrick, D. C. and Lei, D. (1985). Toward an empirical prioritization of contingency variables. *Academy of Management Journal*, **28**(4), 763–788.
Hamel, G. (1991). Competition for competence and inter-partner learning within international strategic alliances. *Strategic Management Journal*, **12**, 83–103.
Hamel, G., Heene, A., eds., (1994), *Competence-Based Competition*, Wiley & Sons.
Hammermesh, R. G. (ed.) (1983). *Strategic Management*. New York, NY: John Wiley & Sons.
Hansen, G. S. and Wernerfelt, B. (1989). Determinants of firm performance: the relative importance of economic and organizational factors. *Strategic Management Journal*, **10**(5), 399–411.
Harrigan, K. R. (1983). Research Methodologies for Contingency Approaches to Business Strategy. *Academy of Management Review*, **8**(3), 398–405.
Hayes, R.H., Abernathy, W.J., (1980), "Managing Our Way to Economic Decline", *Harvard Business Review*, v58n4, 67–77.
Hedberg, B. and Starbuck, W. (1981). *Handbook of Organizational Design*. Oxford: Oxford University Press.
Henderson, B., (1979), *Henderson on Corporate Strategy*, ABT Books.
Herron, L. (1990). *The Effects of Characteristics of the Entrepreneur on New Venture Performance*. Unpublished doctoral dissertation, University of South Carolina.
Hitt, M. A. and Ireland, R. D. (1985). Corporate distinctive competence, strategy, industry and performance. *Management Journal*, **6**(3), 273–293.
Hitt, M. A. and Ireland, R. D. (1986). Relationships among corporate level distinctive competencies, diversification strategy, corporate structure and performance. *Journal of Management Studies*, **23**(4), 401–416.
Hitt, M. A., Hoskisson, R. E. and Harrison, J. S. (1991). Strategic competitiveness in the 1990s: challenges and opportunities for U.S. executives. *The Executive*, **5**(2), 7–22.
Hitt, M. A., Ireland, R. D. and Palia, K. A. (1982). Industrial firms' grand strategy and functional importance: moderating effects of technology and uncertainty. *Academy of Management Journal*, **25**, 265–298.
Hitt, M. A., Ireland, R. D. and Stadter, G. (1982). Functional importance and company performance: moderating effects of grand strategy and industry type. *Strategic Management Journal*, **3**, 315–330.
Hofer, C. W. (1975). Toward a contingency theory of business strategy. *Academy of Management Journal*, **18**, 784–810.
Hofer, C.W., (1980), "Turnaround Strategies", *Journal of Business Strategy*, 1: 19–31.
Hofer, C. W. (1983). ROVA: A new measurement for assessing organizational performance. *Advances in Strategic Management*, **2**, 43–55.
Hofer, C. W. (1990), "How CEO's Set the Strategic Direction of their Organizations: A Comparison of Reginald Jones' and Jack Welsh's Leadership of General Electric", Unpublished Working Paper, University of Georgia.
Hofer, C. W. and Bygrave, W. D. (1992). Researching entrepreneurship. *Working Paper*, University of Georgia and Babson College.
Hofer, C.W., Bygrave, W.D., (1992), "Researching Entrepreneurship", *Entrepreneurship Theory & Practice*, v16n3, 91–100.

Hofer, C. W. and Schendel, D. (1978). *Strategy Formulation: Analytical Concepts*. St Paul, MN: West Publishing Company.

Hrebiniak, L. G. and Joyce, W. F. (1985). Organizational adaptation: strategic choice and environmental determinism. *Administrative Science Quarterly*, **30**, 336–349.

Huff, A. S. and Reger, R. K. (1987). A review of strategic process research. *Journal of Management*, **13**(2), 211–236.

Irvin & Michaels (1989). Skill-Based Competitive Advantage, *McKinsey Quarterly, Fall*, 1–8.

Itami, H. and Roehl, T. W. (1987). *Mobilizing Invisible Assets*. Cambridge, MA: Harvard University Press.

Jelinek, M. (1979). *Institutionalizing Innovation: A Study of Organizational Learning Systems*. New York, NY: Praeger.

Jick, T. D. (1979). Mixing qualitative and quantitative methods: triangulation in action. *Administrative Science Quarterly*, **24**, 602–611.

Katz, J. (1983). A theory of qualitative methodology: the social system of analytical fieldwork. *Contemporary Field Research* (ed. Emerson), 127–155, Waveland Press.

Katz, R. L. (1974). Skills of an Effective Administrator. *Harvard Business Review*, **52**(5), 90–102.

Kazanjian, R. K. and Drazin, R. (1987). Implementing internal diversification: contingency factors for organizational design choices. *Academy of Management Review*, **12**(2), 342–354.

Kirk, J. and Miller, M. L. (1986). *Reliability and Validity in Qualitative Research*. Newbury Park, CA: Sage Publications.

Kirzner, I. M. (1979). *Perception, Opportunity and Profit*. Chicago: Chicago University Press.

Klien, J.A., et.al., (1991), "Skill-Based Competition", *Journal of General Management*, v16n4, 1–15.

Klein, J. E., Edge, G. M., Kass, T. (1991). Skill-based competition. *Journal of General Management*, **16**(4), 1–15.

Kotter, J.P. (1990), *A Force for Change: How Leadership Differs from Management*, Free Press, N.Y.

Lado, A. A., Boyd, N. G. and Wright, P. (1992). A competency-based model of sustainable competitive advantage: toward a conceptual integration. *Journal of Management*, **18**(1), 77–91.

Lakatos, I. and Musgrave, A. (eds.) (1970). *Criticism and The Growth of Knowledge*. Cambridge: Cambridge University Press.

Lane, N. E. (1987). *Skill Acquisition Rates and Patterns: Issues and Training Implications*. New York, NY: Springer-Verlag.

Lawler, E., Ledford, G., (1992), "A Skill-Based Approach to Human Resource Management", *European Management Journal*, v10n4, 383–391.

Lawrence, B. S. (1984). Historical perspective: using the past to study the present. *Academy of Management Review*, **9**(2), 307–312.

Lawrence & Lorsch, (1967), *Organization & Environment*, Harvard Business School Press.

Lecraw, D.J., (1984), "Diversification Strategy & Performance", *Journal of Industrial Economics*, v22n2, 179–198.

Lengnick-Hall, C. A. and Lengnick-Hall, M.L. (1988). Strategic human resources management: a review of the literature and a proposed typology. *Academy of Management Review*, **13**(3), 454–470.

Lenz, R. T. (1980). Strategic capability: a concept and framework for analysis. *Academy of Management Review*, **5**(2), 225–234.

Lenz, R. T. (1981). Determinants of organizational performance: an interdisciplinary review. *Strategic Management Journal*, **2**, 131–154.

Lie, D., Hitt, M., Bettis, R., (1996), "Dynamic Core Competences Through Meta-learning and Strategic Context", *Journal of Management*, v22n4, 549–569.

Lippman, S. A. and Rumelt, R. P. (1982). Uncertain imitability: an analysis of interfirm differences in efficiency under competition. *Bell Journal of Economics*, **13**(2), 418–438.

M.I.T. Commission on Industrial Productivity (1989), *Report on U.S. Productivity*, MIT Press.

MacMillian, K. and Farmer, D. H. (1979). Redefining the boundaries of the firm. *Journal of Industrial Economics*, 277–285.

Magaziner, I. and Patinkin, M. (1989). *The Silent War: Inside the Global Business Battles Shaping America's Future*. New York, NY: Random House.

Maier, N. (1965). *Psychology in Industry*. Boston, MA: Houghton Mifflin.

March, J. G. and Olsen, J. P. (1975). The uncertainty of the past: organizational learning under ambiguity. *European Journal of Political Research*, **3**, 147–171.

Mauri, A.J., Michaels, M.P., (1998), "Firm and Industry Effects with Strategic Management: An Empirical Examination", *Strategic Management Journal*, v19n3, 211–219.

McCleary, R. and Hay, R. A. (1980). *Applied Time Series Analysis for the Social Sciences*. Beverly Hills, CA: Sage Publications.

McDowall, D. M., McCleary, R., Meidinger, E.E. and Hay, R.A. (1980). *Interrupted Time Series Analysis*. Beverly Hills, CA: Sage Publications.

McKelvey, B. (1982). *Organizational Systematics*. Berkeley, CA: University of California Press.

Menard, S. (1991). *Longitudinal Research*. Beverly Hills, CA: Sage Publications.

Meyers, P. W. (1990). Non-linear learning in large technological firms: period four implies chaos. *Research Policy*, **19**(2), 97–115.

Miller, D. (1982). Evolution and revolution: a quantum view of structural change in organizations. *Journal of Management Studies*, **19**, 131–151.

Miller, D. (1986). Configurations of strategy and structure: towards a synthesis. *Strategic Management Journal*, **7**, 233–249.

Miller, D. and Friesen, P. H. (1980). Momentum and Revolution in Organizational Adaptation. *Academy of Management Journal*, **23**(4), 591–614.

Miller, E. J. and Rice, A. K. (1967). *Systems of Organization*. London: Tavistock Institute.

Mintzberg, H. (1987). The strategy concept I: five P's for strategy. *California Management Review*, **30**(1)11–24.

Mintzberg, H. (1987). The strategy concept II: another look at why organizations need strategies. *California Management Review*, **30**(1) 25–32.

Mintzberg, H. (1989). You meet the humblest managers on a Honda. *Across the Board*, **26**(10), 10–12.
Mintzberg, H., (1990), "The Design School: Reconsidering the Basic Premises of Strategic Management, *Strategic Management Journal*, v11n3, 171–195.
Mintzberg, H. and Waters, J. A. (1982). Tracking Strategy. *Academy of Management Journal*, **25**, 465.
Mintzberg, H. and Waters, J. A. (1985). Of strategies, deliberate and emergent. *Strategic Management Journal*, **6**(3), 257–272.
Naugle, D. G. and Davies, G. A. (1987). Strategic skill pools and competitive advantage. *Business Horizons*, **30**(6) 35–42.
Nelson, R. R. and Winter, S. G. (1982). *An Evolutionary Theory of Economic Change*. Boston, MA: Harvard University Press.
Newman, W. (1992). The process of integrating: crucial but neglected feature of managing. *Working Paper*, Columbia Business School.
Nonaka, I. and Johansson, J. K. (1985). Japanese management: what about the "hard" skills? *Academy of Management Review*, **10**, 181–191.
Ohmae, K. (1983). *The Mind of the Strategist*. New York, NY: Penguin.
Osigweh, C. A. B. (1989). Concept fallibility in organizational science. *Academy of Management Review*, **14**(4), 579–594.
Palepu, K. (1985). Diversification strategy, profit performance and the entropy measure. *Strategic Management Journal*, **6**, 239–255.
Panzar, J. C. and Willig, R. D. (1981). Economies of scope. *American Economic Review*, **71**(2), 268–272.
Partridge, E. (1958). *Origins: A Short Etymological Dictionary of Modern English*. London: Routledge & Kegan Paul.
Pascale, R. T. (1991). *Managing on the Edge*. New York, NY: Simon & Schuster.
Pascale, R. T. and Athos, A. G. (1981). *The Art of Japanese Management: Applications for American Executives*. New York, NY: Warner Books.
Patton, M. Q. (1990). *Qualitative Evaluation and Research Methods*. Newbury Park, CA: Sage Publications.
Penrose, E. T. (1959). *The Theory of the Growth of the Firm*. New York, NY: Wiley & Sons.
Peteraf, M.A., (1993), "The Cornerstones of Competitive Advantage: A Resource-Based View", *Strategic Management Journal*, v14n3, 170–191.
Peters, T. J. (1984). Strategy follows structure: developing distinctive skills. *California Management Review*, **26**(3), 111.
Peters, T. and Waterman, R. (1982). *In Search of Excellence*. New York, NY: Harper & Row.
Petts, N., (1997), "Building Growth on Core Competences – A Practical Approach", *Long Range Planning*, v30n4, 551–561.
Pfeffer, J. (1987). Bringing the environment back in, in *The Competitive Challenge* (ed. D. J. Teece). New York, NY: Harper & Row, 119–135.
Pfeffer, J. and Salancik, G. R. (1978). *The External Control of Organizations: A Resource Dependence Perspective*. New York, NY: Harper & Row.
Pinder, C. and Moore, L. (1979). The resurrection of taxonomy to aid in the development of middle-range theories. *Administrative Science Quarterly*, **24**, 99–118.

Porac, J. F. and Thomas, H. (1990). Taxonomic mental models in competitor definition. *Academy of Management Review*, **14**(4), 562–578.
Porter, M. (1985). *Competitive Advantage*. New York, NY: Free Press.
Porter, M.E., (1980), Competitive Strategy: Techniques for Analyzing Industries & Competitiors, Free Press.
Post, H.A., (1997), "Building a Strategy on Competences", *Long Range Planning*, v30n5, 733–740.
Prahalad, C. K. and Bettis, R.A. (1986). The dominant logic: a new linkage between diversity and performance. *Strategic Management Journal*, **7**, 485–501.
Prahalad, C. K. and Hamel, G. (1990). The core competence of the corporation. *Harvard Business Review*, **68**(3), 79–91.
Quinn, J.B., Paquette, P.C., (1990) "Technology in Services: Creating Organizational Revolutions", *Sloan Management Review*, v31n2, 67–78.
Quinn, J.B., et.al., (1990), "Technology in Services: Rethinking Strategic Focus", *Sloan Management Review*, v31n2, 79–87.
Quinn, J.B., Mintzberg, H. (1991), The Strategy Process, Third Edition, Prentice-Hall.
Reed, R. and DeFillippi, R. J. (1990). Causal ambiguity, barriers to imitation, and sustainable competitive advantage. *Academy of Management Review*, **15**(1), 88–102.
Reich, R. B. (1991). *The Work of Nations*. New York, NY: A. Knopf.
Reimann, B. C. (1982). Organizational competence as a predictor of long run survival. *Academy of Management Journal*, **25**(2), 323–334.
Romanelli, E. and Tushman, M. L. (1986). Inertia, environments and strategic choice: a quasi-experimental design for comparative longitudinal research. *Management Science*, **32**, 608–621.
Rosenberg, N. (1985). Perspectives on Technology, Armond Press, N.Y.
Rothwell, R. and Whiston, T. G. (1990). Design, innovation and corporate integration. *R & D Management*, **20**(3),
Rumelt, R., (1991), "How Much does Industry Matter?", *Strategic Management Journal*, V12n3, 167–185.
Rumelt, R. P. (1979). Evaluation of strategy: theory and models, in *Strategic Management: A New View of Business Policy and Planning* (eds. D. Schendel and C. W. Hofer). New York, NY: Little Brown, 189–212.
Rumelt, R. P. (1982). Diversification strategy and profitability. *Strategic Management Journal*, **3**, 359–369.
Rumelt, R. P. (1986). *Strategy, Structure, and Economic Performance*. Harvard, MA: Harvard Business School Press.
Rumelt, R. P. (1987). Theory, strategy and entrepreneurship, in *The Competitive Challenge* (ed. D. J. Teece). New York, NY: Harper & Row, 137–158.
Saint Exupery, A. de (1943). *The Little Prince*. San Diego, CA: Harcourt Brace Jovanovich.
Salter, M. S. and Weinhold, W. A. (1979). *Diversification Through Acquisition*. New York, NY: Free Press.
Sanchez, et.al., (1996), *Dynamics of Competence-Based Competition*, Pergamon.
Schendel, D. and Hofer, C. W. (1979). *Strategic Management: A New View of Business Policy and Planning*. New York, NY: Little Brown.

Schmalensee, R. (1985). Do markets differ much? *American Economic Review*, **75**, 341–351.
Schultz, T. W. (1961). Investment in human capital. *American Economic Review*, (1), 1–17.
Schultz, T. W. (1990). *Restoring Economic Equilibrium: Human Capital in the Modernizing Economy*. Worcester: Billing & Sons.
Schulz, W. C. (1992). Emergence of the real-time computer reservation system in the U.S. airline industry, 1958–1989: a paper on strategic innovation. *Technovation*, **12**(2), 66–74.
Schulz, W. C. and Siffin, W. (1983). Organizational technology: a perspective on organizations as technolgy. *Working Paper*, Indiana University.
Schulze, W. S. (1992). Two resource-based models of the firm?: Definitions and implications for research. *Academy of Management Proceedings*.
Schulze, W.S., (1994) *Environmental & Organizational Determinants of Firm Conduct: A Test of Resource-Based Theory at the Business-Level, Ph.D. Dissertation, University of Colorado, Boulder.*
Schumpeter, J. A. (1934). *The Theory of Economic Development*. Cambridge, MA: Harvard University Press.
Selznick, P. (1957). *Leadership in Administration*. Berkeley, CA: University of California Press.
Siffin, W., (1983) "Notes from a Graduate Seminar on Technology & Strategy, Indiana University, Spring 1983.
Simon, H., (1981), *Sciences of the Artificial*, MIT Press.
Singleton, W. T. and Spurgeon, P. (eds.) (1975). *Measurement of Human Resources*. New York, NY: Halsted Press, Wiley & Sons.
Snow, C. C. and Hrebiniak, L. G. (1980). Strategy, distinctive competence, and organizational performance. *Administrative Science Quarterly*, **25**, 317–335.
Sousa De Vasconcellos, J. A. and Hambrick, D. C. (1989). Key success factors: test of a general theory in the mature industrial product sector. *Strategic Management Journal*, **10**(4), 367–382.
Spender, J. C. (1996) Making knowledge the basis of a dynamic theory of the firm, *Strategic Management Journal*, **17**, 45–62.
Spradley, J. P. (1979). *The Ethnographic Interview*. New York, NY: Holt, Reinhart & Winston.
Stalk, G. and Hout, T. M. (1990). *Competing Against Time: How Time-based Competition is Reshaping Global Markets*. New York, NY: Free Press.
Stalk, G. E. P. and Shulman, L. E. (1992). Competing on capabilities: the new rules of corporate strategy. *Harvard Business Review*, March–April, 57–69.
Stevenson, H. H. (1976). Defining Corporate Strengths and Weaknesses. *Sloan Management Review*, **17**, 51–68.
Stevenson, H. H. and Jarillo, J. C. (1990). A paradigm of entrepreneurship: entrepreneurial management. *Strategic Management Journal*, **11**, 17–27.
Stoner, C. R. (1987). Distinctive competence and competitive advantage. *Journal of Small Business Management*, **25**(2),
Strauss, A. L. (1987). *Qualitative Analysis for Social Scientists*. Cambridge: Cambridge University Press.

Subramanian, R., (1998), "Dynamics of Competence-Based Competition: Theory & Practice in the New Strategic Management", *International Journal of Organizational Analysis*, v6n4, 373–375.
Sutton, Robert, I., (1987), "The Process of Organizational Death: Disbanding and Reconnecting", *Administrative Science Quarterly*, v32 n4, 542–569.
Syrett, M., Hogg, C. (1992), *Frontiers of Leadership*, Oxford Press.
Teece, D. J. (1980). Economies of scope and the scope of the enterprise. *Journal of Econoimc Behavior and Organization*, **1**, 223–247.
Teece, D. J. (1982). Towards an economic theory of the multiproduct firm. *Journal of Economic Behavior and Organization*, **3**, 39–63.
Teece, D. J. (1987). *The Competitive Challenge*. New York, NY: Harper & Row.
Teece, D. J. *et al.* (1997). Dynamic capabilities and strategic management, *Strategic Management Journal*, **18**(7) 509–533.
Teece, D. J., Pisano, G. and Shuen, A. (1992). Dynamic capabilities and strategic management. *Working Paper*, for the Consortium on Competitiveness and Cooperation, Berkeley.
Terreberry, S. (1968). The evolution of organizational environments. *Administrative Science Quarterly*, **12**, 590–613.
Thompson, J. D. (1967). *Organizations in Action*. New York, NY: McGraw-Hill.
Thorelli, H. B. (1977). *Strategy + Structure = Performance*. Bloomington, IN: Indiana University Press.
Tilles, S. (1983). How to evaluate corporate strategy, in *Strategic Management* (ed. R. G. Hammermesh). New York, NY: John Wiley & Sons, 77–94.
Tregoe, B. B. and Zimmerman, J. W. (1980). *Top Management Strategy: What it is and How to Make it Work*. New York, NY: Simon & Schuster.
Tregoe, B., Zimmerman, J. *et al.* (1980, 1988). *Vision in Action*. New York, NY: Simon & Schuster.
Trevelyan, E. (1974). *The Strategic Process in Large Complex Organizations: A Pilot Study of New Business Development*. Unpublished doctoral dissertation, Harvard University, University of Michigan Microfilm.
Tsoukas, H. (1989). The Validity of Idiographic Research Explanations. *Academy of Management Review*, **14**(4), 551–561.
Tushman, M. L., Moore, W. L. (1988), Readings in the Management of Innovation, Cambridge Press.
Ulrich, D. and Lake, D. (1991). Organizational capability: creating competitive advantage. *Academy of Management Executive*, **5**(1), 77–92.
Venkatraman, N. (1989). The concept of fit in strategy research: towards verbal and statistical correspondence. *Academy of Management Review*, **14**(3), 423–444.
Venkatraman, N. (1989). Strategic orientation of business enterprises: the construct dimensionality and measurement. *Management Science*, **35**(8), 942–962.
Venkatraman, N. and Prescott, J. E. (1990). Environment–strategy coalignment: an empirical test of its performance implications. *Strategic Management Journal*, **11**(1), 1–23.
Waddock, S. (1989). Core strategy: end result of restructuring? *Business Horizons*, **32**(3), 49–55.
Walsh, J. P. and Ungson, G. R. (1991). Organizational Memory. *Academy of Management Review*, **16**(1), 57–91.

Warner, M. (1989). Human-resources implications of new technology. *Human Systems Management*, **14**(4), 516–531.
Weick, K. (1987). Substitutes for corporate strategy, in *The Competitive Challenge* (ed. D. J. Teece). New York, NY: Harper & Row, 221–233.
Weick, K. E. (1989). Theory construction as disciplined imagination. *Academy of Management Review*, **14**(4), 516–531.
Wernerfelt, B. (1984). A resource-based view of the firm. *Strategic Management Journal*, **5**, 171–180.
Wernerfelt, B. (1989). From critical resources to corporate strategy. *Journal of General Management*, **14**, 4–12.
Westley, F., Mintzberg, H., (1989), "Visionary Leadership & Strategic Management", *Strategic Management Journal*, v10, 17–32.
White, R. E. and Hammermesh, R. G. (1981). Toward a model of business unit economic performance. *Academy of Management Review*, **6**, 213–223.
Winter, S. G. (1987). Knowlege and competence as strategic assets, in *The Competitive Challenge* (ed. D. J. Teece). New York, NY: Harper & Row, 159–184.
Yin, R. K. (1989). *Case Study Research*. Beverly Hills, CA: Sage.
Yip, G. S. (1982). Diversification entry: internal development vs acquisition. *Strategic Management Journal*, **3**, 331–345.
Yuchtman, E. and Seashore, S. E. (1967). A system resource approach to organizational effectiveness. *American Sociological Review*, **32**, 891–903.

INDEX